U.S. INDUSTRY IN 2000
Studies in Competitive Performance

DAVID C. MOWERY, *Editor*

Board on Science, Technology, and Economic Policy

National Research Council

NATIONAL ACADEMY PRESS
Washington, D.C.   1999

NATIONAL ACADEMY PRESS • 2101 Constitution Ave., N.W. • Washington, D.C. 20418

NOTICE: The conference from which the papers in this publication were drawn was approved by the Governing Board of the National Research Council, whose members come from the councils of the National Academy of Sciences, the National Academy of Engineering, and the Institute of Medicine. The members of the board responsible for the project were chosen for their special competences and with regard for appropriate balance.

This publication was supported by the National Aeronautics and Space Administration, the National Science Foundation, the Office of Industrial Technologies of the U.S. Department of Energy, the Alfred P. Sloan Foundation, Ralph Landau, and the Lockheed Martin Corporation. Any opinions, findings, conclusions, or recommendations expressed in this publication are those of the author(s) and do not necessarily reflect the view of the organizations or agencies that provided supoort for the project.

Library of Congress Cataloging-in-Publication Data

U.S. industry in 2000: studies in competitive performance / David
C. Mowery, editor.
  p. cm.
  Papers presented at a conference held at the National Academy of
Sciences in Washington, D.C., on Dec. 8-9, 1997.
  ISBN 0-309-06179-2 (pbk.)
  1. Industries—United States—Forecasting—Congresses. 2.
Economic forecasting—United States—Congresses. 3. United
States—Economic conditions—1981—Congresses. I. Mowery, David
C. II. National Research Council (U.S.). Board on Science,
Technology, and Economic Policy. III. Title: US industry in 2000
  HC106.82 .U17 1999
  338.0973′09′051—dc21   99-6102

Cover: The emblem appearing on the cover of this publication is an illustration of the bronze medallion in the floor of the Great Hall in the National Academy of Sciences building in Washington, D.C. The medallion is the wellhead placed in the floor when the spectroscopic case over which the Foucault pendulum swings is lowered below floor level. The design is based on a map of the solar system published in 1661 by Andreas Cellarius Palatinus. The array of the planets is the Copernican system as know to Galileo.

Printed in the United States of America

The **National Academy of Sciences** is a private, nonprofit, self-perpetuating society of distinguished scholars engaged in scientific and engineering research, dedicated to the furtherance of science and technology and to their use for the general welfare. On the authority of the charter granted to it by Congress in 1863, the Academy has a working mandate that requires it to advise the federal government on scientific and technical matters. Dr. Bruce M. Alberts is president of the National Academy of Sciences.

The **National Academy of Engineering** was established in 1964, under the charter of the National Academy of Sciences, as a parallel organization of outstanding engineers. It is autonomous in its administration and in the selection of members, sharing with the National Academy of Sciences the responsibility for advising the federal government. The National Academy of Engineering also sponsors engineering programs aimed at meeting national needs, encourages education and research, and recognizes the superior achievements of engineers. Dr. William A. Wulf is president of the National Academy of Engineering.

The **Institute of Medicine** was established in 1970 by the National Academy of Sciences to secure the services of eminent members of appropriate professions in the examination of policy matters pertaining to the health of the public. The institute acts under the responsibility given to the National Academy of Sciences by its congressional charter to be an adviser to the federal government and, upon its own initiative, to identify issues of medical care, research, and education. Dr. Kenneth I. Shine is the president of the Institute of Medicine.

The **National Research Council** was organized by the National Academy of Sciences in 1916 to associate the broad community of science and technology with the Academy's purposes of furthering knowledge and advising the federal government. Functioning in accordance with general policies determined by the Academy, the council has become the principal operating agency of both the National Academy of Sciences and the National Academy of Engineering in providing services to the government, the public, and the scientific and engineering communities. The council is administered jointly by both academies and the Institute of Medicine. Dr. Bruce M. Alberts and Dr. William A. Wulf are chairman and vice chairman, respectively, of the National Research Council.

Preface

In 1991 the National Academies of Sciences and Engineering established the Board on Science, Technology, and Economic Policy as a forum in which economists, technologists, scientists, financial and management experts, and policymakers could broaden and deepen understanding of the relationships between science and technology and economic performance. In its first three years, the Board's activities focused on the adequacy and efficiency of public and private domestic investment in physical and human capital. The Board's first report, ***Investing for Productivity and Prosperity***, underscored the need for higher rates of national saving and investment. Its principal recommendation was to shift the base for taxation from income to consumption.

In the past two years, the Board has turned its attention to more microeconomic concerns—technology policies broadly defined and their relationship to international trade relations, determinants of competitive performance in a wide range of manufacturing and service industries, and changes in patterns of R&D and innovation investments. A series of conferences, workshops, and reports, of which this volume is the third, comprises the latter body of STEP work which we are calling, ***U.S. Industry: Restructuring and Renewal***, because it represents a broad assessment of U.S. industrial performance in an international context at a time of domestic economic confidence and optimism but uncertainty about the consequences of fundamental changes in the composition of the economy, processes of innovation, and economic troubles abroad. Previous publications under this title include ***Industrial Research and Innovation Indicators***, the report of a workshop on measurement of industrial research and innovation, and ***Borderline Case: International Tax Policy, Corporate Research and Development and Investment***, a collection of papers by leading tax scholars, practitioners, and policy analysts. A report of the STEP Board's conclusions from the project, *Securing America's Industrial Strength,* is being published simultaneously. This series of activities would not have been possible without the financial support of the National Aeronautics and Space Administration and National Science Foundation and the personal encouragement of Daniel Goldin, NASA Administrator.

This volume was conceived by Ralph Landau, a founding member of the STEP Board. With the exception of the Introduction, the papers included in it were presented at a conference, "America's Industrial Resurgence: Sources and Prospects," held at the National Academy of Sciences in Washington, D.C., on December 8-9, 1997. The authors are members of multidisciplinary research teams studying economic performance and technological change at the industry level—most of them projects sponsored by the Alfred P. Sloan Foundation. The group of industries examined does not represent a carefully selected sample representative of the economy but rather reflects a decision to capitalize on the work of these groups. Notable omissions are the natural resource extraction industries—petroleum and mining—agriculture and forestry, and automobiles. On the other hand, the selection includes three "service" industries—retail banking, trucking, and food retailing—and thus addresses by far the largest and in many ways most dynamic sector of the economy.

To help integrate this work, the Board asked David Mowery, professor at the Haas School of Business at the University of California at Berkeley, to develop a general framework to analyze the determinants of performance over the past 15 or 20 years. The single exception was an analysis of shifts in comparative advantage in the chemical industry over a 150-year period. Following two workshops in which the investigators shared their analysis with other industry experts, the resulting papers were discussed with representatives of the subject industries, interested government officials, and other scholars at the Washington conference. Commentators included Jeff Burke, Seagate Technology; Tim Cyrus, General Motors Powertrain Group; Linda Dworak, Garment Industry Development Corporation; Mike Eskew, United Parcel Service; Kenneth Flamm, The Brookings Institution; F.G. Jauss, U.S. Steel Corporation; Richard Manning, Pfizer, Inc.; Joy Nicholas, Food Marketing Institute; Dan Schutzer, Citicorp; James Sinnett, The Boeing Company; Larry Sumney, Semiconductor Research Corporation; and Ed Wasserman, E.I. du Pont de Nemours & Co. The labor market implications of structural and technological change in these industries were addressed by a panel including Howard Rosen, Minority Chief of Staff of the congressional Joint Economic Committee; Jared Bernstein, Economic Policy Institute; Craig Olson, University of Wisconsin School of Business; and Thomas I. Palley, AFL-CIO. Additional funds for the conference and this publication were provided by the Office of Industrial Technologies, U.S. Department of Energy, the Sloan Foundation, Ralph Landau, and the Lockheed Martin Corporation. The Board is grateful to all of these participants and sponsors, but especially to David Mowery and Ralph Landau.

The industries studied exhibit great diversity in structure and evolution as well as enormous "churning" that has taken a variety of forms, often negative for some regions and workers—employment downsizing, shifts in location of operation, and changes in the skill requirements of many jobs. Nevertheless, the general picture is one of stronger performance in the 1990s than in the early 1980s, attributable to a variety of factors including supportive public policies, competition and openness to innovation, and changes in supplier and customer relationships—factors that might not be as readily apparent if the analysis were of the macroeconomy or at the level of the firm. Vigorous foreign competition forced changes in manufacturing processes, organization, and strategy but then receded, making the performance of U.S. industries look even better. None of these favorable conditions, least of all the latter, is permanent. The studies persuasively make the point that U.S. industries' superior performance in the past decade is not guaranteed to continue. For that reason among others, incidentally, the studies underscore the importance of maintaining independent centers of industry expertise.

This collection has been reviewed in draft form by individuals chosen for their diverse perspectives and technical expertise, in accordance with procedures approved by the NRC's Report Review Committee. The purpose of this independent review is to provide candid and critical comments that will assist the institution in making the published report as sound as possible and to ensure that the report meets institutional standards for objectivity, evidence, and responsiveness to the study charge. The review comments and draft manuscript remain confidential to protect the integrity of the deliberative process.

We wish to thank the following individuals for their participation in the review of this report: Eileen Appelbaum, Economic Policy Institute; Thomas Bailey, Institute on Education and the Economy, Teachers College, Columbia University; France Cordova, University of California at Santa Barbara (review coordinator); Katherine Hughes, Institute on Education and the Economy, Teachers College, Columbia University; Anita McGahan, Graduate School of Business, Harvard University; and Richard Rosenbloom, Graduate School of Business, Harvard University

While the individuals listed above have provided constructive comments and suggestions, responsibility for the final content of this report rests entirely with the authors, the STEP Board, and the institution.

A. MICHAEL SPENCE
Chairman (until June 30, 1998)

DALE W. JORGENSON
Chairman (from July 1, 1998)

STEPHEN A. MERRILL
Executive Director and
Project Director

Contents

U.S. INDUSTRY IN 2000

Studies in Competitive Performance

America's Industrial Resurgence (?): An Overview

DAVID C. MOWERY
University of California, Berkeley

INTRODUCTION

A series of reports in the late 1980s painted a gloomy picture of U.S. industrial competitiveness. Perhaps the best-known, the report of the M.I.T. "Commission on Industrial Productivity," opined that "...American industry is not producing as well as it ought to produce, or as well as it used to produce, or as well as the industries of some other nations have learned to produce...if the trend cannot be reversed, then sooner or later the American standard of living must pay the penalty" (Dertouzos et al., 1989). The Commission report criticized U.S. industry for failing to translate its research prowess into commercial advantage. Since that report's publication, overall U.S. economic performance has improved markedly. Is this improved performance a result of better performance in the industries analyzed by the M.I.T. Commission?[1] What are the dimensions of change in U.S. industrial or economic performance since the early 1980s, at the level of the economy as a whole or at the level of individual sectors?

Economy-wide measures paint a mixed picture of performance improvements and structural change since the early 1980s. The trade deficit has grown and hit a record high of $166 billion in 1997. Nonfarm business labor productivity growth rates have improved since 1990 but remain below the growth rates achieved during the 1945-1980 period. Unemployment and inflation are significantly lower than was true of the 1970s and 1980s. Measures of household income distribu-

---

[1]Much of this improvement in U.S. competitive performance since 1990 is an improvement in performance relative to that of other nations. The severe problems that have hobbled the Japanese economy for much of the 1990s, for example, have weakened the performance of many of the Japanese firms that were among the strongest competitors of U.S. firms during the 1980s.

tion, however, suggest that households in the lowest quintile of the distribution have fared poorly during the past two decades, while the top quintile of the distribution has done well.

Other indicators suggest that the structure of the U.S. research and development system entered a period of significant change beginning in the early 1980s that has yet to run its course. Among other things, industrially financed R&D has grown (in 1992 dollars) by more than 10 percent annually since 1993, but real industrial spending on basic research declined during 1991-1995. Recent growth in industrially financed R&D is dominated by spending on development.

Aggregate performance indicators thus are mixed, although broadly positive. But the relationship between this improved aggregate performance and trends in individual industries, especially those singled out for criticism by the M.I.T. Commission and other studies, remains unclear. A better understanding this relationship requires analysis of trends in these industries.

This volume provides a disaggregated assessment of recent performance in 11 U.S. manufacturing and nonmanufacturing industries.[2] The papers in this volume were commissioned by the National Research Council's Board on Science, Technology, and Economic Policy (STEP) as part of a study of the changing innovative performance of the U.S. economy. Ten of the industry studies were prepared by researchers from investigations of individual industries sponsored during the 1990s by the Alfred P. Sloan Foundation. The Sloan Foundation industry studies, which were in part a response to the reports of the M.I.T. Commission and other groups, examined trends in the performance of individual industries that often are not captured by official statistics. By supporting extensive fieldwork on managerial and competitive challenges faced by industries and firms, the Foundation also sought to change the graduate education of engineering, economics, and management students and thereby influence future research and teaching in U.S. higher education.

Two additional papers on the chemicals industry are included in this volume. The papers were produced by a study overseen by faculty at Stanford and Carnegie Mellon Universities. The first paper, by Ralph Landau and Ashish Arora, is broader in its coverage of international developments and covers a longer time period than other papers in this volume. This paper provides an overview of the many levels at which competitive performance must be assessed and the numerous factors that affect it. Its historical perspective enables the reader to assess the extent to which the factors singled out in other industry studies are likely to influence performance over the long term. The second paper, by Ashish Arora and Alfonso Gambardella, examines more recent trends in the competitive performance of U.S. firms in the international chemicals industry.

The contributors to this volume have each pursued a different approach to

---

[2]The industries are chemicals, pharmaceuticals, semiconductors, computers, computer disk drives, steel, powdered metallurgy, trucking, financial services, food retailing, and apparel.

analyzing "their" industry, reflecting their interests and background as well as the competitive challenges and issues of greatest urgency for each industry. In drafting their papers, contributors were asked to examine several common issues: (1) the record of competitive performance in their industry since 1980, especially vis-à-vis competition from other industrial nations; (2) the influence on this competitive performance of new approaches to managing and organizing the innovation process in their industry; and (3) the influence of "non-technological" factors, including government technology, trade, and regulatory policies.

CHANGES IN COMPETITIVE PERFORMANCE: A SUMMARY

The first and striking conclusion from these papers is the extraordinary diversity in the performance of these eleven industries since 1980. Some, such as the U.S. semiconductor and steel industries, have staged dramatic comebacks from the brink of competitive collapse. Others, including the U.S. computer disk drive and pharmaceutical industries, have successfully weathered stronger foreign competition throughout this period. Foreign competition has been less salient for the nonmanufacturing industries represented in this volume, although domestic deregulation and changing consumer preferences have created a more competitive domestic environment.

The diversity among these industries is partly a reflection of their contrasting structure. Some, such as powdered metallurgy and apparel, are populated by relatively small firms with modest in-house capabilities in conventionally defined R&D. Other industries, such as pharmaceuticals and chemicals, are highly concentrated, with a small number of global firms dominating capital investment and R&D spending. In semiconductors, pharmaceuticals, computer software, and segments of computer hardware, by contrast, one observes a large number of small and large firms that complement one another, often being linked through collaborative R&D relationships. Similar diversity in structure is apparent within the three nonmanufacturing industries. Although entry barriers appear to be high and growing higher in several of the industries discussed in this volume (e.g., chemicals, computer disk drives), in others a combination of technological developments and regulatory change is promoting the entry of new competitors.

Despite this diversity in structure and performance, virtually all of the contributors to this volume argue that performance in "their" industry has improved during the past two decades. The papers use an array of different measures to measure performance, and not all of them are calibrated against the performance of non-U.S. firms in these industries. Nevertheless, the overall portrait is one of stronger performance, not least in the ability of firms to develop and deploy new products and processes. Importantly, where these chapters discuss improvements in innovative performance, they refer to improvements in the deployment, rather than solely the development, of innovations. As many authors point out, firms have strengthened their ability to exploit their own or externally sourced innovations more effectively, rather than focusing exclusively or even primarily on improvements in their research or development capabilities.

The definition of innovation that is most relevant to understanding the improved performance of U.S. firms in these industries thus must be a broad one that includes the adoption and effective deployment of new technology as well as its creation. The chapter on the computer industry refers to the important role of "co-invention," a process in which the users of the products of computer hardware and software firms contribute to the development and improvement of innovations—similar examples can be drawn from other industries. In other industries, specialized suppliers of logistics services, computer "systems integration," and consulting services have played important roles.

Many of the chapters also stress the importance of the efficient adoption of technologies from other industries, nations, or firms. In many cases (e.g., finance, apparel, pharmaceuticals, computers) the adoption of new technologies (including new approaches to managing innovation) has required significant changes in organizational structure, business processes, or workforce organization within the firm. But the essential investments and activities associated with the broad definition of innovation employed here are captured poorly if at all in public R&D statistics. Indeed, many of these activities are not included in even the broader "innovation surveys" undertaken by the National Science Foundation and other public statistical agencies.

The intersectoral flow of technologies, especially information technology, also has contributed to the competitive performance of these industries. The importance of this factor underscores the fallacy of separating "high-technology" from other industries or sectors in this economy. Mature industries in manufacturing, such as apparel, and in nonmanufacturing, such as trucking, have rejuvenated their performance by adopting technologies developed in other industries. The effects of this intersectoral technology flow are most apparent in the nonmanufacturing industries in this volume (trucking, food retailing, and financial services), all of which have undergone fundamental change as a result of adopting advanced information technologies, but there are numerous other examples of this process and its economic importance. Moreover, the management of the adoption process and the effective "absorption" of technology from other sectors are themselves knowledge-intensive activities that often require considerable investment in experimentation, information collection, and analysis.

These chapters raise a related point concerning the interdependence of technologies emerging from different industries in the U.S. economy. U.S. competitive resurgence in industries such as computers and semiconductors relied on the close proximity of U.S. producers and demanding, innovative users in a large domestic market. In addition, the rapid growth of desktop computing in the United States during the 1980s was aided by the availability of imported desktop systems and components, which kept product prices low and propelled adoption of this technology at a faster pace than in most Western European economies or in Japan, where trade restrictions and other policies kept desktop system prices higher. The rapid domestic adoption of desktop computing contributed to the growth of a large packaged software industry that U.S. firms continue to dominate. In other words, the availability of relatively inexpensive complementary technologies supported a process of adoption that spurred further innovation and economic growth. This virtuous circle was further aided by the restructuring of the U.S. telecommunications industry that began in the 1980s. Such restructuring was associated with the entry of numerous providers of specialized and "value-added" services, providing fertile terrain for the rapid growth of firms supplying hardware, software, and services in computer networking. This trend has benefited the U.S. computer industry, the U.S. semiconductor industry, and the domestic users (both manufacturing and nonmanufacturing firms) of products and services produced by both. These and other intersectoral relationships are of critical importance to understanding U.S. economic and innovative performance at the aggregate and industry-specific levels.

The diffusion of information technology, which has made possible the development and delivery of new or improved products and services in many of these industries, appears to be increasing the skill requirements of many jobs that formerly required minimal basic skills. These technologies place much greater demands on the problem-solving, numeracy, and literacy skills of employees in trucking, steel fabrication, banking, and food retailing, to name only a few examples. Many U.S. entry-level and older workers in these industries face serious challenges in adapting to these new skill requirements because of weaknesses in their basic skills rather than in their job-specific training. But these studies point out that the adoption and effective implementation of new technologies places severe demands on the skills of managers and white-collar workers as well. Not only do managers need new skills and an ability to implement far-reaching organizational change, but in industries as diverse as computing or banking, they face pervasive uncertainty about the future course of evolution of technologies and their applications.

As the chapter by Landau and Arora on long-term growth in the chemicals industry points out, nontechnological factors, such as trade and regulatory policy, the environment for capital formation and corporate governance, and macroeconomic policy all play important roles in competitive performance, especially over the long run. The Landau-Arora chapter's analysis of long-term industrial performance focuses on a "matrix" of factors that operate at the level of the institutional and policy environment within which firms operate as well as at the level of the firm. One of the most important of these factors, which affected the entire U.S. economy and rarely figures prominently in sectoral analyses such as those in this volume, is macroeconomic policy. Both monetary and fiscal policy have been less inflationary and less destabilizing during the 1990s than during the 1980s, as the report of the STEP Board on competitive performance points out (National Research Council, 1998). Although we do not yet have a well-developed empirical model of the precise channels through which the macroeconomic environment influences the investment and strategic decisions of managers in the industries examined in this volume, these links appear to be strong and mean that a stable, noninflationary macroeconomic policy is an indispensable component of improved competitive performance.

Another common element that has strengthened the ability of U.S. firms in many industries to regain or maintain their competitive performance, especially in the face of strong foreign competition, is rapid adaptation in corporate strategy and operations. U.S. firms in several of these industries have restructured their internal operations and existing product lines and have developed entirely new product lines, rather than continuing to compete head-to-head with other U.S. or non-U.S. firms in established lines of business. In some cases, efforts by U.S. firms to reposition their products and strategies in the late 1980s and early 1990s were criticized for "hollowing out" these enterprises, transferring capabilities to foreign competitors, and/or abandoning activities that were essential to the maintenance of these capabilities. To a surprising degree, these prophecies of decline have not been borne out. The shift by U.S. computer disk drive manufacturers of much of their production and related technology to offshore sites has not "hollowed out" their competitive capabilities. Nor has the withdrawal of most U.S. semiconductor manufacturers from domestic production of DRAM components severely weakened their manufacturing capabilities in other product lines. In many U.S. industries, the post-1980 restructuring has been associated with entry by new firms (e.g., specialty chemical firms, fabless semiconductor design firms, package express firms, or steel minimills), and in other cases it has been aided by the entry of specialized intermediaries, such as systems integration firms, consultants, logistics firms, or specialized software producers.

Thus, many of the factors cited by the M.I.T. Commission and other studies as detrimental to U.S. competitiveness in the late 1980s, such as the high levels of entry by new firms into industries such as semiconductors or the pressure from capital markets to meet demanding financial performance targets, contributed to this strategic adaptation by many U.S. firms. It is important to note that the results of such restructuring are not always successful. The study of financial services in this volume concludes that much of the merger and acquisition activity in U.S. banking since 1980 has diminished shareholder value rather than increased it. Nevertheless, in many of these industries, such as steel, disk drives, or semiconductors, European and Japanese firms were slower to respond to new competitive forces, often because their domestic financial markets were less unforgiving than those within which U.S. firms operate. This financial environment also has facilitated the high rates of formation of new firms in U.S. industries such as semiconductors and biotechnology.

In other words, factors that during the late 1980s were described as sources of competitive weakness appear to have contributed to the recovery of several of these industries in the 1990s. This perspective, however, leaves at least two issues unresolved. If U.S. firms' restructuring in the 1990s was an important factor in their improved performance, why did such restructuring take so long to begin? And will such restructuring be an occasional or a continuous process in the future? Moreover, the frequency and nature of such rapid structural change have significant implications for worker skills and employment, an important policy issue that has received little attention in most discussions of "industrial resurgence."

CHANGE IN THE STRUCTURE OF THE INNOVATION PROCESS

Since 1980, the structure and management of the innovation process by firms in all eleven of the industries discussed in this volume have changed considerably. The most common changes include (1) increased reliance on external performers of R&D, such as universities, consortia, and government laboratories; (2) greater collaboration with domestic and foreign competitors, as well as customers, in the development of new products and processes; and (3) slower growth or cuts in research spending.

Beginning in the 1980s, a combination of severe competitive pressure, the perception of disappointing returns from their rapidly expanding investments in internal R&D, and a change in federal antitrust policy contributed to the decision by many U.S. firms to "externalize" a portion of their R&D operations. Large corporate research facilities in pioneers of industrial R&D such as General Electric, AT&T, and Du Pont were sharply reduced in size, and a number of alternative arrangements appeared. U.S. firms formed more than 450 collaborative ventures that focused on joint R&D and product development, as reported in their filings with the Department of Justice under the terms of the National Cooperative Research Act (NCRA), between 1985 and 1994 (Link, 1996). Collaboration has become a much more important part of the innovation process in industries as diverse as semiconductors and food retailing since the early 1980s.

U.S. firms also entered into numerous collaborative ventures with foreign firms during the 1980-1994 period. The majority of these international alliances for which the National Science Foundation has data link U.S. and Western European firms (National Science Board, 1998). Alliances between U.S. and Japanese firms also were widespread. Nevertheless, the formation of "intranational" alliances linking U.S. firms with domestic competitors has outstripped the formation of international alliances, according to National Science Foundation data (National Science Board, 1998). Both intranational and international alliances involving U.S. firms appear to be most numerous in biotechnology and information technology. In contrast to most domestic research consortia, a large proportion of U.S. firms' alliances with foreign firms focused on joint development, manufacture, or marketing of products. In addition to the cost-sharing and technology-access motives that also underpinned the formation of many domestic research joint ventures, the international alliances of U.S. firms have been motivated by concerns over access to foreign markets (Mowery, 1988).

U.S. firms in many of the industries examined in this volume reacted to intensified competitive pressure and/or declining competitive performance by reducing their investments in research. Interestingly, these reductions appear to have accelerated during the period of competitive recovery and significant growth in overall R&D spending. During 1991-1995, total spending on basic research declined at an average rate of almost 1 percent per year in constant dollars. This decline reflected reductions in industry-funded basic research from almost $7.4 billion in 1991 to $6.2 billion in 1995 (in 1992 dollars); real federal spending on basic research increased slightly during this period, from $15.5 to almost $15.7 billion. Industry-funded investments in applied research scarcely grew during this period, while federal spending on applied research declined at an annual rate of nearly 4 percent. In other words, the upturn in real R&D spending that has resulted from more rapid growth in industry-funded R&D investment is almost entirely attributable to increased spending by U.S. industry on development rather than research.[3]

Universities' share of total U.S. R&D performance grew from 7.4 percent in 1960 to nearly 16 percent in 1995, and universities accounted for more than 61 percent of the basic research performed within the United States in 1995 (National Science Foundation, 1998). By 1995, federal funds accounted for 60 percent of total university research, and industry's contribution had tripled to 7 percent of university research. The increased importance of industry in funding university research is reflected in the formation during the 1980s of more than 500 research institutes at U.S. universities seeking to support research on issues of direct interest to industry (Cohen et al., 1994). Nearly 45 percent of these institutes involve one to five firms as members, and more than 46 percent of them rely on government funds for support in addition to support from industry.

The passage of the Bayh-Dole Act in 1980[4] triggered considerable growth in university patent licensing and "technology transfer" offices. The Association of University Technology Managers (AUTM) reports that the number of universities with technology licensing and transfer offices increased from 25 in 1980 to 200 in 1990, and licensing revenues of the AUTM universities increased from $183 million to $318 million in the three years from 1991 to 1994 alone (Cohen et al., 1997). U.S. universities increased their patenting per R&D dollar during a period in which overall patenting per R&D dollar was declining significantly.[5]

---

[3]The National Science Foundation reports that industry-funded real spending on "development" grew by more than 14 percent during 1991-1995, from $65 billion to $74.2 billion. Federal development spending declined during this period, reflecting the cutbacks in defense-related R&D spending.

[4]The Bayh-Dole Patent and Trademark Amendments Act of 1980 clarified and rationalized federal policy governing the patenting and licensing of the results of federally funded research performed by small businesses, universities, and other nonprofit institutions. The act generally allowed these performers to file for patents on the results of such research and to grant licenses for these patents, including exclusive licenses, to other parties.

Another important shift in the structure of the innovation process within U.S. industry during this period is the increased presence of non-U.S. firms in the domestic U.S. R&D system. Investment by U.S. firms in offshore R&D (measured as a share of total industry-finance R&D spending) grew modestly during 1980-1995, from 10.4 percent in 1980 to 12.0 percent in 1995. Nevertheless, this flat or slightly declining trend obscures significant intersectoral differences.

The share of industrial R&D performed within the United States that was financed from foreign sources also grew during this period, from 3.4 percent in 1980 to more than 11.0 percent in 1995. Despite this growth, as of 1993 foreign sources financed a smaller share of industrial R&D performed within the United States than is true of Canada, the United Kingdom, or France. Increased foreign financing of R&D activities in the United States was paralleled by an increase in the share of U.S. patents granted to foreign inventors, from 40.4 percent in 1981 to 47.5 percent in 1989 and 45.9 percent in 1993 (National Science Board, 1998). Foreign firms also participated in the formation of research joint ventures with U.S. firms. According to Link (1996), 32 percent of the research joint venture filings under the terms of the National Cooperative Research Act during 1985-1994 listed foreign firms among their members. Finally, a number of foreign firms operating R&D facilities in the United States pursued collaboration with U.S. universities. More than 50 percent of the Japanese R&D laboratories in the United States, more than 80 percent of the U.S.-sited French R&D laboratories, and almost 75 percent of German corporate R&D laboratories in the United States were involved in such collaborative agreements, according to Florida (1997).

This structural change in the U.S. R&D system is transforming the innovation process in many of the industries reviewed in this volume, giving rise to a very different structure from that which prevailed for much of the postwar period in U.S. industry. In industries such as semiconductors or computers, complex networks of firms and relationships among domestic and foreign firms now play a more important role in developing new products. The importance of "co-invention" in the computer industry has given rise to close collaboration between users and producers of hardware and software. In other industries, such as steel or powdered metallurgy, collaboration with customers has expanded, while the large corporate R&D establishments of integrated steel firms have been drastically reduced. The diversity of institutional actors and relationships in the industrial innovation process has increased considerably, even as the investments by U.S. firms in R&D now appear to focus on shorter time horizons.

The restructured innovation process that has contributed to the resurgence of many of the industries reviewed in this volume emphasizes rapid development,

---

[5]The ratio of patents to R&D spending within universities almost doubled during 1975-1990 (from 57 patents per $1 billion in constant-dollar R&D spending in 1975 to 96 in 1990), while the same indicator for all U.S. patenting displayed a sharp decline (decreasing from 780 in 1975 to 429 in 1990), according to Henderson et al. (1994).

adoption, or deployment of technologies, while placing less weight on the development of the long-term scientific understanding that underpins future generations of these technologies. This shift has produced high private returns, but its long-term consequences are uncertain. We discuss some of the policy implications of this shift in the next section.

The discussion in these papers of the changing structure of the innovation process also highlights the difficulty of collecting and analyzing data that enable managers and policymakers to assess innovative performance or structural change. As noted earlier, many of the activities contributing to innovation in these industries are not captured by conventional definitions of "R&D." They include investments in human resources and training, the hiring of consultants or specialized providers of technology-intensive services, and the reorganization of business processes. All of these activities have contributed to the innovative performance of many of the industries examined in this volume. Indeed, the importance of information technology for innovation in many of these industries means that far-reaching organizational changes and investments in numerous complementary activities are essential to successful technology adoption.

POLICY ISSUES AND IMPLICATIONS

The primary objective of the project that produced these papers was improved public understanding of industry-level changes in competitive performance and the factors that have contributed to them, rather than the development of policy recommendations. Nevertheless, the papers in this volume raise a number of issues for public policy. They include (1) the ability of public statistical data to accurately measure the structure and performance of the innovation process in U.S. industry; (2) the level and sources of investment in long-term R&D within the U.S. economy; (3) the role of federal regulatory, technology, trade, and broader economic policies in these industries' changing performance; (4) the importance and contributions of sector-specific technology policies to industry performance; and (5) the worker adjustment issues posed by structural and technological change.

The data currently published by the National Science Foundation (NSF) provide little information on the changes in the structure of the industrial innovation process that were described in the previous section. The NSF R&D investment data, for example, do not shed much light on the importance or content of the activities and investments that are essential to the intersectoral flow and adoption of information technology-based innovations in many of the industries discussed in this volume. Indeed, the NSF and other public economic data do a poor job of tracking the process of technology adoption throughout the U.S. economy, despite the importance of this process for innovative and competitive performance. Moreover, in many of the nonmanufacturing industries that are essential to the development and diffusion of information technology innovations, "R&D investment" per se is difficult to distinguish from operating, marketing, or materials expenses. For example, these data do not consistently capture the R&D inputs provided by specialized firms to "low-technology" users such as trucking and food retailing firms. The public statistical data on innovative activity that are widely used by scholars, managers, and policymakers thus omit important activities that contribute to innovation. Moreover, their coverage of even conventional R&D-related activities in many of the firms and sectors contributing to innovation in the U.S. economy appears to be imperfect.

Over time, therefore, without substantial change in the content and coverage of statistical data collection, our portrait of innovative activity in the U.S. economy is likely to become less and less accurate. These problems were the subject of another workshop sponsored by the STEP Board as part of its overall assessment of the changing U.S. R&D system, and the report on that workshop contains a more detailed discussion of policy issues and options (Cooper and Merrill, 1997).

As I noted earlier, improvements in the competitive performance of many of the industries examined in this volume have occurred in the face of reductions in industry-funded investments in long-term R&D. The changing time horizon of industry-funded R&D investment raises complex issues for policy. Specifically, how if at all should public R&D investments seek to maintain a balance within the U.S. economy between long- and short-term R&D? Many of the studies in this volume argue for closer public-private R&D partnerships, involving industrial firms, universities, and public laboratories. Yet most recent partnerships of this sort have tended to favor near-term, rather than fundamental, R&D investment. This issue remains an important one for policy, and there are few models of successful partnership in long-term R&D that apply across all industries.

A second issue concerns the treatment of the results of publicly funded R&D in the context of such partnerships. A series of federal statutes, including the Stevenson-Wydler Act of 1980, the Bayh-Dole Act, the Technology Transfer Act of 1986, and others have made it much easier for federal laboratories and universities to patent the results of federally funded research and license these patents to industrial R&D partners. Proponents of licensing argue that the establishment of clearer ownership to the intellectual property resulting from federal R&D will facilitate its commercial application. Patenting per se need not restrict the dissemination of the results of publicly funded R&D, but restrictive or exclusive licensing agreements may do so. As the paper on the U.S. pharmaceuticals industry points out, the "open science" performed in U.S. universities, much of which was funded by the National Institutes of Health during the postwar period, has aided this industry's innovative performance. If new federal policies limit the dissemination of the results of this research, however, the long-term competitive performance of the U.S. pharmaceuticals industry could be impaired. Similar issues appear in other industries. The simultaneous growth in industrial reliance on university and publicly funded R&D for long-term research and the increased resort by universities and federal laboratories to patenting and licensing of the results of this research create complex dilemmas that have received too little attention thus far from industry and government officials.

Although the papers in this volume do not yield a single rank-ordering of the public policies that have been most important to the improved performance of these industries since 1980, federal intellectual property, antitrust, trade, and regulatory policies have affected the competitive resurgence of a number of these industries. They have been most effective where the combined effects of these and other policies have supported high levels of domestic competition and have maintained open U.S. markets to foreign exports and investment. Vigilance must be maintained to ensure that revisions in policy in the intellectual property, trade, and antitrust areas do not inadvertently protect firms from competitive pressure. For example, relatively liberal policies toward inward foreign investment allowed U.S. firms to benefit from close observation of the management practices of foreign-owned production establishments in semiconductors, steel, and automobiles, transferring important management and human resources "technologies" to U.S. firms. In addition, as was noted above, the availability and low prices of computer technologies that foreign imports provided to U.S. consumers through much of the 1980s and 1990s sparked the growth of new applications and new segments of established industries. The restructuring and deregulation of sectors such as telecommunications, trucking, and financial services also has intensified competitive pressure on U.S. firms in these and other industries to improve their performance.

The record of technology policy in these industry studies is less clear. The studies suggest that the most effective technology policies involve stable public investment over long periods of time in "extramural" (i.e., nongovernmental) R&D infrastructure that relies on competition among research performers. U.S. research universities are especially important components of this domestic R&D infrastructure. Their importance reflects their role in research and training, as well as the competitive, decentralized structure of this nation's research university system. In some cases, as in federal support for biomedical research through the National Institutes of Health, these investments in long-term research have had major sectoral effects. But the effects of sector-specific technology support policies, such as the defense-related programs of support for disk drive technologies, or even SEMATECH, appear to be more modest in the small number of industries for which they are relevant. Their lack of dramatic effect reflects the tendency for such policies to be episodic or unstable, the relatively small sums of public funds invested in them, and the extremely complex channels through which any effects of such policies are realized. In light of the importance of federal R&D infrastructure investment, changes in the future structure and size of the federal R&D budget, as well as the policies covering the dissemination of the results of this research (see above), bear close scrutiny.

Finally, the effects of industrial restructuring, technology development and adoption, and competitive resurgence on U.S. workers, especially low-skilled workers, in the manufacturing and nonmanufacturing industries examined in this volume merit attention. As noted earlier, the effects of technology adoption and development continue to raise the skill requirements for entry-level and "shop-floor" employment in these industries, including those in the nonmanufacturing sector. In addition, the very agility of U.S. enterprises that has contributed to improvements in their competitive performance since the 1980s imposes a heavy burden on workers for adjustment. Moreover, the perception that such adjustment burdens are unequitably distributed can have significant political effects, revealed most recently in the 1997 Congressional defeat of "fast-track" legislation to support continued trade liberalization. The United States and most other industrial economies lack policies that can improve the ability of workers to adjust to economic dislocation and compete effectively for more remunerative opportunities without increasing labor market rigidity. The political and social consequences of continuing failure by policymakers to attend to these adjustment issues nevertheless could be serious, and the issue merits public scrutiny and debate.

CONCLUSION

The resurgence of U.S. industry during the 1990s is as welcome as it was unexpected, based on the diagnoses and prescriptions of many reports in the 1980s. Indeed, this recovery was well under way in a number of industries at the very time that the M.I.T. Commission report presented its critique. Moreover, in at least some of the key industries identified by the M.I.T. and other studies as competitively threatened, the factors singled out as sources of weakness in the 1980s appear to have become sources of competitive strength in the 1990s. After all, the competitive resurgence of many if not most of the industries discussed in this volume reflects superiority in product innovation, market repositioning, and responsiveness to changing markets rather than dramatic improvements in manufacturing performance per se. The improvements in manufacturing that have occurred in industries such as steel or semiconductors have been necessary conditions for competitive resurgence, but they were not sufficient.

This argument raises a broader issue of particular importance for policymakers. Particularly when the imperfect nature of the data on innovative performance and processes is taken into account, observers of industrial competitiveness must accept the reality that performance indicators have a very low "signal to noise" ratio—i.e., data are unavailable, unreliable, and often do not highlight the most important trends. Uncertainty is pervasive for managers in industry and for policymakers in the public sector. Government policies designed to address factors identified as crucial to a particular performance problem may prove to be ineffective or even counterproductive, when and if the information on the existence or causes of the problem turns out to be inaccurate. This difficulty is partly due to deficiencies in the data available to policymakers, and improvements in the collection and analysis of these data are essential. But in a dynamic, enormous economy such as that of the United States, these data inevitably will provide an imperfect portrait of trends, causes, and effects. In other words, policy must take into account the importance and pervasiveness of uncertainty. Ideally, policies should be adaptive to long-term trends rather than attempting to meet short-run problems that may or may not be correctly identified in the available data.

If many U.S. industries have in fact enjoyed a competitive resurgence in the 1990s, is this state of grace sustainable or likely to prove permanent? As pointed out in the introduction to this chapter, a portion of the improved performance of many of these U.S. industries reflects significant deterioration in Japan's domestic economy. Recovery in Japan's domestic economy may take time, but it will eventually result in an improved business outlook for many of the firms that were effective foreign competitors of U.S. firms during the 1980s.

Even allowing for the uncertainties that are inherent in any attempt to predict the future, it seems unlikely that U.S. firms have achieved a permanent competitive advantage over their counterparts in other industrial and industrializing economies. The sources of U.S. industrial resurgence are located in ideas, innovations, and practices that can be imitated and improved upon by other enterprises at some cost and investment of technical effort. Global competition in the late twentieth and twenty-first centuries will depend more and more on intellectual and human assets that are relatively mobile across international boundaries. The competitive advantages flowing from any single innovation or technological advance are likely to be more fleeting than in the past; economic change and restructuring are essential complements of a competitive industrial structure.

Nevertheless, some relatively immobile assets within the U.S. economy will continue to aid competitive and innovative performance. The first is the sheer scale of the U.S. domestic market, which, even in the face of impending monetary unification in the European Union, remains the largest high-income region that is so deeply economically unified in markets for goods, capital, technology, and labor. Combined with other factors, such as high levels of new firm formation and entry in many industries, this large market provides a "testbed" for the many economic experiments that are necessary in the development and commercialization of complex new technologies. Faced with pervasive uncertainty, neither managers nor government personnel are able to predict the future with accuracy. An effective method to reduce uncertainty through learning is to run economic experiments, exploring many different approaches to innovation in uncertain markets and technologies. Over the course of the post World War II period, the U.S. economy has provided a very effective venue for these experiments, and the growth of new, high-technology industries has benefited from the tolerance for experimentation and failure that this large market provides.

A second important factor in the process of experimentation that is indispensable to the development of new technologies is an effective domestic mechanism for generating such experiments. Here, the postwar U.S. economy also has proven to be remarkably effective. Success has been influenced by large-scale federal funding of R&D in universities and industry as well as a policy structure, including the financial and corporate-governance systems, intellectual property rights, and competition policies that support the generation of ideas as well as attempts at their commercialization and supply the trained scientists and engineers to undertake such efforts.

Both of these assets are longer lived and far less internationally mobile than the ideas or innovations they generate. They contribute to high levels of economic and structural change that are beneficial to the economy overall while imposing the costs of employment dislocation or displacement on some groups and individuals.

The current environment of intensified international and domestic competition and innovation is a legacy of an extraordinary policy success in the postwar period for which the United States and other industrial-economy governments should claim credit. Trade liberalization, economic reconstruction, and economic development have reduced the importance of immobile assets, such as natural resources, in determining competitive advantage. These developments have lifted tens of millions of people from poverty during the past 50 years and are unambiguously positive for economic welfare and global political stability. Nevertheless, these successes mean that competitive challenges and perhaps recurrent "crises" in U.S. industrial performance will be staples of political discussion and debate for years to come. This economy needs robust policies to support economic adjustment and a world-class R&D infrastructure for the indefinite future.

REFERENCES

Cohen, W., R. Florida, and R. Goe. (1994). "University-Industry Research Centers in the United States," technical report, Center for Economic Development, Carnegie Mellon University.

Cohen, W., R. Florida, L. Randazzese, and J. Walsh (1997). "Industry and the Academy: Uneasy Partners in the Cause of Technological Advance," in *Challenges to Research Universities*, R. Noll, ed., Washington, DC: Brookings Institution.

Cooper, R.S., and S. Merrill, eds. (1997). *Industrial Research and Innovation Indicators: Report of a Workshop*, Washington, DC: National Academy Press.

Dertouzos, M., R. Lester, and R. Solow. (1989). *Made in America: The Report of the MIT Commission on Industrial Productivity*, Cambridge, MA: MIT Press.

Florida, R. (1997). "The Globalization of R&D: Results of a Survey of Foreign-Affiliated R&D Laboratories in the United States," *Research Policy* 26:85-103.

Henderson, R., A.B. Jaffe, and M. Trajtenberg. (1994). "Numbers Up, Quality Down? Trends in University Patenting, 1965-1992," presented at the CEPR conference on "University Goals, Institutional Mechanisms, and the 'Industrial Transferability' of Research," Stanford University, March 18-20, 1994.

Link, A.N. (1996). "Research Joint Ventures: Patterns from *Federal Register* Filings," *Review of Industrial Organization*.

Mowery, D.C. (1988). *International Collaborative Ventures in U.S. Manufacturing,* Cambridge, MA: Ballinger.

National Science Board. (1998). *Science and Engineering Indicators: 1998,* Washington, DC: National Science Foundation.

Science, Technology, and Economic Policy Board, National Research Council. (1999). *Securing America's Industrial Strength*, Washington, DC: National Academy Press.

The Dynamics of Long-Term Growth: Gaining and Losing Advantage in the Chemical Industry[1]

RALPH LANDAU
Stanford University
ASHISH ARORA
Carnegie Mellon University

What factors support the long-term growth of industrial societies? This chapter examines the innovative and competitive performance of the chemicals industry over the past 150 years. It draws on a recent book analyzing factors that contribute to the flourishing or failure of companies in the industry in the United States, the United Kingdom, Germany, and Japan—the four countries with the most extensive available data and case histories (Arora et al., 1998). It links this history with the external factors, the "climate" in which these companies operated during different periods and probes the interrelationship. By focusing on the "long run," we hope to bring out some of the underlying factors, such as change in the institutional landscape, that influence industrial and economic performance but that are not easily captured in analyses covering shorter time periods. This analysis of the long-run factors supporting industrial growth and competitiveness complements the discussion of short-term trends in performance in the other chapters of this volume.

The chapter relies on an analytical framework developed elsewhere, a framework that highlights the multiple sources of comparative advantage (Landau et al., 1996) (Figure 1). As Krugman (1996) notes, countries do not compete, firms do. Nonetheless, countries can establish a more or less favorable climate for their firms to compete, helping them to gain comparative advantage for their industry and producing benefits for their home country.

---

[1]The authors are grateful for the invaluable assistance of Johann Peter Murmann and for continuing advice from Nathan Rosenberg and Paul Romer.

National Governance
Socio-Political Climate
Macro Policies
    Fiscal
    Monetary
    Trade
    Tax
Institutional Setting
    Financial
    Legal (including torts, antitrust, and intellectual property)
    Corporate governance
    Professional bodies
    Intermediating institutions
Structural and Supportive Policies
    Education (including university-industry relations)
    Labor
    Science and technology (including the role of engineers and scientists)
    Regulatory and environmental
The Industry Collectively
Companies Within the Industry

FIGURE 1 Levels of sources of comparative advantage.

We conclude that a well-functioning and growing economy depends on

• a complex mix of institutions and policies that extend beyond the legal system and fiscal and monetary authorities to include entities such as national systems of higher education, industry-specific and economy-wide regulation, and trade policy;
• market-based policies that support the interaction of social institutions and policies to generate higher economic growth within a relatively stable macroeconomic environment but avoid unwarranted intervention; and
• the size of the market, the historical development, and the political and social environment of the country in question.

Our long-term view highlights the central importance of technological innovation for the growth of the chemical industry and for industrial societies as a whole. But like other chapters in this collection, we stress that a narrow definition of "technological innovation" is inadequate for this analysis. Instead, technological innovation must be defined to include the broader constellation of risk-taking activities that commercialize the technology and underlying science. These activities are influenced by social institutions and policies.

SOME PERSPECTIVES ON GROWTH

The conventional neoclassical views of the causes of growth neglect other aspects of economic structure, policy, and society that affect the growth or decline in comparative advantage. Most neoclassical models of economic growth assumed a world of perfect competition, in which the only institutions needed to achieve good economic performance are a legal system, a specification of property rights, and an antitrust authority that can prevent the emergence of monopoly power. Experience has taught economists to add to this list government institutions that seek to ensure a stable monetary policy and avoid macroeconomic disruption.

"Growth accounting" studies inspired by neoclassical models have consistently found a significant unexplained residual, labeled "technology" or, more accurately, multifactor productivity. The factors contributing to this residual encompass a substantially broader and more diverse list than those included in narrow definitions of technological change—indeed, they encompass many of the factors included in Figure 1. Moses Abramovitz (1956) coined the words "social capability" to describe the complex of institutions and policies embedded in these levels. The factors influencing economic growth in this conceptual framework are more numerous and complex than the parsimonious list associated with the neoclassical model and its applications in growth accounting.[2]

Newer work on endogenous growth theory has introduced concepts to models of economic growth that fit reality more closely (Romer, 1994, for example). This theory recognizes that the underlying assumptions of neoclassical economists, such as perfect competition, emphasized the central role of new technologies and simultaneously denied the possibility that economic analysis could have anything to say about the processes that affect their creation, improvement, or adoption. These neoclassical models allowed little scope or significance for invention or innovation, learning by doing, technology transfer from abroad, or systematic research and development, all of which can produce new and improved products, processes, and services and greatly enhance the growth process. Although endogenous growth theory has recognized many of these important factors, neither it nor its neoclassical predecessor takes into account historical effects, such as path dependence (Arora et al., 1998).

A richer model of long-term economic growth requires examination of how commercialization of technology actually takes place at the firm level and an understanding of the forces external to the firm that influence that commercialization. Internal factors that are well known from the business literature include management recruiting, research and development, and manufacturing and marketing and need no further detailing here. But external factors also influence the evolution of management strategies, and the nature and channels of this influence

---

[2]See, for example, Lau (1996).

remain poorly understood. The technology of a firm depends in part on the performance of institutions of learning, as well as scientific and engineering research conducted by various public and private institutions. A firm's investment strategies depend on the cost and availability of capital, the division of profits between the owners of the company and the other stakeholders, and the efficient functioning of the labor market. Capital supply in turn depends on the functioning of the external, and now largely international, capital markets, the intermediating institutions such as banks that allocate capital from savers to investors, and the competition for capital by other firms and governments.

Government policy is essential in several areas. Government tax policies affect the net returns to investors for the employment of their capital, which in turn guides future investment. Government budgetary and monetary policies affect national welfare and aggregate domestic demand as well as domestic savings and the cost of capital. Governments set trade policies. Governments must maintain a legal order so that firms know what they and their competitors can and cannot do. Governments provide for much of the education of the labor force and promulgate a variety of regulations to control many aspects of the economy.

These factors are individually complex, and their interrelationships and interactions, intended and otherwise, are even more so. We cannot hope to provide a definitive description of these individual factors, let alone their interaction, for such a lengthy period in four large industrial economies. The historical and analytic discussion that follows instead should alert economic theorists, policymakers, and managers to the complexity of the factors and forces that support long-term growth. This discussion also should give pause to those who proclaim the arrival of a "new paradigm" or the onset of an indefinite period of U.S. economic dominance in chemicals or other industries. An exclusive focus on the near term in such analyses will result in myopic conclusions and prescriptions.

THE CHEMICAL INDUSTRY

The history of the chemical industry offers clear illustrations of the interdependence between the strategies of individual firms and the environment of economic policy and institutions that their home-economy governments create. Three characteristics illustrate the economic and technological significance of the chemical industry as well as its long history.

First, chemicals was the first science-based, high-technology industry. Moreover, with the exception of this century's two world wars, this industry's research and development has been financed almost entirely by private investment. Figure 2 gives the most recent data available in this form.

An estimation of 1997 expenditures is given in Figure 3, which also shows federal funding of R&D by industry for 1996. It is evident that R&D in the chemicals industry, which, along with transportation equipment, is one of the two largest R&D performers in the U.S. economy, is virtually all privately financed.

INDUSTRY	1991	1992	PRIVATELY FINANCED	PERCENTAGE PRIVATELY FINANCED
1. Aerospace	16.63	16.12	6.25	39%
2. Electrical Machinery & Communications	13.42	13.55	9.69	72%
3. Machinery	14.78	15.14	14.07	93%
4. Chemicals	14.65	16.71	16.42	98%
5. Autos, Trucks, Transportation	10.80	10.37	9.48	91%
6. Professional & Scientific Instruments	8.71	9.65	7.43	77%
7. Computer Software & Services	5.77	6.66	3.89	58%
8. Petroleum	2.50	2.34	2.33	99%
TOTAL	87.26	90.54	69.56	

Note: Total R&D in 1992 was $154.5 billion, of which R&D performed by industry was $107.6 billion so that the above are the bulk of R&D performers. Battelle estimates these figures for 1995 at $182 billion and $130.6 billion.

Source: National Science Foundation, Division of Science Resources Studies, "Selected Data on Research and Development in Industry: 1992" and "National Paterns of R&D Resources."

FIGURE 2 The Major R&D Industries, 1991 & 1992 R&D Expenditures (billions of current dollars)

INDUSTRY	ESTIMATED 1997 R&D (billions)
1. Chemicals and Pharmaceuticals	31.4
2. Transportation	30.4
3. Telecommunications	29.0
4. Computers	22.5
5. Electronics	15.2
6. Software	9.9
7. Semiconductors	6.8
TOTAL	145.2

Note: Total R&D for 1997 is estimated by Battelle at $192 billion, of which 62.8 percent ($120.6 billion) will be financed by industry, 32.4 percent ($62.2 billion) by government, and 4.9 percent ($9.4 billion) by others (such as non-profits, universities, research institutions). Numbers have been rounded and may not add to 100.

Source: "1997 R&D Funding Forecast" by Battelle.

Transportation Equipment	52%
Professional & Scientific Instruments	17%
Electric Equipment	9%
Non-manufacturers	15%
Other Manufacturers	7%
	100%

Source: National Science Foundation.

FIGURE 3 The Major R&D Industries for 1997 (top) (billions of current dollars) and Federal Funding of R&D by Industry for 1996 (bottom).

Second, the chemicals industry has generated technological innovations for other industries, such as automobiles, rubber, textiles, consumer products, agriculture, petroleum refining, pulp and paper, health services, construction, publishing, entertainment, and metals. In this regard, the chemical industry illustrates the general tendency for the benefits of internationally competitive industries to spill over to other industries.

Third, the chemicals industry is a U.S. success story. The chemicals industry is one of only two major high-technology industries (aerospace being the other) in which the United States has maintained its competitive lead in international trade. Its growth rate has exceeded that of the overall U.S. economy since World War II.

The output of the modern chemicals industry conveys some sense of the diversity of activity. It includes paints and coatings, pharmaceuticals, soaps and detergents, perfumes and cosmetics, fertilizers, pesticides, herbicides and other agricultural chemicals, solvents, packaging materials, composites, plastics, synthetic fibers and rubbers, dyestuffs, inks, photographic supplies, explosives, antifreeze, and many other kinds of chemicals—more than 70,000 products. It is the leading U.S. export industry, with a long-term favorable balance of trade. Very few industries have the complexity of the chemical industry. The enormous size of this industry in 1996 is shown in Tables 1 and 2 (Figures 4, 5, and 6 contain other comparative data on the U.S. chemicals industry and other U.S. high-technology industries). In Europe, it is second only to the food, drink, and tobacco industries in size and value added and has a consistently positive balance of trade.

TABLE 1 GDP by Industry (1996)

U.S. Manufacturing Sector	$1332 billion
(17.4% of total GDP)	$7636 billion
Chemicals and allied products	$157.8
Industrial machines and equipment	150.2
Electronic and electric equipment	143.8
Food and kindred products	122.6
Fabricated metal products	98.2
Printing and publishing	90.4
Motor vehicles and parts	85.1
Paper products	57.1
Instruments	52.3
Other transportation	49.7
Petroleum and coal	30.1
Other	294.9
Total	$1332.2

Source: U.S. Bureau of Economic Analysis, Survey of Current Business, November 1997.

TABLE 2 The World Chemical Industry in 1996

	Sales volume	
	$ billion	% of total
Western Europe (includes EFTA)	445[a]	28
EFTA =	(36)	(2)
United States	372	24
Japan	216	14
All others	533	34
Total	1566	100

[a] Of which German sales value is about $117 billion and the U.K. is $56 billion.

The most important class of chemicals is the organic compounds, which are much more varied and pervasive than the inorganic compounds (e.g., derived from salt and minerals). Organic inputs like oil and natural gas contain hydro-carbons, which form the backbone of final organic chemical outputs. In the first stage of processing, chemicals such as chlorine and oxygen are added to the hydrocarbon backbones to give the compounds certain desired characteristics. The final output may be nylon or polyester fiber, plastic, a pharmaceutical product, or other products that are rarely considered to be chemical industry outputs.

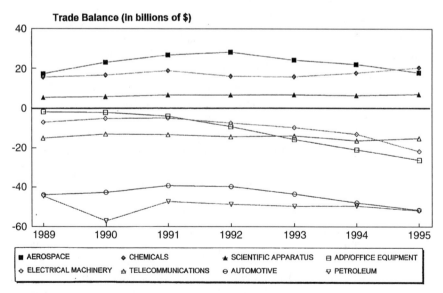

FIGURE 4 U.S. Industries with heavy R&D.
Source: U.S. Bureau of Census.

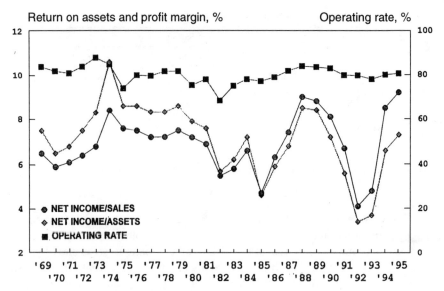

Return on assets and profit margin, %   Operating rate, %

- ● NET INCOME/SALES
- ◆ NET INCOME/ASSETS
- ■ OPERATING RATE

FIGURE 5 Return on assets and profit margin (%) and operating rate (%).
Source: Chemical Manufacturers Association.

A BRIEF HISTORY OF DYNAMIC COMPARATIVE ADVANTAGE IN THE CHEMICAL INDUSTRY

To understand the development of comparative advantage in the chemical industry, it is useful to summarize the essential historical facts before offering a more detailed analysis. England already dominated inorganic chemicals when William Henry Perkin discovered the first synthetic dye (mauve) in 1856 and launched the modern organic chemical industry. England in the mid-1800s was wealthy; it had the know-how, the largest customer base (textiles), and the largest supply of raw material (coal). But the chemical industry let its advantages slip away, and by the end of the 1880s the Germans dominated the organic chemical industry. By 1913 German companies produced 140,000 tons of dyes, Switzerland produced 10,000 tons, and Britain produced only 4,400 tons. The American industry depended mainly on German dyestuff and other chemical imports, although it was a large producer of basic inorganic chemicals.

World War I brought a change in the relative position of the four countries. The United States built its own organic chemical industry, and the German industry fell on hard times. With the tacit support of the German government, their competitive difficulties contributed to the merger of the leading German chemical companies to form the I.G. Farben company. Britain and the United States took advantage of the military defeat of Germany, refusing to give back prewar patents to German firms. Further, by sanctioning the merger that created Impe-

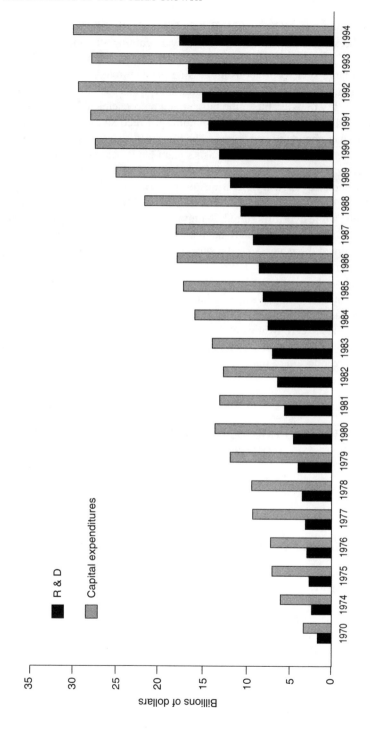

FIGURE 6 Chemical industry R&D versus capital expenditures.
Source: Chemical Manufacturers Association.

rial Chemical Industries (ICI) in 1926, Britain avoided falling further behind Germany. At the same time, the United States was gaining strength through the development of a large petroleum refining industry and was creating new skills in the design of large-scale continuous processing plants through the use of chemical engineering. These skills, largely in the hands of specialized engineering firms, were readily transferable to the burgeoning petrochemical industry, which was based on the cheap petroleum and natural gas feed stocks with which the United States was abundantly endowed. The European chemical industry continued to use coal, rather than petroleum, as its main feedstock through the 1940s.

World War II resulted in the physical destruction of a significant portion of the German chemical industry. During the postwar period, the U.S. industry developed uses for petrochemicals in the production of fibers, plastics, and many other products, while dyestuffs shrank in importance. America's chemical industry grew enormously and dominated the market at least until the 1970s. As world prosperity returned, however, so did a successful chemical industry in Germany and in Europe more generally. Petrochemical industries were soon well-established in Asia, in the oil-exporting countries, and elsewhere. No longer did one country dominate; the industry's growth had made it a truly global industry. Competitive advantage at the firm level came to the fore, with different companies in different countries excelling at what they did best. Japan was the one exception. Although the Japanese chemical industry grew to become the second largest in the world, it never became a major player in international markets for products or technology.

TRACING THE DEVELOPMENT OF COMPARATIVE ADVANTAGE THROUGH THE MATRIX

We now discuss briefly how each level of the matrix in Figure 1 has affected comparative advantage and growth in the chemical industry over the last 150 years.

National Governance and Socio-Political Climate

How do factors related to national political and social factors help to explain the shift in comparative advantage from Britain to Germany after the 1870s until 1914? To begin with, the national governmental structures were very different in the two countries. Britain had a parliamentary system of government; Germany was a collection of 39 political entities that had a customs union but otherwise differed widely in their governmental structures and policies. The competition among the various states contributed to the rise of many German dyestuff companies. Germany's political unification in 1871, under Chancellor Otto von Bismarck, a Prussian, not only created a common market and an investment boom but also produced a foundation for a unified patent system that proved to be very important.

Much has been written about the reluctance of British investors to undertake the higher-risk organic chemical investments that their counterparts in Germany did. Britain had many opportunities for low-risk investments throughout its own empire, the United States, and South America. Germany in contrast had no empire and, even with the creation of a larger domestic market, had a limited market. German senior industrial managers, supported by their investors who were receiving rich dividends, accordingly took bigger risks, including investments in scientific and technological developments of the second (electrical and chemical) industrial revolution.

Political and social factors also influenced the relative positions of the Japanese and U.S. chemical companies, particularly during and after the two world wars. In Japan, the feudal regime was replaced in 1868 by the Meiji restoration, yielding a somewhat more democratic society. But the Japanese military's political influence expanded in the early twentieth century, partly as a result of victories in wars with Korea and Russia, and the military assumed control of Japan's government by 1931 with the invasion of Manchuria.

The two world wars led to major governmental changes in Germany and Japan and influenced the subsequent direction of their chemical industries. The war also led to changes in the U.S. chemical industry. The two world wars, however, drained Britain of much of its economic strength and, as the history of ICI, its largest firm shows, had an influence on its chemical industry. The rise of consumerism and the substitution of natural materials as a result of the deprivations of the war in the West led to the creation of new products and an enlarged demand for the newly developed plastics and other synthetic materials, some of which had been discovered in the interwar years. These developments allowed the chemical industry to grow for many decades much faster than gross domestic product.

Macroeconomic Factors: Monetary and Fiscal Policies

Prevailing government and sociopolitical conditions profoundly influence macroeconmic policies, which have affected the growth pattern of our four national chemical industries in important ways. Macroeconomic policies during the period before the World War I favored British capital exports. Most major industrialized countries sooner or later adhered to the gold standard before World War I. Britain had control of much of the world's gold supply and therefore was able to maintain clear leadership in the international flow of capital. British investors preferred low-risk foreign investment opportunities to the riskier domestic investment options offered by the nascent chemical industry. German investment overseas was constrained by these British policies, and so Germany was forced to export goods instead of capital.

In the nineteenth century, Britain imposed the first modern income tax, which, while low, nevertheless provided a flexible source of government revenues to sustain the costs of its empire. Thus, the average British taxpayer supported the very low-risk investments in the empire, an advantage not available to the Germans.[3]

After its defeat in World War I, Germany faced serious economic problems resulting from the reparations imposed by the Versailles Treaty and a postwar recession. To cope with the effects of the war reparations payments, the German government printed large quantities of money, which produced runaway inflation. The resulting economic turbulence discouraged new investment and contributed to massive unemployment, the political crises that led to Hitler's Nazi government, and eventually war.

Macroeconomic policies in all of these nations changed after World War II. In Germany, Ludwig Erhard's free-market economic policies created the *"Wirtschaftswunder"* (economic miracle), characterized by minimal government interference with the private sector. By contrast, Great Britain after losing its empire pursued the creation of a welfare state that led to price controls, inflationary policies, frequent labor unrest, and recurrent currency crises. Productivity growth in Britain lagged behind that of Germany during most of the post-1945 period, reducing demand growth for chemicals. The Japanese government systematically established its chemical industry through a series of government plans and decrees, developing a petrochemical industry that grew rapidly from the mid-1950s onward.

Another area of difference in national macroeconomic policies was the management of exchange rates in the postwar era.[4] Undervaluation of the yen and the mark during the 1950s and 1960s helped Japanese and German exports and the recovery of both nations from wartime destruction. The "hard dollar" in the early 1980s had a significant and unfavorable effect on U.S. exports; the hard yen of the later 1980s had a similar effect on Japanese exports. In both Japan and the United States, however, overvalued exchange rates intensified competitive pressures on domestic exporters to improve their innovative performance and efficiency.

Macroeconomic Factors: Trade Policies

Between 1879 and 1882 Germany imposed higher tariffs on heavy chemicals, which hurt the British soda exporters and helped to enhance the competitive position of the German soda industry. The German organic chemical industry,

---

[3]A much fuller discussion of more recent developments in tax policy and their effects on industrial investment may be found in Jorgenson and Landau (1993).

[4]A discussion of the exchange-rate policies of leading governments during the interwar period, along with some consideration of the effects of these policies on the chemicals industries of the United States, Japan, Great Britain, and Germany, may be found in Arora et al. (1998).

however, succeeded in keeping dyes tariff free, reflecting the dominant position of German firms in the global dyestuffs trade. Throughout Britain adhered to a free trade policy. The American inorganic chemical industry flourished during this period but only after the government raised tariff barriers, which further damaged British exports of these products.

In response to the World War I, the United States and Great Britain enacted strong tariff protection for their infant organic chemical industries. Subsequent increases in tariffs throughout the world dramatically reduced Germany's world market share in this industry. Japan was essentially a closed economy and devoted an increasing share of national investment to military preparations after the military seized power in the 1930s.

As the dominant political and economic power in the West after 1945, the United States was able to reverse the trend of the trade barriers and cartels of the interwar years. A number of trade agreements were launched under the leadership of the United States including the formation of the General Agreement on Tariffs and Trade, the Kennedy Round, the Uruguay Round, and finally the World Trade Organization. Step by step other barriers have been lowered in the postwar years, and trade as a whole has stimulated the growth of many economies and industries, including the chemical industry. Worldwide trade in the postwar era has grown about sevenfold, outstripping a quadrupling of overall gross domestic product within the global economy during this period.

Institutional Setting: Legal Institutions

A profound difference arose at the end of the nineteenth century between the legal systems of the United States and the European countries. Whereas the United States passed the Sherman Antitrust Act of 1890 and the Clayton Antitrust Act of 1914 to discourage trust formation, Germany in 1897 declared cartels to be legal, making it possible for German firms to eliminate competition among themselves in many branches of the chemical industry. Germany and Britain participated in a large number of international cartels that led to fixed prices and market shares in the most important segments of the world chemical industry during the interwar period. The cartelization of the interwar German chemicals industry in I.G. Farben, which was dominated in many aspects by BASF, contributed to the ultimately unsuccessful policy of developing high-pressure synthetic fuels technologies. In the United States, by contrast, brisk competition among domestic chemicals and petroleum firms led to the development of the petrochemical industry and to the rapid growth of a number of companies that in 1920 were still quite small, such as Dow.[5] Some U.S. firms entered into technical exchange

---

[5]Still another important legal influence on corporate innovation and performance, especially in the postwar U.S. chemicals industry, is product liability. High levels of liability litigation and costly court judgments may have discouraged the introduction of some new products by U.S. chemicals firms, especially pharmaceutical products.

agreements with European firms. Japan had no antitrust policy until after World War II.

Germany had no national patent system until 1877. In a certain sense the newly rising entrepreneurial chemical firms, the most prominent being BASF, Bayer, and Hoechst, were lucky because the absence of a unified patent system allowed them to copy with impunity the technologies developed abroad. Following political unification, when chemical science and the German chemical industry had advanced sufficiently to make R&D investment lucrative if its results were protected by patents, German chemical companies joined in the lobbying campaign for the creation of a unified domestic patent system. Although Britain had a unified patent system, the excessive breadth of patents and the limited scientific understanding of the field of organic dyestuffs weakened British organic chemicals firms, which endured long-lasting patent conflicts. The United States also had a patent system, but by the late 1800s its inorganic chemical industry had reached a level of maturity where patents were not nearly so important. The patent system became important in the American chemical industry only after the organic chemical industry began to develop around the time of World War I.

As mentioned earlier, German chemical firms lost their patents in Britain and the United States as a result of both wars, which put them at a significant disadvantage in these two important markets. Access to these German patents eventually enabled British and American firms to enter markets that German firms previously had dominated. Patents are still very important in pharmaceuticals and the newer field of biotechnology.

Institutional Settings: Financial Institutions

One difference across the four countries that was already apparent before World War I concerns the relationship of chemical companies to their domestic financial systems. Britain at that time had the most advanced financial services industry in the world, located in London, but young firms in the chemical industry had great trouble raising money because the British bank system largely took a hands-off attitude toward its clients' welfare. Firms had to prove profitability before they could qualify for loans, which was always difficult in a risky science-based industry. The Germans, however, had a strong investment bank system and very soon developed a relationship banking system whereby the principal banks took a direct interest in assisting companies to which they lent money. Banks not only helped companies manage their affairs but frequently took ownership interest in them. In this sense, the German industrial banks resembled the American venture capitalists of the present day. Indeed, the largest of the nineteenth century German chemical companies, BASF, from the beginning had its investment bankers from Ladenburg & Sons on its board and listed among its shareholders. Rapid expansion of the German chemical companies required large amounts of finance, even though the companies rapidly became very profitable, and their relationships with large banks were indispensable.

The American financial system was dominated by firms such as J.P. Morgan, with extensive experience in railroad financing; but Morgan's investment banking efforts were centered on the big basic industries such as oil and steel. The Mellon group in Pittsburgh likewise helped finance many growing companies, such as U.S. Steel, Pittsburgh Plate Glass, the Koppers Company, and Alcoa, but devoted little attention to the nascent chemicals sector. Other U.S. banks followed much the same policy as the British financial institutions.

In Japan after the Meji restoration of 1868, large industrial holding companies (*zaibutsu*) appeared around large banks. Their attentions, however, were focused primarily on military requirements as Japan entered into a series of wars including the conquest of Korea and the Russo-Japanese War early in the twentieth century. Chemicals, with some exceptions such as fertilizers and explosives, were not an important part of this development.

During the interwar period, the financial markets in Germany remained much the same as they had before World War I, with close bank-industry relationships. The American companies had much greater difficulty in growing rapidly because of the hands-off banking style, reinforced by the passage of the Glass-Steagall Act; but the size and flexibility of the American financial system as a whole produced a greater reliance by U.S. chemicals firms on equity finance than was true of their counterparts in Western Europe or Japan for much of the postwar period. After World War II Japan's *zaibatsu* were converted to *keiretsu*, but the banking relationships persisted and cross stockholding among the groups continued.

Institutional Setting: Corporate Governance

Issues of corporate governance did not arise in any of these countries, although the U.S. capital markets put greater pressure on American firms to pay out profits. Gottfried Plúmpe (1990) shows that the large American firms achieved consistently higher returns on investment than did the British and German firms during the transwar period. Since World War II, their relationship banking system has essentially insulated German firms from shareholder pressure. In the 1990s, however, corporate governance has become an important issue within German management and in the financial industry, as globalization of German industry has proceeded rapidly and capital markets have become internationalized. We return to this issue later.

Structural and Supportive Policies: Environmental Regulation

The chemicals industry raised environmental concerns in Germany, the United States, and Great Britain at an early stage. Nevertheless, until relatively recently industry and jobs were so important that environmental concerns had little effect on the operation of these firms. Regulatory and environmental considerations became much more important after World War II, even though the problem had been perceived a hundred years earlier. The publication of Rachel Carson's *Silent Spring* in 1962 galvanized a rapidly growing environmental movement throughout the free world, which spurred adoption of accompanying regulation. These regulations have imposed substantial costs on the manufacturers of chemicals in many industrial countries, especially Germany, which enforces its regulatory laws vigorously. Many European chemicals firms devote as much as 15 percent of average annual capital spending to environmental remediation, a cost that has led some firms to move operations out of the industrialized countries of Europe into some of the newly industrializing countries of Asia, where regulatory policies are less stringent.

Structural and Supportive Policies: Labor

The Communist movement that ignited in 1848 produced tense labor relations in Germany and Britain. The Bismarck government passed strict laws to control the labor unions, which kept an adequate labor peace. But the government also pioneered in creating a social safety net for workers, and many German chemical companies erected company housing and provided other benefits to their work force. The Communist movement in Britain touched off a long history of adversarial labor relations. Management and owners largely controlled the House of Commons and were able to politically subordinate labor until well into the twentieth century.

Owners and managers also dominated U.S. labor policy during this period, and the American Federation of Labor did not wield much power until after World War I. During the interwar period, regulatory and environmental policies were subordinated to the greater needs of a faltering economy in the four countries of our study, and the urgency of job formation also modulated the demands of labor. Only in the late 1930s were strong unions formed in the United States.

Following World War II, Germany established a domestic "social contract" among employers, the government, and unions that led the country along a high investment, low labor conflict path; Germany had labor peace and high taxes. In Britain, despite the arrival of the "fair shares for all" economics, labor conflicts and more general adversarial relations between management and unions hurt the introduction of efficient production methods until the Thatcher period in the 1980s. In the United States strikes and other labor problems were relatively minor during the postwar era; as the power of labor unions diminished, their political strength weakened. The same is true in Japan where labor negotiations were almost pro forma and were dictated primarily by industry. After a very short period of violent labor strife, management and labor in Japan developed life-time employment in the large companies as an effective peace-agreement that has lasted to the present day, despite growing problems brought about by the recession of the 1990s.

Structural and Supportive Policies: Education, Science, and Technology

The rise of the German chemical industry in the nineteenth century was supported by the most advanced research capability in organic chemistry at its leading universities. Justus von Liebig established the first academic research institute for chemicals at the university in Giessen in the late 1820s. Liebig's approach closely associated systematic research with teaching. Both before and after political unification, the governments of the various German states invested heavily in universities to increase their intellectual capital. The young German chemical companies maintained close contacts between their own chemical researchers and the universities, which enabled the German firms to develop the new technologies faster and more fully than did their British and American counterparts. The American research university followed much later and did not materially affect American innovation until virtually the end of the nineteenth century. Japan had no important institutions of science and learning during this period.

Both the German state and its firms and the U.S. private foundations and firms supported science and technology research earlier and to a much greater degree than did Britain. Few English universities focused on chemistry and technology. The tradition in Britain was set by the Oxford-Cambridge model of education, which emphasized the classics and prepared students for careers in the church, the diplomatic service, the armed forces, and the government.

In the United States, in contrast, the rise of the research university in science and engineering gave a strong boost to the American chemical industry. Many chemistry teachers in the United States had undertaken their graduate studies in Germany, and they helped build advanced research and training capabilities in many American universities in the early part of the twentieth century. Unlike either Germany or Britain, an American tradition of engineering also arose that proved beneficial because of the large, homogeneous domestic market of the United States, which favored the extensive application of large-scale mass production technologies.

The United States was the first nation to develop a system of manufactures by utilizing standardized parts. As early as 1851 in the London Crystal Palace Exhibition, American exhibitors displayed firearms made by these methods that were much lower in cost than those shown by other countries. In 1862 the United States passed the Morrill Act, which created the land grant colleges for training students in agricultural, engineering, and other mechanic arts. The Massachusetts Institute of Technology (M.I.T.), the first technologically based land grant college, was established in 1865. It remained entirely an undergraduate engineering school until the beginning of the twentieth century, a pattern that was typical elsewhere in the United States. Britain had some engineering curricula but lacked one specifically adapted to the chemical industry. In nineteenth century Germany, engineers were treated like second-class citizens. Engineering was taught not in universities but in technical institutes (*technische hochschulen*), which were not allowed to grant advanced degrees until the early part of the twentieth century. Consequently, German chemists dominated the German chemical companies in this period and used engineers, primarily mechanical engineers, to help them design plants, relying on fairly direct scale-up from laboratory apparatus.

This U.S. engineering tradition in the early part of this century led directly to the development of the chemical engineering profession, which came to fruition in 1920 when M.I.T. established the first independent department of chemical engineering and developed a new engineering model called unit operations (Walker et al., 1923). At about the same time, as a consequence of their close inspection of the German chemical industry after World War I, American universities became much more concerned with improving chemistry education. Britain, Germany, and Japan were much later in developing the chemical engineering profession, and so the United States was able to gain an important lead in designing and operating large, continuous process chemical plants.

Another important factor in the rise of chemical engineering in the United States was the growth in consumer demand for low-cost gasoline to fuel the automobiles that more and more people were buying. U.S. chemicals and petroleum firms were compelled to develop new techniques for refining unprecedented quantities of petroleum. The oil companies had anticipated the growth in automobiles. The first major petroleum industry R&D organization, Esso Research and Engineering, was established by Standard Oil of New Jersey in 1919. ESSO worked closely with M.I.T. faculty to establish chemical engineering as a genuine intellectual field at the university level, and many institutions adopted it. The pioneers in the creation of petrochemicals were two oil companies, Standard Oil of New Jersey (now Exxon) and Shell, and two chemical companies, Union Carbide and Dow (Spitz, 1988).

At the same time the German universities, even technical universities, began to withdraw from close contact with industry. Although BASF had developed the Haber-Bosch process to produce synthetic ammonia in 1913, a remarkable feat of chemical engineering, this accomplishment was never shared with the educational institutions, and as a result no chemical engineering discipline arose in Germany before World War II. The absence of a strong engineering tradition in Germany contributed to a lag when the new petrochemical era dawned after the war. Japan's first chemical engineering department was established in 1940; but it had little influence for several years and remains less advanced than Japanese electrical engineering, materials science, and other fields.

A major post-war change in university and government relationships took place in the United States, following the publication of Vannevar Bush's *Science: The Endless Frontier* in 1945. For the first time the federal government became the main source of undergraduate scholarships through the GI Bill and granted fellowships and research grants to faculty in the sciences. Compared with the United States and Britain, German spending on universities has declined since the war, particularly in the last quarter century. The German states support higher education, but the relationship between universities and industry has cooled. Although German chemical firms appear to get all the chemists they need in Germany, many of the brightest students now go to the United States to receive their postdoctoral training, and German universities have lost their leading position in chemistry to American universities.

The Industry and Firms

As noted earlier, Britain dominated inorganic chemicals by the mid-1800s. The British industry, with many firms competing intensely, was efficient and able to hold down prices. The incumbent firms, however, failed to make investments in the newer alkali technologies, such as the Solvay process, because of thin profit margins and their reluctance to scrap investments in older processes. The rapidly growing U.S. domestic market gave American firms the incentive to build large and efficient inorganic chemical plants, eliminating the competitive advantage of foreign firms in this segment. The German firms also had more efficient technologies than did the British and also freed themselves from British imports in the late nineteenth century. The organic chemical industry, however, showed a completely different pattern, as German dyestuff companies proved to be extraordinarily innovative and very profitable and soon dominated world markets. In the late 1880s the Germans drew on their dyestuff technology to enter the pharmaceuticals industry and became strong players in this area well before the United States and Great Britain. Japan had no strength in chemicals in this period.

Three major firms—BASF, Bayer, and Hoechst—soon dominated the German industry. Companies like BASF, which was the largest in the world in the last part of the nineteenth century, were enormously profitable, encouraging even further growth. These German firms maintained close relationships with universities and proceeded in developing a new corporate function—the R&D laboratory. After the passage of the unified patent law in 1877, German firms further cemented their lead in organic chemicals by organizing systematic, large-scale efforts to create new chemical products. In contrast to British firms, German chemical firms developed strong marketing capabilities, which aided their penetration of export markets. Neither Britain nor the United States replicated this industrialization of innovation until after World War I, giving German firms a strong comparative advantage. Britain's slower growth rate in the chemical industry was to a large extent the result of its inability to compete against the constantly innovating German firms.

The economic difficulties of the interwar years, such as protectionism and macroeconomic turbulence, triggered a wave of mergers that created large chemical firms in Germany (I.G. Farben) and Britain (ICI) and several large firms in the United States. The American chemical firms had made large profits during World War I, which enabled them to diversify and invest in R&D and acquisitions. Du Pont, for example, purchased a 25 percent interest in General Motors that subsequently gave the chemical company substantial profits both from dividends and the ultimate sale of that stock under an antitrust decree after World War II. Union Carbide was formed by such mergers, as was the Allied Chemical Company. Japan still had very little new activity in chemicals, except in fertilizers.

British and American companies profited from the tariff protection they had sought after World War I. The technical excellence and the long traditions of the German companies, however, permitted I.G. Farben to continue extensive research and investment in new fields such as polymers, even though they were still based on coal and would prove to be ultimately noncompetitive. Despite I.G. Farben's virtual monopoly of the domestic market, innovation in the German chemicals industry did not die out, in part because of the need for substantial exports to make up for the weak domestic market.

Dow Chemical Company, started in 1920, grew rapidly during the interwar period as did Union Carbide. Du Pont invested in the research and development that produced nylon, its most profitable polymer discovery, and its numerous acquisitions greatly strengthened the company. I.G. Farben was strong in several chemical sectors but under Carl Bosch, who had been head of BASF before World War I, I.G. Farben pursued an ambitious synthetic-fuels development program that consumed enormous amounts of capital. Dyestuffs and pharmaceuticals provided most of the profits of I.G. Farben until the Nazi government came to power. From that point on, the firm focused more and more on the creation of synthetic fuel from coal as a part of Hitler's autarchy policies, and I.G. Farben was no longer able to dictate its own policies as it had before (Plúmpe, 1990).

After World War II, the big three German chemical companies, profiting from their long tradition and the favorable circumstances created by the German government, grew very rapidly to become the three largest chemical companies in the world today. Only in the last few years has it become obvious that some parts of their businesses are unprofitable, and the first steps are being taken toward divestiture and acquisitions in order to produce a better overall profit picture. The entry of the major U.S. oil companies, including Exxon, Amoco, and Mobil, created stiff competition in the basic commodity. Britain's two oil companies, British Petroleum and Shell, also entered the petrochemical industry after World War II.

The Recent Influence of Financial Markets on Corporate Governance

The conditions prevailing in Britain in the early 1990s created intense stockholder pressure on ICI, which split at the beginning of 1993 into the pharmaceuticals firm Zeneca, and ICI, which retained the traditional chemicals products.

The American firms, especially Du Pont and Dow Chemical, remain among the leaders in the global chemicals industry. After a serious plant explosion in India in 1984, Union Carbide was the object of a hostile takeover that resulted in a significant diminution of its presence in the chemical industry. Likewise, Allied Chemical, which in 1920 was the largest chemical company in the world, has virtually exited the industry as a result of short-sighted policies in the interwar years when its managers paid high dividends while virtually neglecting R&D. Aside from the new startup companies that have been responsible for the first commercial biotechnology products, established pharmaceutical companies in the United States continue to dominate the industry because of their strong marketing capabilities, whereas in Europe the conventional chemical companies are still very large in pharmaceuticals. In fact, the influence of the shareholder value philosophy can be seen in the consistently higher profitability, return on assets, and market capitalization of the major American chemical companies compared with their European competitors, where management control has persisted until recently.

The strong Japanese chemicals industry has not been influential in international markets. The reasons are not only the absence of shareholder control but also an intrusive government bureaucracy, protected markets, and capital controls in the earlier decades; a weak domestic university system; and the fierce rivalries of the many *keiretsu* groups, which led to too many similar plants of less than world scale. In a society traditionally resistant to change, the international weaknesses of Japan's chemical firms also reflect the nation's historic lack of participation in international markets.

The historical and expected future performance of these large chemicals firms, like industrial firms in other sectors, is reflected in their market valuation, especially for those firms whose shares are widely held and traded in liquid, efficient equities markets. This valuation also reflects investor expectations of management's commitment to increase shareholder value, and international comparisons of the relationship between sales and market capitalization reflect national differences in the power of "mass shareholder" (as opposed to management, "main bank," or *keiretsu* shareholder) influence on management.

The comparisons in Figure 7 illustrate how changes in corporate governance have changed the strategies of the major chemical companies, particularly in Europe.[6] In a well-rounded chemical company, sales volume should approximately equal the firm's total market capitalization. As the figure shows, however, the ratio of sales to market capitalization differs widely among five leading chemicals firms. The figure shows the market value of ICI before its 1993 spinoff of Zeneca as well as the combined market value of the two firms since 1993. The market obviously values the new versions of ICI much more favorably than it did

---

[6]The Appendix to this chapter compares the relationship between sales and market valuation for a number of U.S. firms in various industries.

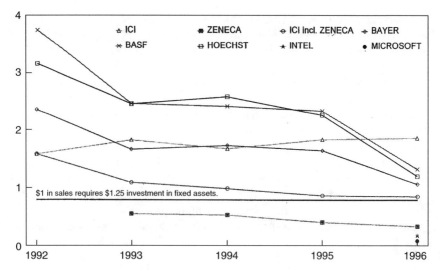

FIGURE 7 Sales/market capitalization.
Note: Sales for Microsoft are estimated based on results for the six months ended 12/31/96.
Source: Chemical Manufacturers Association.

the original unified ICI. Zeneca has joined the ranks of high-technology companies with a high price-earnings multiple, whereas the new ICI has drifted toward
a lower level that is comparable to some of the German firms today. The new ICI
is less internally balanced than the German companies, a factor that contributed
to ICI's 1997 purchase of the specialty chemical businesses of Unilever and its
divestiture of remaining commodity businesses. What the markets, which have
valued Zeneca favorably, will think of the remainder of the old ICI that is focused
on chemicals remains to be seen.

In contrast the stock market was treating two of the three German firms with
caution, because their total sales exceed their capitalization by almost 2:1. Bayer
is exceptional in this regard, because its heavy emphasis on health care and pharmaceuticals has led the market to value the company's prospects somewhat more
favorably than either Hoechst or BASF. Other factors are involved, however.
Pharmaceutical companies have higher earnings per dollar of sales revenue, and
therefore price/earnings ratios will be higher despite the lower sales to market
capitalization ratios. German chemicals firms now are attempting to increase
shareholder value amid discussion of methods to realize the underlying values of
the different businesses by various devices that will not add heavy tax burdens.

Corporate governance issues have now become a major factor in establishing
firm strategies. In Japan, where the trend toward mergers of rather small companies by international standards has been very slow to develop, these issues are
gathering steam. Mitsubishi has finally succeeded in uniting its two chemical
firms, and Mitsui has announced a comparable move. Such rationalization is

proceeding at a much more rapid pace in the European industry, where there has been a great deal of divestiture and acquisition of divisions and businesses and the formation of many joint ventures or alliances. The industry is in the midst of a major shakeup but the traditional trajectories of the major companies, influenced by their corporate capabilities and history, will continue to exert a dominant influence over their strategies. Nevertheless, these strategies differ from firm to firm, and the recent wave of mergers, alliances, and restructuring reflects the varying ways managements are striving to increase their profitability and shareholder values.

CONCLUSIONS

This historical survey of the chemicals industry yields conclusions that largely complement, rather than conflict with, the findings of other chapters in this volume. Competitive strength in the long run, as in the two or at most three decades covered in the other chapters, rests on a robust institutional infrastructure and supportive government policies that are general, rather than highly sector specific, in their target and intent. A stable, predictable, macroeconomic environment and a pro-investment policy lead to higher long-term growth. The case of the chemical industry over 150 years suggests that high-technology industries develop better in a country when they can draw on strong national research and teaching universities in science and engineering. For this industry at least targeted government science and technology policies do not matter very much compared with the constellation of institutions and policies incorporated in the matrix.

Although abundant natural resources may help a domestic industry get started, they do not afford a lasting lead. In a peaceful world where natural resources can be shipped all around the world, know-how and economies of scale are decisive factors in maintaining competitive advantages. But the climate maintained by the national government institutions and policies is probably of equal importance. Despite the widespread diffusion of technology and capital, national interests are not always the same, and the constellation of these policies and institutions contributes to or detracts from comparative advantage and growth. Thus, for example, the cost of capital in recent decades has differed in the four countries considered here, with Japan's low cost contributing to its investment boom—and subsequent bust.

The development of a large and sophisticated home demand in the beginning of an industry life cycle has several advantages. It allows more than one national firm to develop competitive skills and to build large-scale plants that give them a cost advantage over producers with smaller plants. Of course, readily accessible foreign markets can make up at least in part for a relatively small home market.

Competition among firms has generally led to higher levels of innovation in the chemical industry. A high-technology industry seems to be most innovative when there is just enough competition to spur the creation or improvement of products and processes and yet allow firms to make sufficient profits to provide the ability and incentives to invest in R&D.

In an industry with increasing returns to scale and no tariff protection, early entry into the business is an important factor for long-term competitive success. Many of the firms that became early leaders in the chemical industry continue to dominate it today, despite a dramatic proliferation of new products and processes. Expansion outside the home country to newly developing countries serves to provide competitive advantages to large profitable companies in the industrialized nations but is not enough to maintain dominance without continuing developments in technology.

As capital markets have been internationalized, corporate governance issues have become a prime concern to shareholders and managements. Thus, technical specialists, who have historically dominated the senior management of most chemical firms, may give way to more finance and business managers. Nevertheless, technology remains a prime driving force for gaining and retaining competitive advantage in this industry.

REFERENCES

Abramovitz, M. (1956). "Resource and Output Trends in the United States since 1870." *American Economic Review* 46(2):5-23.

Arora, A., R. Landau, and N. Rosenberg, eds. (1998). *Chemicals and Long-Term Growth: Insights from the Chemical Industry.* New York: John Wiley & Sons.

Bush, V. (1945). *Science, The Endless Frontier.* New Hampshire: Ayer.

Carson, R. (1962). *Silent Spring.* Greenwich, CT: Fawcet Crest.

Jorgenson, D. and R. Landau, eds. (1993). *Tax Reform and the Cost of Capital.* Washington, DC: The Brookings Institution.

Krugman, P. (1996). "A Country is not a Company." *Harvard Business Review* 74(1): 40-51.

Landau, R., T. Taylor, and G. Wright. (1996). *The Mosaic of Economic Growth*, Stanford: Stanford University Press.

Lau, L. (1996). "The Sources of Long-Term Economic Growth: Observations from the Experience of Developed and Developing Countries." In *The Mosaic of Economic Growth*, R. Landau, T. Taylor, and G. Wright, eds. Stanford: Stanford University Press, 63-91.

Plümpe, G. (1990). *Die I.G. Farbenindustrie AG: Wirtschaft, Technik und Politik 1904-1945.* Berlin: Duncker & Humblot.

Romer, P. (1994). "The Origins of Endogenous Growth." *Journal of Economic Perspectives* 8(1):20.

Spitz, P. (1988). *Petrochemicals.* New York: John Wiley & Sons.

Walker W., W. Lewis, and W. McAdams. (1923). *Principles of Chemical Engineering.* New York: McGraw-Hill.

APPENDIX

The analysis of market capitalization and sales for selected companies, together with industry averages provides insights into corporate performance in other U.S. industries. A preliminary assessment reveals similar interfirm and interindustry differences in investors' assessments of the prospects for future growth. The data discussed in this appendix are taken from the *Business Week* issue of March 30, 1998, and are shown in Appendix Table 1. Complete financial analysis is, of course, more sophisticated than this simplified approach, but these results are still useful. As mentioned, in some industries such as pharmaceuticals, profits are higher per dollar of sales than in others. Although the stock markets have risen substantially since March 1998, the relative positions have been maintained and this may continue to be the case when the markets decline. Of course, the wave of mergers and acquisitions will affect the relative positions of some of these companies.

What can be deduced from these data? One must bear in mind the observations of Albert D. Richards in our volume on *Chemicals and Long-Term Economic Growth: Insights from the Chemical Industry*—markets do a pretty good job of forecasting the future of companies.

1. Some chemical companies, such as DuPont and Monsanto, are seen as not having reached maturity, despite their large size. In both cases, these favorable valuations reflect the firms' increased emphasis on R&D and product development in the life sciences. By contrast, Dow is more heavily committed to basic commodity chemicals and is seen as average and relatively mature, while Union Carbide's prospects are not as brilliant.

2. Among pharmaceutical companies, which clearly are the favorites of investors for the reasons stated earlier, Merck and Pfizer shine. Among the other firms in this industry, particular note should be taken of Amgen, the most successful of the biotechnology companies. Its successes are seen as giving it the potential to rank eventually with the major pharmaceutical companies, but it is at an earlier stage of development.

3. Investors see good prospects in the financial services industry and are watching smaller, well-managed banks, such as Bank of New York.

4. In the steel industry, the decline of the former colossus USX-US Steel illustrates the diminishing outlook for a mature company in an industry where technological innovation is modest and the focus is on cost cutting. The newer minimill Nucor comes out as much better in the eyes of investors. Alcoa is an example of a large metals company that continues to innovate and manage well.

5. The extraordinary records of Microsoft and Intel are noted in Figure 7. Nonetheless, the more traditional computer companies such as IBM and Hewlett-Packard still find much favor in investors' eyes. The improved valuation of IBM is a particularly dramatic example of the effects of a management shakeup.

6. The retailing industry presents a wholly different picture; the large groups are unable to improve their market capitalizations greatly, with the exception of specialty retailers, such as the Gap.

APPENDIX TABLE 1 Capitalization Ratios

	Ratio	Capitalization $ billion Feb. 27, 1998	Sales $ Billion 1997
Chemical companies			
DuPont	1.51	69,373	45,079
Monsanto	4.03	30,317	7,514
Dow	1.04	20,748	20,018
UCC	0.88	5,722	6,502
Industry average	**1.62**	**12,667**	**7,801**
Pharmaceutical companies (health care)			
Merck	6.48	153,374	23,670
Pfizer	9.15	114,453	12,504
Bristol-Meyers-Squibb	5.96	99,615	16,701
Johnson & Johnson	4.46	100,813	22,629
American Home Products	4.29	60,869	14,196
Amgen	5.82	13,983	2,401
Industry average	**3.76**	**27,539**	**7,329**
Financial services: Banks			
Citicorp	1.98	60,062	30,300
Chase Manhattan	1.97	52,231	27,365
Bank America	2.50	53,324	21,318
Bank of New York	4.28	21,955	5,124
J.P. Morgan	1.71	21,070	12,353
Industry average	**2.73**	**20,559**	**7,533**
Financial Services: Non-banks			
Morgan-Dean Witter	1.53	41,518	27,132
Merrill-Lynch	1.69	63,696	37,609
Industry average	**1.74**	**16,378**	**9,394**
Metals and Mining			
USX-U.S. Steel	0.44	3,026	6,871
Nucor	1.08	4,527	4,185
Alcoa	0.95	12,655	13,319
Industry average	0.91	3,436	3,782
Computers and Software and Office Software			
Microsoft	15.67	205,265	13,098
IBM	1.29	101,532	78,508
Hewlett-Packard	1.57	69,750	44,416
Intel	5.85	146,730	25,070
Industry average	2.63	23,311	8,875
Retailing			
Walmart	0.88	104,015	117,958
Gap	2.71	17,659	6,508
Federated Dept.	0.63	9,835	15,608
Sears Roebuck	0.50	20,776	41,469
Industry average	**0.80**	**14,573**	**18,182**
Aerospace			
Boeing	1.15	52,810	45,800
Industry average	**0.99**	**22,134**	**22,359**
Transportation			
FedEx	1.86	7,335	12,571
Ryder Systems	0.56	2,865	4,894
Industry average	**1.00**	**9,137**	**9,097**
Miscellaneous			
GE	2.80	254,455	90,840
Exxon	1.29	157,201	122,089

7. Aerospace, the other major manufacturing industry with a significant positive balance of payments in the United States, is led by Boeing whose performance is comparable to some of the large chemical companies that continue to innovate.

8. The transportation stocks are too few in number to be very meaningful, but, with the exception of Fedex, they fall into an average that suggests no great promise for growth.

9. A few exceptional companies are shown for purposes of comparison. The case of GE is especially interesting. A relatively old, highly diversified manufacturing company, GE has become the largest company on Wall Street in terms of its market capitalization, with a capitalization/sales ratio that is remarkable for such a giant. Exxon, another large capitalization company, is more normal, perhaps, but still shines by comparison with some of the others in the table.

The overall conclusion from this table is quite clear. Investors are rewarding those companies and industries that they perceive to have a technological and managerial capability for growth. This table does not present foreign companies because the data are more difficult to locate, but there is little doubt that the same general trends prevail there too. The influence of the financial markets is spreading, and so is shareholder value. As capital becomes more and more global, this trend is inevitable.

Chemicals[1]

ASHISH ARORA
Carnegie Mellon University
ALFONSO GAMBARDELLA
University of Urbino

Unlike several of the other manufacturing industries discussed in this volume, the U.S. chemical industry began to experience its competitive "crisis" in the late 1960s, well before most other U.S. industries encountered growing competitive pressure from foreign sources and well before the onset of the "oil shocks" of the 1970s that affected prices for the industry's key raw material. Slower growth for its dominant products, including polymers, along with the growth of production capacity offshore, led many leading U.S. chemical firms to pursue diversification programs during the 1970s, with mixed results. In the 1980s, a far-reaching restructuring in the industry, consisting of divestitures and actions to focus firms on a narrower line of products and processes, contributed to improved results in many U.S. chemical firms. This restructuring process began earlier and has proceeded further in the U.S. chemical industry than in those of continental Europe and Japan.

The development during the 1940s and 1950s of a group of independent developers and sellers of process technology, known as specialized engineering firms (SEFs) accelerated the international transfer of technologies and planted the seeds of the competitive challenges faced by the U.S. and European firms in the 1960s and 1970s. But the technological response of these firms, especially that of U.S. firms in the 1980s and 1990s, to these competitive challenges has involved the development of new variants of existing products, customized to the needs of specific users, and greater integration between products and process

[1]The authors are grateful to Ralph Landau for helpful comments and to Marco Ceccagnoli for excellent research assistance. This version has benefited from David Mowery's comments and suggestions. We alone are responsible for all remaining errors and omissions.

45

innovation. These steps have not only eroded the markets for the services and technologies of SEFs but also reduced investment in basic research.

EVOLUTION OF INDUSTRY STRUCTURE SINCE WORLD WAR II[2]

With sales of more than $1 trillion a year, the chemical industry is one of the largest manufacturing industries in the world, as well as one of the oldest and most complex. The modern chemical industry comprises a myriad of products from sulfuric acid to fragrances and perfumes. But there have been two principal driving forces behind the industry's growth in the last half century—first, polymer science, which developed the synthetic fibers, plastics, resins, adhesives, paints, and coatings that virtually define "modern" materials; second, chemical engineering, which made it possible to produce these materials at costs low enough to ensure their success.

Petrochemicals—chemicals produced using oil and natural gas as inputs—are the base of most of these modern materials and are perhaps the most important component of the post-World War II chemical industry. The United States, which has abundant oil and natural gas reserves, was the first country to develop a petrochemicals industry, beginning early in the century. World War II had a major impact on technology and the industry's structure. As part of the war effort, the U.S. government funded large programs for research and production of synthetic rubber and created massive demand for oil for aviation fuel.[3] After the war the demand for cars and gasoline skyrocketed, and by 1950 half of the total U.S. production of organic chemicals was based on natural gas and oil. By 1960 the proportion was nearly 90 percent (Chapman, 1991). Several oil companies, most notably Shell, Exxon, Amoco, and Arco, become major producers of basic and intermediate chemicals derived from petroleum feedstocks.

The United States was the dominant chemical-producing nation at the end of World War II. The German and British industries had been devastated by the war, either directly by bombing or indirectly through damage to the economic infrastructure. But the chemical industry in both countries rebuilt and grew rapidly, shifting its organic chemical production to petrochemicals nearly as quickly as had the United States.[4] During the 1950s and the 1960s Japan made an astonishingly rapid entry into petrochemicals, leading to a rapid growth of the chemical industry. Apart from the three main *keiretsu*—Mitsui, Mitsubishi, and

---

[2]This and the next section draw upon some of our earlier work, including Arora and Gambardella (1998) and Arora and Rosenberg (1998).

[3]Morris (1994) provides a detailed account of the synthetic rubber case. For aviation fuel, see for instance, Spitz (1988) and Aftalion (1989).

[4]In 1949, 9 percent of the United Kingdom's total organic chemical production was based on oil and natural gas; the proportion rose to 63 percent by 1962 (Chapman, 1991). In Germany, the first petrochemical plant was set up in the mid-1950s, and by 1973 German companies derived 90 percent of their chemical feedstocks from oil (see also Stokes, 1994).

TABLE 1 Chemical Production, 1913-1993: Selected Countries

	USA	Britain	Japan	Germany	Rest of the world	World
1913	1.53 (16)	1.1 (11)	0.15 (2)	2.4 (24)	4.82 (47)	10
1927	9.45 (42)	2.3 (10)	0.55 (2)	3.6 (16)	6.6 (30)	22.5
1938	8 (30)	2.3 (9)	1.5 (6)	5.9 (22)	9.2 (33)	26.9
1951	71.8 (43)	14.7 (9)	6.5 (4)	9.7 (8)	63.3 (32)	166
1970	49.2 (29)	7.6 (4.5)	15.3 (9)	13.5 (8)	85.4 (49.5)	171
1980	168.3 (23)	31.8 (4.5)	79.2 (11)	59.3 (8)	380.4 (53.5)	719
1993	313.5 (25)	49.6 (4)	208 (17)	98.6 (8)	580.3 (46)	1250

Note: Sales given in billions of Reichsmark up to 1938, in billions of Deutchmark for 1951, and in billions of U.S. dollars thereafter. Percentage share in total world output given in parentheses.
Source: Eichengreen (1998).

Sumitomo—several other companies, such as Asahi Chemical, Maruzen Oil, and Idemitsu, made considerable investments in petrochemical plants (Hikino et al., 1998).

The technological lead of U.S. chemical producers in petrochemicals was eroded as oil companies and engineering design firms diffused the technology internationally. Technology for producing a variety of important products—from the basic petrochemical inputs such as ethylene to materials such as polyethylene, polypropylene, and polyester—became more widely available. Moreover, the development of a world market in oil meant that the oil and natural gas endowments of the United States did not prove to be an overwhelming source of comparative advantage.[5]

By the end of the 1960s, European countries and Japan had succeeded in closing much of the gap with the United States. Since then, relative shares in world output have largely remained constant, with the exception of a small decline in the U.S. share and a rise in the Japanese share (Table 1). In the 1980s a cheaper dollar and declining growth opportunities in their home markets prompted European firms and, to a lesser extent, Japanese firms to expand heavily into the U.S. market. The expansion, accomplished through direct investments, as well as acquisitions and alliances, underlined both the globalization of the industry, as well as the declining U.S. dominance. Since most of the leading companies in the world are highly globalized, one must be cautious in linking the performance of

---

[5]Government regulation of oil imports in the United States also played an important part. Since the late 1930s, the oil industry had been regulated by the government. Among other things, production of individual companies was regulated to prop up the domestic price of oil. After World War II, the regulations were extended to restrict imports of oil. The net effect, according to Chapman (1991), was that the crude oil acquisition costs for U.S. refineries was 60-80 percent higher than the landed costs in Western Europe through the late 1950s and 1960s. However, one should note that many U.S. firms have access to another cheap source of light hydrocarbons (such as ethane, propane, and butane)—namely, natural gas.

firms with their so-called national industries. With this proviso, we also note that the leading U.S. chemical firms have performed better than their European counterparts in the last decade or so, suggesting that perhaps the decline in U.S. dominance has been stemmed (see Figures 1 and 2 and discussion below).

In 1992, seven European countries—Germany, the United Kingdom, Italy, France, Netherlands, Belgium, and Denmark—taken as a group produced about 20 percent more than the U.S. chemical industry output; as indicated earlier, these figures have been fairly stable over the 1980s and early 1990s. Japan, as Table 1 also suggests, has been increasing its share of world chemical production, producing a little less than two-thirds of the U.S. level in 1992. Even so, Japan's share of world chemical production is smaller than its share in world manufacturing. By contrast, Europe as a whole has a higher share of world chemical output than it does of world manufacturing output, while the U.S. share in chemicals is about the same as its share in manufacturing output.

Although the relative shares of the leading industrial countries in terms of output or exports have been largely constant over the past thirty years, there are some clear patterns that emerge at a more disaggregated level. First and foremost, within most subsectors, relative output and export figures have been fairly stable over the 1980s and early 1990s. European production of pharmaceuticals in 1991 was about 90 percent that of the U.S., with Germany and the United Kingdom. each producing about 30 percent of the U.S. output, and these figures were largely constant over the 1980s. For the same period, Japanese production had been increasing slightly and by the end of the 1980s was about 70 percent of

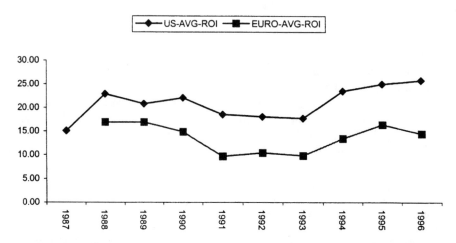

FIGURE 1 Pretax return on investment for leading U.S. and European chemical companies, 1987-1996.

Source: Global Vantage database from Standard and Poor. See text for details of sample.

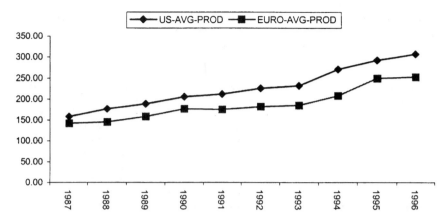

FIGURE 2 Labor productivity: total revenue per employee, for leading U.S. and European chemical companies, 1987-1996.
Source: Global Vantage database from Standard and Poor.

the U.S. output. In terms of exports, Germany alone shipped 20 percent more than U.S. exports, increasing slightly over the 1980s, while the United Kingdom exported about as much as the United States. Japanese exports, on the other hand, were only about 20 percent of the U.S. level, albeit increasing slightly over the 1980s. Similarly, in industrial chemicals (ISIC 351-352), German exports are about the same level as U.S. exports while Japan and the United Kingdom each has exports of about 40 percent of the U.S. level.[6]

Table 2, which provides an output index for selected chemical subsectors, shows that drugs and pharmaceuticals have grown the fastest of all subsectors in the United States, followed by synthetic polymer-based sectors such as rubber and plastic products and plastic materials. Although data for Japan and Europe cannot be compared directly with each other or with the United States, tentative conclusions can nonetheless be drawn. In Japan, basic petrochemicals and aromatic products (such as benzene) remain the major growth sectors. By contrast, pharmaceuticals, basic chemicals, and specialty chemicals have shown the greatest growth in Europe. These patterns are borne out by patent statistics. U.S. chemical patenting is relatively specialized in drugs, and the degree of this specialization increased between 1973 and 1996. Chemical patenting in the United Kingdom is specialized in drugs and agricultural chemicals and away from plastics and fibers. Relative to their share in overall chemical patents, the Japanese patent heavily in plastics and fibers and away from drugs and agro-chemicals. As one might expect, German chemical patenting is relatively specialized in industrial organic chemicals.

---

[6]All figures from the CMA (1997).

TABLE 2 Index of Chemical Production in the United States, European Union, and Japan 1990 = 100

Japan	1992	1993	1994	1995	1996
All chemicals	102.0	100.1	105.5	112.4	113.9
Petrochemicals	104.5	102.2	109.4	120.3	122.6
Aromatics	119.3	127.1	132.9	149.3	144.3
Alkalis and chlorine	97.0	94.2	93.8	98.6	97.1
Inorganic chemicals and pigments	96.2	90.8	90.9	95.6	93.4
Organic chemicals	105.7	102.0	105.4	116.5	119.7
Cyclic intermediates and dyes	105.9	104.8	116.3	126.0	126.2
Plastics	99.4	96.3	102.2	110.4	114.5
Synthetic fibers	106.3	106.9	108.8	116.1	112.7
Synthetic rubber	97.5	91.9	94.7	105	106.6
Fertilizers	95.5	92.9	91.4	91.7	89.4

United States	1986	1988	1990	1992	1994	1996
Chemicals and products	84.69	94.76	100.00	102.77	106.99	111.92
Basic chemicals	80.83	88.34	100.00	98.81	90.22	86.36
Alkalis and chlorine	85.94	97.84	100.00	98.33	106.49	110.72
Industrial organic chemicals	79.60	93.52	100.00	95.33	96.76	97.33
Synthetic materials	86.44	98.54	100.00	104.28	113.45	122.42
Plastic materials	82.58	97.48	100.00	104.93	116.37	128.33
Synthetic fibers	91.14	100.72	100.00	102.99	107.11	109.37
Drugs and medicines	85.84	93.09	100.00	113.25	119.93	132.39
Soaps and toiletries	88.39	97.30	100.00	100.10	106.11	104.20
Paints	100.77	106.41	100.00	95.69	109.38	119.81
Agricultural chemicals	74.50	89.64	100.00	99.60	99.40	102.69
Rubber and plastic products	84.82	95.66	100.00	108.46	125.92	130.69

Source of U.S. data: Chemical & Engineering News, June 23, 1997.

European Union	1991	1992	1993	1994	1995	1996	1997[a]
Chemical industry (NACE 24)	100.1	102.7	102.3	109.1	113.4	115.5	119.1
Basic chemicals (NACE 241) (excluding fertilizers and nitrogen compounds)	97.6	97.7	96.2	104.7	107.8	108.5	111.3
Pesticides and other agro-chemicals (NACE 242)(including fertilizers and nitrogen compounds)	97.7	85.9	89.2	90.9	100.5	104.0	109.7
Paints, inks, and varnishes (NACE 243)	97.4	98.3	97.8	105.3	105.3	106.6	110.1
Pharmaceuticals (NACE 244)	106.8	113.6	114.7	120.4	129.4	130.9	133.4
Soap and toiletries (NACE 245)	99.2	103.1	102.8	109.0	108.1	108.2	112.1
Other chemical products (NACE 246)	100.3	103.5	104.2	108.4	109.0	112.1	116.1

[a]Only January-April.

Source of E.U. data: Our calculation based on ESCIMO database (European Chemical Industry Council).

Overall, however, the relative positions of the major industrial producers has not changed much in the last 25 years.[7] Indeed, the major change has been the increase in chemical production outside the leading industrial countries (see Table 1). In 1959 the United States, Japan, and Western Europe accounted for virtually all chemical exports. By 1993 their combined share had fallen to two-thirds, with the rest coming from Asia, particularly South Korea and Taiwan, Eastern Europe, and the Middle East. Much of the capacity addition took place during the 1970s and especially the 1980s.

Following the big technology push in the industry during the 1950s and 1960s, technology diffused more widely than it ever had before. Specialized engineering firms played a key role in creating a global market for process technologies for a large number of basic and intermediate chemicals. The maturing technology, along with increasing competition and slower demand growth, lowered the payoffs to traditional types of innovations. Commercialization became more expensive and required ever more sophisticated knowledge of customers and the market. Faced with overcapacity, the industry restructured, beginning in the 1980s in the United States, and a few years later in Western Europe. The drive to reduce cost dominated the initial restructuring phase, driven in part by the relentless pressure from shareholders and their representatives. Major realignments of the product portfolios of many firms followed, with many mergers and acquisitions and the rise of entirely new firms in the industry.

During this phase, many firms cut down on R&D and refocused R&D expenditures on short-term projects and away from more fundamental research. In the past couple of years, there are some indications that the industry may be entering a new phase of technological change and R&D spending appears to be picking up as well. Nonetheless, the restructured firm portfolios beg the question of who will perform the basic research that continues to be very important for the future of the industry. The current situation points to the possible need for increased government support for R&D in an industry that has hitherto largely financed its research by itself.

POLYMER CHEMISTRY: "MATERIALS BY DESIGN"

With synthetic dyestuffs as its engine of growth, Germany dominated the chemical industry from the 1870s until World War I.[8] Advances in organic chemistry clarified how carbon atoms are linked to hydrogen and other atoms to form

---

[7]The production and export figures are based on STATCON, while the patenting figures are based on U.S. patents. Details of the analysis are available from the authors on request.

[8]The German chemical industry was strong in other fields such as inorganics and high-pressure chemical processing from coal. BASF for instance had developed the contact process for sulfuric acid, and it was responsible for many process innovations, which culminated in the development of the Haber-Bosch process. BASF (within I.G. Farben) also pioneered research in the 1920s and 1930s on coal hydrogenation to produce synthetic gasoline.

more complex molecules. Over several decades German firms developed this knowledge into a general purpose technology for producing new dyes such as alizarin and indigo. Moreover, they soon discovered that the organic chemistry behind the creation of synthetic dyestuffs could also be harnessed for other applications such as pharmaceuticals and photographic materials.

If synthetic dyestuffs represented the start of "materials by design," polymer chemistry represents its maturation. The theoretical work of Herman Staudinger and other German scientists in the 1920s, postulating that many natural and synthetic materials such as cotton, silk, and rubber consist of long chains of the same molecule linked by chemical bonds, pointed to ways of developing a series of new products by using different building-block molecules and changing the way in which these molecules were connected. Long and systematic experimentation was still needed to produce commercially successful products, but over time the advances in the scientific understanding of the relationship between molecular structure and physical properties made the research much more productive.

A key challenge in producing a polymer is controlling the length and physical structure of the macromolecule. Catalysts are the main instrument used for this purpose because they permit control over the rate and manner in which monomers connect to each other. The discovery of the Ziegler-Natta catalysts for the production of linear polyethylene and polypropylene is probably the most successful case. Indeed, research into new catalysts remains the focus of research efforts involving existing polymers, and the recently developed metallocene catalysts are viewed by many as a major breakthrough for plastics (Thayer, 1995).

As with synthetic dyestuffs, polymer science was marked by knowledge-based economies of scope. By establishing relationships between properties of materials and their molecular structures, polymer chemistry provided a systematic basis for product innovations in several downstream sectors. For example, the macro properties of the polymer material, such as its strength or malleability, can be changed by varying the physical orientation of the molecules in a polymer chain. In addition, by applying heat and pressure, or by controlling density or melt indexes, many polymers can be made into any desired shape. The same basic material can then be used as a fiber, sheet, or film or molded to form a component or product of a specific shape.[9] The product may be further fine-tuned in other ways. For example, chemists learned that engineering resins could be enhanced and extended by adding fillers and reinforcements such as glass or carbon to the polymer (Seymour and Kirshenbaum, 1986). By varying the amounts of these materials, one could produce different grades of the engineering

---

[9]For instance, nylon with less than 15 percent crystallinity can be used to produce soft shopping bags, women's underwear with 20-30 percent crystallinity, sweaters with 15-35 percent crystallinity, stockings with 60-65 percent crystallinity, tire cords with 75-90 percent, and fishing lines with more than 90 percent crystallinity (Mark, 1994).

resin. Ultimately, polymer chemistry supplied a common technological base in five distinct areas—plastics, fibers, rubbers and elastomers, surface coatings and paints, and adhesives.[10]

The striking feature of these examples is that the underlying technological base generated opportunities for linking product markets that, to a final user, would appear to have nothing in common with each other. Because the technology had to be adapted for specific uses, commercializing a new polymer product required knowledge of the use of the material. In other words, there were economies of scope but, in order to realize those economies, firms had to become knowledgeable about downstream users in a wide variety of markets.

As with synthetic dyestuffs, the rise of polymer chemistry opened up vast new opportunities in the industry. But unlike dyestuffs, these opportunities were exploited by a much larger number of firms with comparable commercial and technological capabilities and from many countries. Consequently, even small information leaks allowed very rapid imitation. Thus, many chemical companies and some oil producers found themselves operating and competing in very similar markets. For example, Union Carbide, Goodrich, General Electric, I.G. Farben, and ICI performed research on polyvinyl chloride (PVC) and produced the polymer from the very beginning. Similarly, Dow, I.G. Farben, and Monsanto were all involved in polystyrene from very early on. Du Pont, ICI, BCC, Monsanto, Kodak, and many others invested in various kinds of polyamides, acrylics, and polyesters (Spitz, 1988; Aftalion, 1989). The net result was an increase in competition in virtually every market segment.

Thus, ironically enough, the diffusion of polymer science meant that the crucial problem in innovation shifted from *how* to produce different products to *what* to produce.[11] Companies had to decide which applications were to be developed among the many that could be produced. This increased importance of "what to produce" increased the relative importance of marketing and downstream links with users to find out how to tailor products for their needs (Hounshell, 1995).

These trends are well illustrated by Keller's (1996) case studies of innovation in polymers. For instance, Keller notes that Quantum Chemicals, one of the largest producers of polyethylene, does little fundamental research in polyethylene processes. Instead, it focuses on process improvement and optimization of processes licensed from other firms including Du Pont, Union Carbide, and BP. It competes by providing "customer service" rather than lower prices or superior

---

[10]See Landau (1998) for a detailed discussion of two major innovations—polypropylene, and purified terephthalic acid for polyester. Spitz (1988) discusses how different the activities involved in the development of synthetic fibers were from those for plastics.

[11]As noted earlier, the only exception to this was the research into catalysis and the improvement of existing catalysts. However, it is worth noting that cost reduction was not the only motive here. Rather, catalysts make possible greater control over the properties of the final product and thus catalyst research can be seen as an integral part of product innovation.

properties of the product itself. Its research staff helps customers use various grades of polyethylene more efficiently in their processing equipment. To this end, Quantum has a large collection of commercial- and semicommercial-scale polymer processing equipment, including film lines and injection molding equipment, which it uses to demonstrate how its products would work. This and other examples show that polymers are increasingly seen as part of a system, rather than a product itself. Thus the desired properties will vary according to the various uses to which the polymer will be put. This implies that the effective economies of scope are more limited than those implied by the technology itself.

As noted earlier, the opportunities opened up by polymers induced a number of firms to enter, with the inevitable result that profits were sometimes below expectations. Even though individual firms may have been disappointed by the returns to their investments in synthetic polymers, there is no denying that synthetic polymers and chemical products based on these polymers were very successful, accounting for nearly 20 percent of the total value of shipments of the U.S. industry in 1970. The success of synthetic polymers owes a great deal to a steep drop in the cost of basic petrochemicals, which are the building blocks for synthetic polymers. This cost reduction was realized through process innovation, both radical and incremental, in petrochemicals and polymers. In turn, the development of chemical engineering was key to the progress in chemical processing technologies.

CHEMICAL ENGINEERING: THE SCIENCE OF THE CHEMICAL PROCESS

If polymer chemistry is the science of chemical products, chemical engineering is the science of the chemical process. The job of the chemical engineer is to develop manufacturing processes for chemical products that emerged from laboratories, using commercially available equipment and inputs instead of the glass beakers and expensive reagents used in laboratories. The objective is to produce at unit costs that are low enough to make the product commercially viable, typically by scaling up production to produce large quantities of output in a continuous flow plant.[12]

Beginning with the concept of unit processes, chemical engineering was an attempt to abstract the essential and common features of chemical processes for a wide variety of products. The systematic isolation, categorization, and analysis of the basic processes (unit processes and, later, unit operations) common to all chemical industries meant that an engineer trained in terms of unit operations

---

[12]Scale up has therefore been a traditional focus of chemical engineering. As we discuss below, with slower growth and increasing product differentiation, the focus may be changing to emphasize flexibility and reduction in the cost of small scale production (see for instance Shinnar, 1991).

could mix and match these operations as necessary to produce a wide variety of distinct final products. The separation between product and process innovation also made possible a division of labor that was to affect the competitive position of the leading chemical producers, a theme explored in greater detail later in this chapter. The general purpose nature of chemical engineering made it possible for university research and training to play an important role in applying engineering science to the practical problem of designing large-scale processes.

Chemical engineering was a distinctly American achievement, which testifies to the unique nature of the university-industry interface in the United States.[13] The large size of the market had introduced American firms to the problems involved in large-scale production of basic products, such as chlorine, caustic soda, soda ash, and sulfuric acid as early as the beginning of this century. This ability to deal with a large volume of output, and eventually to do so with continuous process technology, was to become a central feature of the chemical industry in the twentieth century.

This focus on large-scale production had additional benefits when it turned out that the new petrochemical technologies had strong plant-level economies of scale, with capital costs rising by less than two-thirds when production capacity was doubled. Because "scaling up" output was not a simple matter, and involved considerable learning, early experience with process technologies gained American firms a head start when petrochemicals became the dominant feedstock after World War II. In an earlier era, this head start might have been expected to last for a long time. In petrochemicals, however, the rise of a new market—for engineering and construction services, and eventually for process technology itself—allowed other countries to catch up quickly.

EXTENT OF MARKET AND DIVISION OF LABOR

The rise of this new market involved a new division of labor and involved a new type of firm—specialized process design and engineering contractors, hereafter the SEFs. In addition to supplying proprietary processes, some SEFs also acted as licensers on behalf of chemical firms and provided design and engineering know-how. During the past ten or fifteen years, SEFs may have declined in importance but in the post-World War II period as a whole they have played an important role in developing new and improved processes and a crucial one in diffusing new technologies.

As one might expect, given the comparative emphasis on large-scale production, the United States enjoyed an early lead in chemical engineering of plants. The first SEFs were formed in the early part of this century, and their clients were

[13]See Landau and Rosenberg (1992) for a discussion of the role of M.I.T. in the development of chemical engineering as a discipline. The discussion here is based on Rosenberg (1998).

typically oil companies. Prominent among the early SEFs are companies such as Kellogg, Badger, Stone and Webster, UOP, and Scientific Design.[14] Steeped in a tradition of secrecy, chemical firms were initially reluctant to outsource plant design and engineering. Established chemical companies were especially reluctant to buy technology from SEFs or to enter into alliances with SEFs to develop new technology. But World War II, with its leveling effect, changed this as well.[15] The rise of synthetic polymers united what had been a number of disparate markets such as fibers, plastics, rubber, and films, thereby increasing the number of potential entrants and the demand for the services of SEFs. After the war American SEFs combined their know-how about process design with proprietary technologies and offered technology packages to customers overseas.[16]

Initially European and Japanese firms, and later firms in the Middle East and East Asia, benefited greatly from the technology transfer by the SEFs. Between 1960 and 1990 roughly three-fourths of the petrochemical plants built all over the world were engineered by SEFs (Freeman, 1968; Arora and Gambardella, 1998). By providing technology licenses to firms the world over, SEFs played a major role in the diffusion of chemical, especially petrochemical, technologies. As independent developers of technology, SEFs were similar in some respects to today's biotechnology companies, often partnering with several different chemical firms in developing new technologies.[17]

SEFs had a major impact on industry structure, both in the United States and abroad. A number of firms entered the industry, especially in petrochemicals, shown in Table 3a and 3b. Often, these were not "new" firms but rather firms that had operated in other sectors of the chemical industry and wished to exploit some real or perceived competitive advantage. These firms typically entered on the

---

[14]However, SEFs also started operating in some bulk chemicals such as sulfuric acid, and ammonia. The Chemical Construction Corporation built sulfuric acid and other plants, while the Chemical Engineering Corporation targeted synthetic ammonia and methanol processes, both attaining some success in project exports to Europe as well (Haynes, 1948, vol. IV).

[15]Landau (1966), writing two decades after the end of the war, noted that the "... the partial breakdown of secrecy barriers in the chemical industry is increasing ... the trend toward more licensing of processes."

[16]The technology transfers were not all from the United States to Europe. Spitz (1988) points out that many European chemical firms, particularly technology-rich but cash-poor German firms, were willing to license their technologies for revenue in the 1950s. For much of this period, however, Japan remained a net importer of technology.

[17]For instance, Badger used its fluidized bed catalytic process to develop processes for phthalic anhydride with Sherwin Williams, ethylene dichloride with BF Goodrich, and acrylonitrile with Std. of Ohio. UOP similarly had a number of strategic partnerships with Dow (Udex—benzene extraction), Shell (sulfonale—benzene extraction), Ashland oil (Hydeal—dealkylation of toluene), Toray (Tatoray—disproportionation of toluene and C4 to benzene and xylene), and BP (Cyclar—reforming of LPG into aromatics). Scientific Design, an innovative SEF, followed a different strategy, developing technology without recourse to joint research.

TABLE 3a Producers, Capacities, and Concentration Ratios for Selected Petrochemicals in the United States, 1957, 1964, 1972, and 1990

	Ethylene	Polyethylene[a]	Ethylene oxide	Ethylene dichloride	Styrene
1957					
Producers	16.0	13.0	7.0	8.0	8
1964					
Producers	20.0	17.0	10.0	10.0	9.0
Capacity	4.9	1.2	0.8	0.9	1.0
Concentration ratio	52.1	47.8	70.5	42.1	70.3
1972					
Producers	25.0	21.0	13.0	11.0	12.0
Capacity	9.8	3.7	1.9	4.7	2.6
Concentration ratio	40.5	40.8	72.0	60.8	54.1
1990					
Producers	22.0	16.0	12.0	11.0	8.0
Capacity	18.4	10.3	3.3	8.4	4.0
Concentration ratio	30.8	37.9	54.2	57.5	52.1

[a]Includes LDPE and HDPE.

Notes: Capacity is measured in millions of tons. Concentration ratio is measured as the share of the largest three producers, expressed as a percentage.

Source: Chapman (1991), table 5.2, page 104.

TABLE 3b Producers, Capacities, and Concentration Ratios for Selected Petrochemicals in Western Europe, 1955, 1964, 1973, and 1990

	Ethylene	Polyethylene[a]	Ethylene oxide	Styrene
1955				
Producers	13.0	N/A	N/A	N/A
Capacity	0.2	N/A	N/A	N/A
Concentration ratio	52.7	N/A	N/A	N/A
1964				
Producers	25.0	23.0	15.0	10.0
Capacity	2.0	0.8	0.3	0.5
Concentration ratio	27.6	40.3	33.6	57.7
1973				
Producers	38.0	40.0	16.0	15.0
Capacity	11.6	5.2	1.4	>2.8
Concentration ratio	24.3	18.7	41.5	40.2
1990				
Producers	29.0	26.0	10.0	12.0
Capacity	16.0	9.3	1.7	4.0
Concentration ratio	25.9	27.2	41.4	47.6

[a]Includes LDPE and HDPE.

Notes: Capacity is measured in millions of tons. Concentration ratio is measured as the share of the largest three producers, expressed as a percentage.

Source: Chapman (1991), table 5.3, page 105.

basis of licensed rather than internally developed technology.[18] Indeed, as we discuss below, many chemical producers also licensed their technology to others. Thus, a major consequence of SEFs was, paradoxically enough, to reduce the strategic importance of process technology, in essence by helping to develop and supply a market for technology (Arora et al., 1998). The large number of potential licensees and the possibility of competing innovation made it difficult for a chemical firm to gain long-term advantage from a single innovation. Only by continual improvements and innovation could a company hope to derive a long-term advantage, and in some cases even that was not sufficient.[19]

In addition to inducing entry and creating competition on a global scale, the development of a market in technology licenses brought to the fore the importance of other factors influencing competitive success—availability of raw materials and capital, proximity to market, and other idiosyncratic factors such as severity of environmental regulation and macroeconomic instability. The important point is that although initially the benefits of the division of labor between chemical producers and SEFs accrued to U.S. chemical firms, over time these benefits became available to chemical producers in other countries as well. The very factors that underpin the U.S. success also enabled other countries to catch up.

SEFs provide a vivid illustration of economies of scale—economies of specialization—that operate at the level of the industry rather than of the plant or the individual firm. The growth of SEFs was directly linked to the rapid growth of a large, relatively homogenous market. As demand growth has faltered, the SEF sector has itself shrunk and consolidated. SEFs have drastically reduced their own R&D expenditures, and some have left the industry or been acquired by others. In the future, the division of labor may arise in new arenas, such as the emergence of firms specializing in providing software tools for process design and for simulating how the designed chemical process works. These software tools may include modeling of distillation and reaction columns, surface chemistry, and computational models of fluid dynamics. Some elements of this system

---

[18]In a study of 39 commodity chemicals in the United States in a period from the mid-1950s to the mid-1970s, Lieberman (1989) found that controlling for demand conditions, experience accumulated by incumbents did not act to deter new entry. Given the importance of learning by doing, this suggests that entrants had access to other sources of know-how, most likely from SEFs. This interpretation is further supported by Lieberman's findings that entry into concentrated markets, which were also marked by low rates of patenting by non-producers (both foreign firms and SEFs), usually required that the entrant develop its own technology. By contrast, less concentrated markets were associated with high rates of patenting by non-producers and high rates of licensing to entrants. In a related study of a subset of 24 chemicals Lieberman (1987) found that high rates of patenting by non-producers were also associated with faster rates of decline in prices.

[19]Spitz (1988) describes how a number of companies entered into vinyl chloride and PVC based on new technologies. However, since a number of new technologies became available, few firms managed to get sustained profits unless they enjoyed some other advantage such as a cheap source of raw materials.

are already in place. Firms that supply these tools may be U.S.-based, but as with SEFs, chemical producers elsewhere are also likely to benefit.

RESTRUCTURING

The rise of the SEFs was an important factor in increasing competition for chemical producers in the United States. The oil shocks of the 1970s perhaps only hastened the inevitable consolidation in the industry. For an industry whose growth was closely tied to manufacturing growth, the oil shock meant a decline in demand precisely at a time when its costs were rising and when opportunities to innovate were becoming rarer. The combination of increasing entry, slower demand growth, and diminishing opportunities for major product innovations on the scale of nylon or polyester forced a consolidation of industrial structure.

The adjustment to the new equilibrium was slow and painful. Economies of large-scale production meant that existing producers had sunk large investments in capacity, especially in the basic intermediates. The problem was magnified because many chemical and petrochemical operations were highly integrated both vertically and horizontally. Typically, a reduction in capacity at one plant left excess capacity at others. Thus, a reduction in the output of one product could reduce the manufacturing efficiency of an entire production complex. To make matters worse, some firms failed to foresee the slow growth of demand and continued to invest. A comprehensive realignment of expectations was completed only during the 1981-1982 recession (Aftalion, 1989; Chapman, 1991; Lane, 1993).

The process of restructuring illuminates how the strategic importance of different types of investments, together with the nature of corporate strategy and corporate governance, changes as a high technology industry matures. During the first phase of restructuring, the industry rationalized capacity by phasing out older and less efficient capacity. This phase of the restructuring appears to have been accomplished in the United States by the mid-1980s. It was followed by a restructuring in the corporate sector that reached its peak in the United States in the late 1980s. Restructuring in Western Europe appears to have lagged that in the United States by about five years or so. Unlike the United States, where the restructuring has been largely market driven, in Europe, especially in Italy and France where the state had large ownership stakes, government intervention has also driven the process (Martinelli, 1991). In Japan, MITI has played a major role by coordinating capacity rationalizations through its Industrial Structure Council (Bower, 1986).

The slower pace of reform in Europe is not without its costs. The available evidence shows that the leading U.S. chemical firms have emerged from the restructuring in better shape than their European counterparts. Figure 1 shows that the average return on investment for the leading U.S. firms has been consistently higher than for European counterparts over the ten-year period 1987-1996. In terms of output per employee, a traditional measure of performance, the leading U.S. firms have outperformed the leading European chemical firms.[20]

The process of restructuring has been most marked in the basic and intermediate petrochemicals, the sectors with the strongest competition. Several traditional chemical companies in the United States and Europe are exiting from some of their commodity chemical businesses and moving downstream, focusing on businesses where product differentiation based on quality and performance allows for higher margins. In their place, oil companies such as Shell, BP, Exxon, Arco, and Amoco and other firms such as Vista, Quantum, Cain, Sterling, and Huntsman have stepped in. Many of the latter are new firms that have taken over the existing commodity chemicals businesses of firms such as Conoco, Texaco, Monsanto, and USX. In Europe as well, new focused firms such as Borealis, Clariant, and Montel have been formed by merging businesses of existing companies. The new companies seem to be separating into those that produce high value added, specialty chemicals and those that manufacture larger-volume commodity chemicals. Thus many of the synergies and economies of scope that were characteristic of the industry are seen as less important, most markedly in the apparent separation between chemicals and life sciences.

From a corporate viewpoint, firms are becoming narrower but deeper. They are reducing the number of business they are in, reversing a long trend of diversification based on economies of scope.[21] Through a series of divestitures, acquisitions, mergers, and alliances, they are attempting to increase both the absolute size as well as market share of their remaining businesses. Global expansion of existing businesses is a key part of this strategy. Although globalization has become something of a cant word, there is no doubt that chemical companies in the developed countries are becoming increasingly globalized in their outlook and operations. In other words, as firms are being driven to specialize more narrowly in business areas in accordance with comparative advantage or core competence, they seek a broader geographical platform on which to operate. Freer movement of both goods and capital have doubtless played a very important role in this respect.

The search for size could be driven by some efficiency motives, such as the desire to spread fixed costs—process research, sales, and management—over larger volumes (see for example, Cohen and Klepper, 1992). Alternatively, the search for size could be an attempt to control price competition by exercising price leadership. Consolidation can enable a firm to manage inter-brand compe-

---

[20]The data are taken from the *Global Vantage* database published by Standard & Poor's. The sample of U.S. firms consists of the 27 firms listed in Table 4a. The European sample consists of 20 of the 24 firms listed in Table 4b, with Enichem, Borealis, SKW Trostberg, and Clariant excluded because of lack of data.

[21]To some extent, these divestments are a correction of the earlier tendencies toward conglomeration and unrelated diversification of the 1960s. However, the restructuring has gone well beyond getting rid of the unrelated businesses.

tition more effectively, especially when faced with a large "competitive fringe" of producers. This may be an important motive behind the division swaps where a large firm will exchange one business division for a different business division owned by another firm.[22] Both forces are probably at work although industry sources point to efficiency as the dominant motive.

The timing and pattern of restructuring points to the role of capital markets. Restructuring began in the United States and has taken place far more slowly in Europe, and has been even slower in Japan (Hikino et al., 1998). The increasing importance of mutual funds and pension funds and the greater attention to "shareholder value" have pressured managements to improve financial performance. The social welfare implications of the restructuring are unclear, and the debate has been closely tied up with the broader debate on the virtues and vices of the Anglo-Saxon system of finance versus the bank-based systems of Germany and Japan (Richards, 1998; Da Rin 1998). Stock markets appear to disfavor diversified firms with portfolios that include both commodities and specialties. The reasons perhaps lie in the greater difficulties of managing such firms as well as the greater difficulties in evaluating the performance of the management of a diversified company, particularly when the company has a mix of research intensive and less research intensive businesses.[23]

WHAT IS HAPPENING TO R&D, WHY, AND WHAT ARE THE CONSEQUENCES?

The restructuring of the chemical industry initially focused on cost reduction and, among other actions, managements of firms in both the United States and Europe cut back spending on R&D. Total R&D spending for the leading American chemical firms between 1985 and 1995 increased at a little more than 3 percent a year in *nominal* terms and even declined from 1993 to 1995. Table 4a shows that for a sample of the leading U.S. chemical companies, growth in real R&D spending has been concentrated, growing at more than 5 percent annually

---

[22]Consider the restructuring in PVC in Europe. Such agreements played a critical role in the restructuring of the PVC market. First, BP and ICI signed a deal in 1981 which led to the consolidation of their businesses. BP ceded its PVC operations to ICI, which pulled out of polyethylene by relinquishing its activities to BP. This concentrated PVC in ICI and polyethylene in BP. Then, in 1985, ICI formed a joint-venture, European Vinyls, with the Italian company Enichem, which merged the PVC businesses of the two firms. The new company became the major European PVC producer. Similarly in polypropylene, Statoil and Neste have merged their petrochemical operations to form Borealis (sales $2.3 billion), Europe's largest and the world's fifth largest polyelefin producer.

[23]The breakup of ICI is a case in point. In 1992, before the demerger, ICI's sales were about $21 billion, and market capitalization was about $13 billion. By 1995, while sales increased by less than 10 percent to about $23 billion, its market capitalization *doubled* to about $26 billion. By contrast, in 1992 Bayer had sales of $26 billion and a market capitalization of about $11 billion. The corresponding figures for 1995 are $31 billion and $18.5 billion.

TABLE 4a Real R&D Spending by U.S. Chemical Companies, 1986-1995 ($ Millions)

Company	1986	1987	1988	1989	1990	1991	1992	1993	1994
Dow Chemicals[a]	751	807	897	973	1213	1191	1289	1224	1201
Monsanto	740	740	753	666	654	644	651	610	580
Rohm and Haas	165	171	181	195	190	188	199	200	192
Union Carbide	184	191	185	202	204	195	155	135	130
Air Products & Chemicals[b]	76	69	84	79	77	82	85	90	92
W.R. Grace[c]	117	129	138	139	158	154	151	132	126
International Flavors	48	54	58	58	61	64	71	73	77
Lubrizol	63	75	75	78	79	82	90	87	87
Morton International[d]	36	37	41	42	51	61	61	67	63
Witco	27	28	28	28	28	29	29	39	39
Hercules	88	89	86	88	98	88	70	74	62
Ethyl	58	60	69	68	69	71	73	44	48
Nalco Chemical	41	42	43	46	48	48	48	49	44
Olin	69	75	67	74	70	42	39	40	33
Petrolite[e]	15	14	14	13	13	11	12	14	12
Total	2478	2581	2718	2750	3014	2977	3085	2963	2868

Pharmaceutical

	1986	1987	1988	1989	1990	1991	1992	1993	1994
Pfizer	336	401	473	531	640	757	863	974	1139
Merck & Co.	480	566	669	751	854	998	1112	1173	1231
American Home Products[f]	227	247	328	345	369	431	552	663	817
Bristol-Myers Squibb	474	563	688	789	881	993	1083	1128	1108
Eli Lilly[g]	427	466	512	605	703	767	925	955	839
Schering-Plough	212	251	298	327	380	426	522	578	620
Total	2156	2494	2968	3348	3827	4372	5057	5471	5754

Diversified

	1986	1987	1988	1989	1990	1991	1992	1993	1994
Procter & Gamble[d]	479	549	615	628	693	786	861	956	1059
DuPont	1156	1223	1319	1387	1428	1298	1277	1132	1047
3M	586	650	721	784	865	914	1007	1030	1054
AlliedSignal	459	400	415	381	426	381	320	313	318
PPG Industries	204	227	232	233	218	220	203	201	218
FMC	146	132	144	150	158	135	145	149	167
Total	3030	3181	3446	3563	3788	3734	3813	3781	3863

[a]Sold Marion Merrell Dow drug operations to Hoechst in 1995.

[b]Fiscal year ends Sept. 30.

[c]Sold off water treatment and process chemicals, health care, and agro-biotech businesses in 1996.

[d]Fiscal year ends June 30.

[e]Inactive date: July 2, 1997. Merged into Baker Hughes Inc. in July 1997.

[f]Acquired American Cyanamid in December 1994.

[g]Spun off and sold medical devices businesses in 1994.

Source: Our calculations based on *Chemical & Engineering News, Facts and Figures.*

in pharmaceutical companies, at a mere 1 percent annually in diversified companies such as Du Pont and 3M, and not at all in chemical companies. There are some signs that R&D spending may have turned the corner in the late 1990s as the focus shifts from cost reduction to growth. Whether this shift is merely a cyclical phenomenon or represents a new phase remains to be seen, but a recent industry-supported study has identified the key science and technology areas for research (Vision 2020, 1996). Table 4b, which presents data for a sample of the

TABLE 4b Real R&D Spending by European Chemical Companies, 1992-1996, in 1992 dollars ($ millions)

Company	1992	1993	1994	1995	1996
BASF	1362.60	1262.16	1224.82	1312.64	1406.41
Hoechst	1931.80	1983.35	2153.38	2187.23	2079.57
Bayer	2059.60	2060.39	2030.66	2048.98	2219.85
Rhone-Poulenc	1164.10	1224.66	1254.64	1296.33	1451.63
ICI[a]	1176.00	559.57	360.88	272.45	275.30
Akzo Nobel[b]	551.20	539.40	639.21	622.04	622.40
Norsk Hydro	110.00	106.22	89.93	102.79	109.73
Roche Group	1607.80	1713.16	1808.90	1747.30	1826.42
Degussa[c]	320.60	305.14	259.35	271.32	289.24
Solvay	393.70	381.10	355.80	359.68	366.76
Air Liquide	183.50	89.08	87.92	146.55	144.53
Enichem	98.00	137.05	123.68	97.13	74.20
BOC Internat.[c]	128.20	130.40	131.73	134.76	141.94
DSM	252.20	226.43	199.04	192.95	213.47
Courtaulds[d]	60.30	59.91	67.40	58.94	59.07
Kemira	N/A	N/A	48.03	49.51	56.30
Novo Nordisk	262.30	294.57	313.52	317.15	357.16
Laporte	9.50	14.98	15.53	16.60	15.87
Dyno	15.90	16.45	N/A	17.26	18.18
Novartis[e]	N/A	N/A	N/A	2691.1[f]	2729.85
Borealis	N/A	N/A	N/A	61.77	56.94
SKW Trostberg[g]	N/A	N/A	N/A	97.13	95.89
Clariant[h]	N/A	N/A	N/A	51.87	47.07
Albright&Wilson[i]	N/A	N/A	N/A	11.98	12.27
Total	11,687.3	11,104.0	11,164.4	14,165.5	14,670.1

[a]Spun off life-sciences business as Zeneca.
[b]Acquired Nobel Industries on November 8, 1993.
[c]Fiscal year ends September 30.
[d]Fiscal year ends March 31.
[e]Formed by merger of Ciba and Sandoz on December 20, 1996.
[f]Pro forma results.
[g]Launched on stock market on May 24, 1995.
[h]Spun off from Sandoz on July 1, 1995.
[i]Launched on stock market on March 8, 1995.
Source: Our calculation based on *Chemical & Engineering News*, June 23, 1997.

leading European countries, similarly shows R&D spending has grown there, primarily in the life sciences area.[24]

The slowdown in R&D spending also reflects a growing international parity in technological capabilities. Recent estimates suggest that Japanese firms are now at par with their European and U.S. counterparts in R&D spending as a percentage of sales. R&D intensity in the Japanese chemical industry has been rising steadily during the last ten years, from less than 4 percent in 1985 to well over 5 percent in 1995, making it more research-intensive than the European and U.S. industries (Figure 3). However, this trend may not be sustained in view of the recent, well-publicized, recession in Japan.

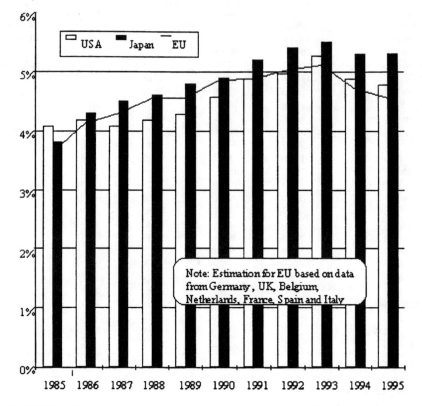

FIGURE 3 Chemical industry R&D spending as a percent of sales: International comparisons.

Source: R&TD and Innovation in the EU - Economic Bulletin - June 1997 Article 2 chart 4, http://www.cefic.be/Eco/eb9706b.htm

---

[24]We lack comparable data for Japanese chemical companies but as Figure 3 below suggests, R&D spending by Japanese firms is likely to have gone up during this period.

Nonetheless, other trends point to the increasing international parity as well. Royalties and license fees paid by U.S. subsidiaries grew at an annual rate of 23 percent between 1986 and 1996 and stood at about $1.5 billion a year in the mid-1990s. Royalties and license fees paid to American firms grew at 15.7 percent annually during the same period and stood at about $3 billion a year in the mid-1990s. Thus, the United States is still a net seller of chemical technology, but its outflows are growing faster than its inflows. This is consistent with the fall in the relative share of U.S.-based companies in U.S. chemical patents, a figure that has fallen from 66 percent in 1970 to 55 percent in 1980 to about 52 percent in 1995. During the same time, Japan's share increased from 6 percent to 12 percent to nearly 17 percent.

Further examination of the structure of R&D reveals other trends consistent with this story. In real terms, chemical R&D has grown, but much of that growth has been in the drugs and medicines sectors (Figure 4).[25] R&D spending in real terms has remained constant at best in industrial chemicals and sectors such as paints and inorganics. Indeed, the share of industrial chemicals (inorganic, organic chemicals and plastics and synthetics) in total chemical R&D spending has declined from 43 percent in 1970 to 30 percent in 1995.

To some extent, these trends reflect the increasing importance of drugs in the chemical sector and the concomitant decline in the share of plastics, fibers, and

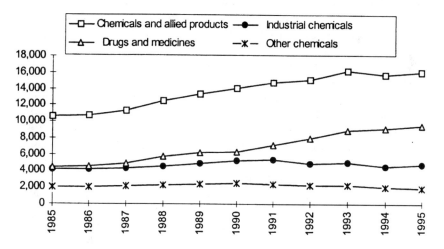

FIGURE 4 Company finance R&D expenditures ($ millions in 1992 constant dollars). Source: Our calculations from CMA, 1997 data.

---

[25]Figure 4 shows company financed R&D only but this accounts for the vast bulk of R&D spending in chemicals.

organic and inorganic intermediates. For instance, the share of industrial and others in shipments, declined from 60 percent in 1980 to around 50 percent in 1995. But there is more to the story. With the maturation of the technology in polymers and chemical engineering, R&D intensity in the industry has declined, if pharmaceutical related R&D is excluded. In industrial chemicals it has hovered between 4.4 and 4.1 percent, while in the other non-drugs sectors it has declined steadily from 3.3 percent in 1986 to close to 2 percent in 1995. For drugs, the figure has risen steadily from about 8.4 percent in 1986 to well over 10 percent for the same time period, approaching 12 percent in barely two years.

For the U.S. chemical industry as a whole, National Science Foundation (NSF) figures show that the "D" of R&D accounts for about 53 percent of the total. In industrial chemicals, however, the share of development tends to be higher and appears to have risen over time. Although changes in definition and coverage make precise comparisons difficult, the available data show that for company-financed R&D, the share of development increased from about 53 percent in 1989 to 62 percent in 1995.

These figures point to the changing role of technology in the industry. Simply put, there is a market for certain types of technologies. As noted in the discussion of SEFs, chemical firms are much more willing than before to license their technology for profit. Moreover, many firms explicitly consider licensing revenues to be a part of the overall return from investing in technology. In turn, this readiness to license technology implies that generic or basic research will not be replicated as widely as it used to be. Instead, firms will license the generic process technology and, as illustrated by the example of Quantum, focus on adapting and improving the technology to best suit their needs and those of their customers.

Perhaps the most vivid example is the metallocene catalysts, which have been used in many of the most significant process innovations in recent years because they provide greater impact strength and toughness, melt characteristics, and clarity in films than do existing catalysts. Total investment worldwide in metallocene research has been estimated at close to $4 billion (Thayer, 1995). Commercially first used in the production of polyethylene in 1991, metallocene catalysts are being applied to a wide variety of polymers, exemplifying the inherent economies of scope in the technology.

Several firms are active in this research area. Dow and Exxon are regarded as being ahead of the rest in polyethylene, while BASF, Hoechst, Mitsui Toatsu, Fina, and Exxon are also active in polypropylene, and Du Pont and Nova are developing alternative catalyst systems. Both Dow and Exxon have allied with other process innovators to combine the catalyst system with processing technologies largely specific to the major polymers like polyethylene and polypropylene. For instance, Exxon has formed a technology joint venture, Univation, with Union Carbide, combining its Unipol technology with Exxon's catalyst (Chemical Week, 1997a). Dow and BP have a similar arrangement.[26] What is noteworthy is that both groups are actively trying to license their technology for commodity-grade but not specialty-grade polyethylene. This has encouraged other firms to develop complementary technologies. For example, BASF, Phillips Petroleum, and BP are developing ways of using metallocene catalysts in existing slurry processes, many of which will be licensed.[27] Interestingly enough, although SEFs are not among the major innovators, their role in improving processes in the past suggests they are likely to develop improvements and modifications in metallocene catalyst-based processes in the future. They are therefore likely once again to play an important role as diffusers of new process technology.

This willingness to license demonstrates that firms consider technology to be valuable but not necessarily the key source of competitive advantage. As documented in Arora (1997), the metallocene licensing is only the continuation of the trend that began in the 1970s. In addition to companies such as Union Carbide, Amoco, Montedison (later through Montell, its joint venture with Shell), Phillips, Exxon, and BP that have been licensing their technologies for quite some time, a number of leading chemical producers such as Dow, Monsanto, Du Pont, and Hoechst are actively rethinking their traditional reluctance to license. For instance, Dow expects to earn $100 million in licensing revenues by 2000, while Du Pont hopes to reach the same target by 2005. Monsanto has licensed its acrylonitrile technology and is looking to license its acrylic fiber and detergent technology (Chemical Week, 1997c.). Even Hoechst is reported to be contemplating a reorganization of its R&D structure, with an explicit emphasis on licensing technologies developed in-house (Chemical Week, 1996). As a result, technological capability is more evenly distributed than ever before.

A complementary trend is that R&D itself is being globalized. As Table 5 shows, American companies are directing a substantial fraction of their R&D spending overseas; in 1995 American firms spent $4.2 billion—nearly a quarter of their total R&D budget—on research overseas. We lack comparable figures for foreign firms but a recent survey found that, excluding biotechnology and pharmaceuticals, there were 42 foreign-owned chemical research laboratories in the United States, accounting for about $400 million in R&D spending and employing more than 11,000 people. Counting biotechnology and pharmaceuticals, the aggregate figures increase to nearly $3 billion in annual R&D spending, or about 15 percent of the industry total, and more than 30,000 employees (Florida, 1997).[28]

---

[26]There are other technology sharing alliances as well in this area, including Dow-Idemetsu and Exxon-Mitusi Petrochemicals in polyethylene, and Dow-Montell, Hoechst-Exxon, Hoechst-Mitsui Petrochemicals, and Fina-Mitsui Toatsu in polypropylene.

[27]For instance, a spokesman for Phillips is quoted as saying that the company is likely to offer its proprietary metallocene LLDPE slurry technology for license (Chemical Week, 1997b).

[28]According to a recent news report, in 1995 Hoechst spent a majority of its R&D budget outside Germany for the first time in its history.

TABLE 5 Company Financed R&D Performed Outside the U.S. by U.S. R&D
Performing Domestic Companies and their Foreign Subsidiaries

	All chemicals ($ millions)	Industrial and others ($ millions)	Pharmaceuticals, ($ millions)	All chemicals (percent)	Industrial and others (percent)	Pharmaceutical (percent)
1985	843	444	399			
1986	1071	579	492	12.36	11.56	13.45
1987	1243	625	618	13.16	11.68	15.09
1988	1548	855	693	14.30	14.42	14.14
1989	1532	609	923	12.83	9.47	16.75
1990	2007	720	1287	15.24	9.93	21.75
1991	2401	1009	1392	16.63	13.47	20.04
1992	2676	1045	1631	17.73	14.60	20.56
1993	2833	1318	1516	17.13	17.80	16.60
1994	2456	917	1539	14.83	13.22	15.99
1995	4194	1632	2562	24.19	22.87	25.11

Source: NSF/SRS, *Survey of Industrial Research and Development* (1995).

Internationalization of R&D is not entirely new to the industry. The famous
technology cooperation agreement in the 1930s between Standard Oil and I.G.
Farben was based on bringing together U.S. expertise in refining technologies
and German expertise in organic chemistry and coal gassification and liquefac-
tion. The patents and processes agreement between Du Pont and the British ICI
also had a patents and processes agreement that lasted for more than a decade
until the end of World War II. Similarly, in 1928 Shell Chemicals set up its
petrochemical R&D unit in Emeryville, California, rather than in the Netherlands or Britain. The international technology cartels that Solvay and Nobel put
together in the nineteenth century in alkali and dynamite, respectively, show quite
clearly that international technology cooperation has a long history in chemicals.
This internationalization has gained strength in recent decades, reversing the fragmenting effect of World War II.

Several forces are driving the current internationalization. Products now
have to be customized to meet the needs of local customers. Different regions of
the world appear to be becoming technologically specialized. Both factors reflect
a division between basic or general purpose research not tied to specific applications and downstream application and development research, which is decentralized and globally dispersed. Fundamental research, on the other hand, is
becoming geographically concentrated. Thus, many leading German firms view
Germany as the best suited for fundamental research in organic synthesis, Japan
for electronics chemicals, and the United States for life sciences.

This separation between general-purpose research and research specifically tailored to products and applications has important public policy implications. Traditionally, R&D, including basic chemical research, has been privately financed. Firms were larger, had more diversified portfolios, and hence could better appropriate the benefits of the new knowledge they produced. Under the current competitive pressures in the output and capital markets, companies are moving their R&D toward more applied, product-oriented research. Yet, many industry analysts argue that a key challenge facing the chemical industry is the development of a number of new methods and generic technologies that can be used as critical tools and knowledge bases for innovation in many chemical industry segments. For example, areas such as chemical synthesis or catalysis, as well as broad fields, such environmentally friendly products or processes require in-depth understanding of products, processes, and related phenomena before more specific problems and applications can be solved effectively. Similarly, although the life sciences have already witnessed several important scientific advances, there is still a significant need for fundamental research to enhance new product development opportunities in areas such as drugs, biocatalysts, and bioprocessors. The development of computerized modeling techniques has also become a key challenge. Although molecular models have been used for some time in the pharmaceutical sector, their use is now expanding in areas such as organic chemicals, as well as new materials and processes. Chemical modeling techniques require greater understanding of the fundamental aspects of the phenomena that have to be modeled as well as the creation of simulators, which are inevitably general in nature (Vision 2020, 1996).

In fact, the need for more general tools and methods is not confined to the newest technologies and scientific disciplines. Basic research will also be key in the development of new processes, in the design and engineering of new plants, and in enhancing the manufacturing efficiency of chemical production. The plant design and engineering tools in use today have not changed much from those of chemical engineering in the 1960s. A more systematic use of computerized software engineering tools in the design and operation of plants is in its early stages. On many occasions new plants are still built almost entirely "from scratch," with little re-use of concepts, tools, or even equipment used in similar plants. Thus, another R&D goal is to develop "modularized" equipment or process structures that can be employed repeatedly in many plants of similar type or nature.

A related issue is how to reduce the engineering costs of designing and constructing new plants as well as the unit production costs in plants that produce at smaller scales than has been the tradition. As noted earlier, unit production costs in chemicals have been lowered typically by increasing the scale of plants. But today's overcapacity problems imply that companies have to devise ways to obtain low unit costs of production even with plants of smaller scale. Advances in this area can come about only through a better understanding of process science and engineering to standardize the underlying structure of the processes as well as equipment and plant components.[29]

The questions that arise are who will carry out this basic R&D, and who will pay for it? The problem is serious because basic R&D is marked, much more than product-specific R&D, by high fixed costs and economies of scale. These costs can be amortized only if they can be spread over a very large downstream market or a large number of different markets. The chemical industry does have large and diversified companies, but in recent years these companies have been narrowing their product portfolios at the same time as the knowledge required is growing more complex and interdisciplinary. Reduced market opportunities for many chemical companies, caused by slower demand growth and increased competition, have intensified the problem (see, for instance, Lenz and Lafrance, 1996).

In short, today the chemical industry is facing the classic problem of market failure in R&D. Basic and long-term R&D offer the potential for significant long-term benefits, but individual firms may not have enough incentives to explore that potential. The solution will almost certainly involve industry-wide research projects in areas such as environmental technologies. Such agreements are already in place in the United States under the auspices of the Chemical Manufacturer's Association, and there are similar pressures in Europe, in environmental and many other chemical fields, coming from the European Chemical Industry Council (CEFIC) and the European Union itself (AllChemE, 1997). Inter-firm joint ventures and university-industry collaborations will also be important parts of the solution (Vision 2020, 1996; Lenz and Lafrance, 1996). Thus, wider collaborative arrangements will be key factors in providing new, critical enabling technologies that encompass many distinct product applications. Governments and public research agencies will play an important role in this respect, both in funding pre-competitive research and in improving cooperation in the upstream research.

One possible future outcome is that the major investments in basic R&D will be concentrated in a few regions or countries, with the new knowledge then being diffused or transferred more broadly. In some cases, the concentration is likely to depend on the location of the first movers, those that are the first to make the investments in a particular research area. We argued earlier that different countries or areas may become increasingly specialized in different research fields. It is likely that the United States will play a key role in this respect. The U.S. chemical industry has been the first to raise concerns about the need for fundamental chemical research, a problem that Europeans are only recently beginning to discuss (AllChemE, 1997). As a result, the United States as a nation may well bear a large fraction of the fixed cost that is needed to advance basic research in

---

[29]For instance, Shinnar (1991) reports the example of ICI, which succeeded in building a new 500-ton ammonia plant that was competitive with traditional world class ammonia plants producing 2000 tons of ammonia daily.

chemistry, particularly in key areas such as the life sciences, environment, and computerized modeling of products and processes, including software. Other countries will get access to this knowledge at marginal rather than average cost, but with a lag. This phenomenon has already been observed in the life sciences, where much of the basic research is being conducted in the United States, some of it by foreign-owned companies, with the results being used by companies from other countries.

As a final note, one should not forget the lessons of the SEFs in the 1960s. The knowledge that flows from one country to the rest of the world is not just "in the air." The transfer typically requires intermediating institutions that carry the burden of moving the knowledge across locations. The SEFs had the right incentives to "sell the technologies" because they had no stake in the product markets and hence were not restrained by the fear of creating greater downstream competition. Other intermediating institutions will probably play a very similar role in the next few decades. Small and medium sized research-intensive biotechnology companies have already acted as intermediating institutions in the early rise of the biotechnology industry, and the SEFs themselves are quite likely play this role again once new chemical process technologies become more standardized. Similarly, many independent software vendors specialized in commercial software for molecular modeling, process simulation, and the like are increasingly diffusing the computerized tools for more efficient automated chemical research and engineering processes.

CONCLUSIONS

The evolution of the chemical industry has been driven by advances in technology and by the institutions that have facilitated the growth of new markets. In addition to the conventional market growth in the form of demand from developing countries, the evolution of the chemical industry has also been profoundly affected by the growth of a market for technology and a market for capital. When technology becomes widely available, albeit at a price, it ceases to be a decisive source of competitive advantage, be it for firms or for countries. Instead, competitive advantage must be sought elsewhere, in cheaper inputs or in closeness to markets. Similarly, a global market for capital gives shareholders the opportunity to look for the best returns, putting managements under pressure to cut costs and improve shareholder value.

In the chemical industry, technological superiority was often a key component of competitive advantage, and the clear relationship between advances in chemistry and chemical engineering had led the market leaders to fund a substantial amount of basic research. Developments in the industry in recent years have weakened this incentive. These developments have tended to raise the payoff to applied, business-driven research relative to more basic and fundamental research. The industry has responded by forming industry-wide research initiatives in specific areas and by inter-firm joint ventures. However, there appears to be an important role for the public sector because basic research remains vital for the future of the industry. An equally important role may exist for international cooperation, at least in areas such as environmentally benign technologies and products.

REFERENCES

Aftalion, F. (1989). *History of the International Chemical Industry*. Philadelphia: University of Pennsylvania Press.

AllChemE. (1997). "Chemistry: Europe and the Future," Special Report, *http://www.cefic.be/allcheme/ indextext.html*.

Arora, A. (1997). "Patent, Licensing and Market Structure in the Chemical Industry," *Research Policy* (December).

Arora, A., and A. Gambardella. (1997). "Domestic Markets and International Competitiveness," *Strategic Management Journal* Special Issue 18:53-74.

Arora, A., and A. Gambardella. (1998). "Evolution of Industry Structure in the Chemical Industry." In *Chemicals and Long-Term Economic Growth*, Arora, Landau, and Rosenberg, eds. New York: Wiley.

Arora, A., and N. Rosenberg. (1998). "Chemicals: A U.S. Success Story." In *Chemicals and Long-Term Economic Growth*, Arora, Landau, and Rosenberg, eds. New York: Wiley.

Arora, A., A. Fosfuri, and A. Gambardella. (1998). "Division of Labor and International Technology Spillovers." mimeo. Pittsburgh: Heinz School, Carnegie Mellon University.

Bower, J.L. (1986). *When Markets Quake: The Management Challenge of Restructuring Industry*. Boston: Harvard Business School Press.

Chandler, A., T. Hikino, and D. Mowery. (1998). "The Evolution of Corporate Capability and Corporate Strategies within the World's Largest Chemical Firms." In *Chemicals and Long-Term Economic Growth*, Arora, Landau, and Rosenberg, eds. New York: Wiley.

Chapman, K. (1991). *The International Petrochemical Industry*. Oxford: Basil Blackwell.

CMA. (1997). "Statistical Yearbook." Washington, DC: Chemical Manufacturer's Association.

Chemical Week. (1996). "Hoechst Studies Venture Model For Research." *Chemical Week* (October 30):41.

Chemical Week. (1997a). "Univation's New Metallocenes." *Chemical Week* (November 12).

Chemical Week. (1997b). "Phillips Enters LLDPE Market with Metallocene Process." *Chemical Week* (February 5):8.

Chemical Week. (1997c). "Turning Process Know-How Into Profits." *Chemical Week* (July 23):45.

Cohen, W. M., and S. Klepper. (1992). "The Anatomy of Firm R&D Distributions." *American Economic Review* 82(4):773-799.

Da Rin, M. (1998). "Finance in the Chemical Industry." In *Chemicals and Long-Term Economic Growth*, Arora, Landau, and Rosenberg eds. New York: Wiley.

David, P. A., and G. Wright. (1991). "Resource Abundance and American Economic Leadership." CEPR publication number 267, Stanford University.

Eichengreen, B. (1998), "Monetary, Fiscal, and Trade Policies in the Development of the Chemical Industry." In *Chemicals and Long-Term Economic Growth*, Arora, Landau, and Rosenberg eds. New York: Wiley.

Florida, R. (1997). "The Globalisation of R&D: Results of a Survey of Foreign-Affiliated R&D Laboratories in the USA." *Research Policy* 26:85-103.

Freeman. (1968). "Chemical Process Plant: Innovation and the World Market." NIER, 45(August):29-51.

Haber, L.F. (1958). *The Chemical Industry during the Nineteenth Century.* Oxford: Oxford University Press.

Haber, L.F. (1971). *The Chemical Industry, 1900-1930.* Oxford: Clarendon Press.

Haynes, W. (1948). *American Chemical Industry.* Vol 1-6. New York: D. Van Nostrand Company.

Hikino, T., et al. (1998). "The Japanese Puzzle: The Development of the Japanese Chemical Industry." In *Chemicals and Long-Term Economic Growth,* Arora, Landau, and Rosenberg, eds. New York: Wiley.

Hounshell, D. (1995). "Strategies of Growth and Innovation in the Decentralized Du Pont Company." In *Innovations in the European Economy between the Wars,* Caron, F., P. Erker, and W. Fischer, eds. Berlin: Walter de Gruyter.

Hounshell, D., and J. K. Smith. (1988). *Science and Strategy: Du Pont R&D, 1902-80.* Cambridge: Cambridge University Press.

Keller, J. (1996). "A Historical Perspective on Polymers: The Science, Technology, Markets, and Institutions." Unpublished manuscript. Dept. Of Engineering and Public Policy, Carnegie Mellon University.

Landau, R. (1966). *The Chemical Plant: From Process Selection to Commercial Operation.* New York: Reinhold Publishing Co.

Landau, R. (1998). "The Process of Innovation." In *Chemicals and Long-Term Economic Growth,* Arora, Landau, and Rosenberg, eds. New York: Wiley.

Landau, R. (1997). "Education, Moving from Chemistry to Chemical Engineering and Beyond." *Chemical Engineering Progress* (January):52-65.

Landau, R., and N. Rosenberg. (1992). "Successful Commercialization in the Chemical Process Industries." In *Technology and the Wealth of Nations,* Rosenberg, Landau, and Mowery, eds. Stanford: Stanford University Press.

Lane, S. (1993). "Corporate Restructuring in the Chemical Industry." In *The Deal Decade: What Takeovers and Leveraged Buyouts Mean for Corporate Governance,* Margaret Blair, ed. Washington, DC: Brookings Institution.

Lenz, A., and J. Lafrance. (1996). *Meeting the Challenge: U.S. Industry Faces the 21st Century.* Washington, DC: The Chemical Industry, Office of Technology Policy, U.S. Department of Commerce.

Lieberman, M. (1987). "Patents, Learning by Doing, and Market Structure in the Chemical Processing Industries." *International Journal of Industrial Organization* 5:257-276.

Lieberman, M. (1989). "The Learning Curve, Technological Barriers to Entry, and Competitive Survival in the Chemical Processing Industries." *Strategic Management Journal* 10.

Manufacturing Chemists Association. 1971. *Chemical Statistics Handbook.* 7th Ed., Washington, DC: 432.

Mark, H. (1994). "The Development of Plastics." *American Scientist* 72:156-162.

Martinelli, A., ed. (1991). *International Markets and Global Firms.* London: Sage Publications.

Morris, P. J. T. (1994). "Synthetic Rubber: Autarky and War." In *The Development of Plastics,* Mossman and Morris, eds. Cambridge, U.K.: Royal Society of Chemistry.

Nelson, R. R. and G. Wright. (1992). "The Rise and Fall of American Technological Leadership: The Postwar Era in Historical Perspective." *Journal of Economic Literature* (December):1931-1964.

Richards, A. (1998). "Governance of Chemical Firms" In *Chemicals and Long-Term Economic Growth,* Arora, Landau, and Rosenberg, eds. New York: Wiley.

Rosenberg, N. (1972). *Technology And American Economic Growth.* New York: Harper Row.

Rosenberg, N. (1998). "Technological Change in Chemicals: The Role of University-Industry Relations." In *Chemicals and Long-Term Economic Growth,* Arora, Landau, and Rosenberg, eds. New York: Wiley.

Seymour, R. and G. Kirshenbaum, eds. (1986). *High Performance Polymers: Their Origin and Development.* New York: Elsevier Science Publishing.

Shinnar, Reuel. (1991). "The Future of Chemical Industries." *Chemtech* (January):58-64.

Spitz, P.H. (1988). *Petrochemicals: The Rise of an Industry.* New York: Wiley.

Stokes, R. G. (1994). *Opting for Oil.* Cambridge, U.K.: Cambridge University Press.

Thayer, A. (1995). "Metallocene Catalysts Initiate New Era in Polymer Synthesis." *Chemical & Engineering News* (Sept 11):1-8.

Vision 2020. (1996). *Technology Vision 2020.* The U.S. Chemical Industry, American Chemical Society.

Wright, G. (1990). "The Origins of American Industrial Success, 1879-1940." *American Economic Review* 80(4):656-668.

Steel

RICHARD J. FRUEHAN
DANY A. CHEIJ
DAVID M. VISLOSKY
Carnegie Mellon University

The U.S. steel industry is an interesting case study of competitiveness and innovation because of its recent history of near-economic collapse, followed by a rebirth fueled by a continuous drive for improvement (Ahlbrandt et al., 1996). By the 1980s, global competition, domestic labor disputes, and other factors had seriously undermined the foundations of the U.S. steel industry (Hoerr, 1988). In response to these competitive and economic pressures, many of the industry's large, integrated steel producers successfully restructured their organizations and operations. Today, the industry is highly competitive and profitable.

From the standpoint of R&D activities, the outcome has been somewhat surprising. Whereas R&D resources have decreased dramatically in the drive to cut costs, the U.S. steel industry's technology innovation performance as a whole has improved. Factors such as the effective management of R&D and technological resources; the acquisition of technology and innovative ideas from suppliers, customers, and competing steel producers; and collaborative research efforts have all created an environment that fosters improvements in production efficiency, technological developments, economic prosperity, and global competitiveness.

For the U.S. steel industry as a whole, R&D resources have been more effectively utilized in collaborative research efforts involving a number of companies, both domestic and foreign, their suppliers, and to a lesser degree, universities. These collaborations have contributed to the industry's innovative and economic performance, especially in the last decade.

Still subject to global and domestic competitive pressures, the U.S. steel industry is undergoing rapid changes even though research capabilities in the industry have been greatly reduced (Fruehan and Uljon, 1995; Ahlbrandt et al.,

1996).[1] Thus there is a motivation within the industry to explore R&D activity and project management practices in the wake of economic downsizing and against the backdrop of competitive, economic, and technological challenges.

This chapter documents the changes in the U.S. steel industry's production, productivity, profits, and R&D activities before, after, and during the industry's restructuring period in the 1980s. It also examines the industry's development and acquisition of technology, and the various sources of innovations and technology—including in-house R&D, relationships with suppliers and customers, government funded R&D, and collaborative research with various partners. In addition, it discusses various facets of the industry's R&D activity, as well as other factors that may influence the industry's competitiveness.

CHANGES IN INDUSTRY PERFORMANCE

By the late 1980s, the U.S. steel industry seemed to be in irreversible decline. In the previous decade, half the workforce employed in the U.S. steel industry— some 250,000 workers—lost their jobs, production of unfinished steel in the United States declined by more than 12 percent, and plant closures and downsizing brought the U.S. industry's production capacity down 25 percent (Ahlbrandt et al., 1996). Under severe financial pressures, research staff budgets for industry's internal R&D operations decreased by up to 75 percent throughout the 1980s (Dennis, 1991; Fruehan, 1996).

A National Academy of Engineering steel industry study conducted in 1985 concluded that the steel industry was no longer technologically progressive (Hannay and Steele, 1986). The study found that of 28 process advances under development, only two—direct reduction and continuous casting—were likely to be adopted in the next five years. The lack of R&D activity in new process development was attributed to the high capital cost of the research and low estimates of return on investment. The study concluded that leadership in technology alone would not rescue the domestic steel industry from its economic slump. Other factors, such as foreign pressures on price, labor productivity, cost of raw materials, energy, labor, plant location in relation to markets, and future estimates of production overcapacity would be equally and in some cases more important determinants of future performance. The following section examines how the U.S. steel industry has responded and restructured itself in terms of production, productivity, and financial performance during the last two decades.

---

[1]As evidenced by the recent explosion of electric arc furnace-thin slab casting plants, and other recent technological advances including massive coal injection in the blast furnace, and the large production of ultra clean and interstitial free steels (Albrandt et al., 1996).

Production and Market Share

By the early 1980s, foreign competitors—primarily in Europe and Japan—had made serious inroads into U.S. market share from sales of high-quality steel products. From a position of world dominance, U.S. steelmakers' share of the world steel market fell to approximately 10 percent because of foreign competitors' expanded capacity and their implementation of new and improved technologies. By 1983, Japan's share of the world steel market had grown to 16 percent, making it the new world leader. Since then, Japanese growth has slowed and its market share has decreased. Meanwhile, U.S. producers have made a partial comeback—thanks to the downsizing and restructuring of the integrated mills and the strong entrance of U.S. minimill operators.

For the last 20 years, U.S. production capacity has exceeded the actual production of raw steel (see Figure 1). This gap was largest in the early 1980s, when imports of raw steel also reached their highest point: 25 percent or more of the U.S. steel supply. In 1983, the gap between capacity and production was about 75 million tons. Recently, this gap has narrowed significantly; in 1996, it was less than 10 million tons. In comparison, world capacity has exceeded world production by more than 200 million tons for the last 15 years. The production gap has narrowed because the U.S. steel industry, especially the integrated producers, has improved its efficiency compared to a decade ago; now U.S. integrated producers are one of the lowest cost producers for their market. In addition, a larger ratio of capital investment per worker-hour has increased productivity.

In steel product markets where minimills have competed with integrated producers, minimills have gained market share because their costs, and thus their prices, have been lower. Minimills' ability to produce many types of steel products efficiently still exerts a constant pressure on the integrated producers. To-

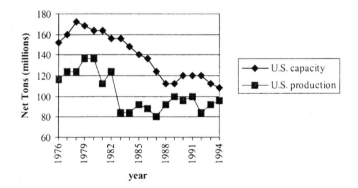

FIGURE 1 U.S. raw steel production and capacity.
Source: Cyert and Fruehan, 1996.

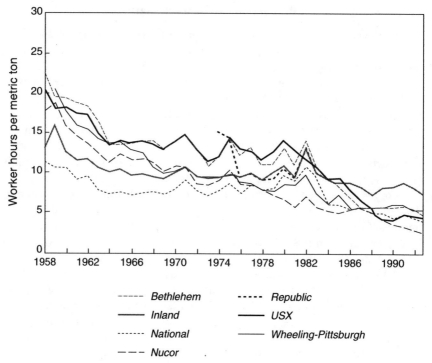

FIGURE 2 Labor-hours per ton produced: U.S. steel firms.
Source: Lieberman and Johnson (1995).

day, U.S. minimill producers such as Nucor rank among the most efficient steelmakers in the world (Fruehan et al., 1997).

Recently, integrated steel firms in developing countries such as Korea have become leaders in production efficiency. In fact, in 1996, the Korean firm POSCO was the world's most profitable integrated steelmaker and arguably the most efficient, at least until the recent economic crisis in Korea (Lieberman and Johnson, 1995). Another developing country, Brazil, has also improved in production efficiency. With its low labor costs, it may soon become a major factor in the global steel market.

Productivity

The U.S. steel industry has made remarkable improvements in productivity in the past 15 years. The following section discusses the changes in three measures of productivity—labor, capital, and total factor productivity. The section is based heavily on a study of productivity in the steel industry performed by Lieberman and Johnson (1995).

Labor Productivity

U.S. integrated producers have lagged behind their foreign competitors in terms of labor productivity since the 1960s. For almost two decades after 1964, U.S. integrated firms' labor productivity remained stagnant. However, labor productivity has been steadily improving among U.S. steelmakers. Notably, a standard measure of labor productivity in the steel industry—labor-hours per ton produced—shows U.S. performance increasing from a range of 7 to 14 labor-hours per ton in the early 1980s to approximately 5 labor-hours per ton a decade later (see Figure 2). In contrast, the labor productivity of Japanese steelmakers has remained steady at about the current U.S. level of 5 labor-hours per ton since the early 1970s, with only small incremental gains.

Improvements continued throughout the 1980s and the 1990s. This is illustrated in Figure 3, which shows the labor-hours per tonne for two major integrated producers, Bethlehem and U.S. Steel, and the largest scrap-based producer, Nucor.

Although a standard measure, labor-hours per ton fails to account for differences in the extent of diversification and vertical integration of firms; nor does the measure account for differences in steel "quality" and the extent of finishing operations. Also, different companies measure labor-hours differently, and issues such as contracting and outsourcing bias the statistics. For example, in Japan over half of the nonprofessionals in a plant are contract workers, while in the United States this figure has increased as much as 25 percent in some plants. The total number of labor-hours per ton may be 20 percent higher in Japan and 10 percent higher in some U.S. plants. In non-union plants, the percentage of contract workers is generally lower, and in some cases zero.

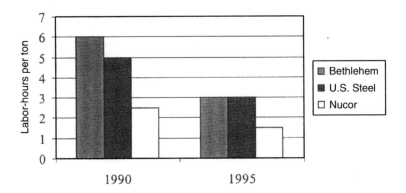

FIGURE 3 Labor productivity at leading U.S. steel firms.
Note: Labor-hours per ton based on the metric tonne (1000kg).
Source: Cyert and Fruehan (1996).

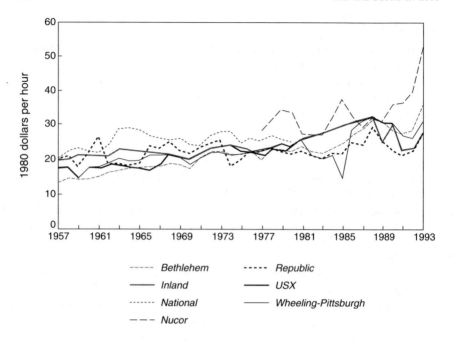

FIGURE 4 Value-added per worker-hour: U.S. steel firms.

To account for this bias, labor productivity may also be calculated in terms of value-added per worker-hour (Lieberman and Johnson, 1995). This measure accounts for employee effort and the use of capital.[2] Using the value-added metric, Figure 4 shows that labor productivity for U.S. steel firms remained stagnant between 20 and 25 value-added dollars (1980 U.S. dollars) per worker-hour until the early 1980s and began rising through the early 1990s to between 28 and 38 value-added dollars per worker-hour. In comparison with the labor-hours per ton, the trends for labor productivity show steady improvement since the early 1980s.

In contrast, Japanese steelmakers show a dramatic increase in value-added per worker-hour, increasing almost ten-fold since the late 1950s, and ending at 38 to 48 value-added dollars per work-hour in the 1990s. However, this dramatic increase is due in large part to the exclusion of workers who were dispatched to unconsolidated subsidiaries and the heavy outsourcing initiated by Japanese steel firms, both of which were common practices in Japan in the 1980s.

---

[2]Value-added is the difference between a firm's total sales and its purchases of raw materials and contracted services.

Capital Productivity

Before 1980, the capital intensity of U.S. steel firms, measured by capital investment per worker, grew very little. Because integrated steelmaking is a capital-intensive and competitive global industry, U.S. producers found it difficult to earn the rates of return necessary to justify substantial new investment. In fact, no new integrated steel plants have been built in the United States in the last 35 years, and only recently has the industry invested in additional production capacity (Fruehan et al., 1997). However, primarily because of the massive downsizing at U.S. steel firms in the 1980s, U.S. capital intensity grew substantially, from a fixed investment per worker that was below $70,000 in 1980 to over $100,000 in 1993 at all surviving U.S. firms except for Inland (see Figure 5). Yet the U.S. investment per employee is less than half that invested by Japan and four times less than Korean firms in that same time period. These differences reflect slightly leaner staffing by Japan and Korean firms and also higher rates of plant and equipment investment—in the case of Korea attributable in large part to heavy government subsidies to its steel industry, and in the Japanese case attributable in part to encouragement from the banking system (Fruehan et al., 1997).

Total Factor Productivity

Total factor productivity, which is regarded as a more appropriate measure of overall efficiency in production plants, is a weighted average of labor productivity and capital productivity. From the late 1950s to the 1990s, the total factor

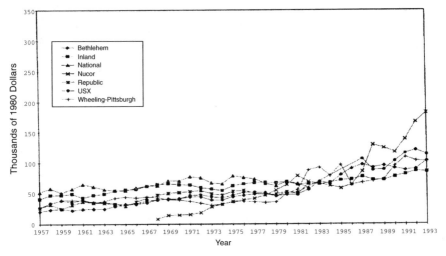

FIGURE 5 Fixed capital per employee: U.S. steel firms.

productivity of U.S. steel firms rose about 50 percent (see Figure 6). Currently the total factor productivity for Japan, Korea, and the United States is roughly equivalent. Interestingly, the steel industries in all three countries have shown very different trends in their capital and labor input, but they have each arrived at comparable efficiency levels in the last decade.

Quality Improvement

In addition to improving productivity, U.S. steel makers have also dramatically improved quality during the last fifteen years. Much of the improvement has stemmed from technological advances, such as secondary refining and continuous casting. But "working smarter," through training, continuing education, and quality control, has also been critical.

One measure of quality improvements is customer acceptance. The U.S. steel industry's most critical customer has been the automotive industry. A decade ago, rejection rates for steel of poor quality at automotive companies were typically three to six percent. Today, the rejection rates are about 0.5 percent—a tenfold improvement. Other examples of quality improvements are the new steel grades and types, such as corrosion-resistant steels. Before these new grades existed, automobiles in the northern United States suffered extensive corrosion or

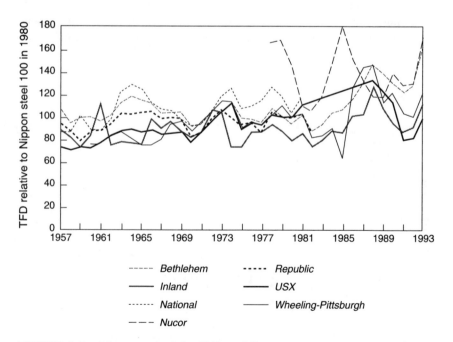

FIGURE 6 Total factor productivity: U.S. steel firms.

rust within five years. Today, automobiles that incorporate these new steel grades last fifteen years or longer before extensive corrosion and rust occurs. Furthermore, steels are stronger, lighter, and more formable for specific applications and the production of complex products.

A significant impetus for many of the quality improvements was the Japanese automobile producers located in the United States. They demanded higher quality steels than were previously produced, and they also required extensive quality control within steel production plants. Once it became clear that these high-quality steels could be produced, U.S. automotive firms and other industrial manufacturers soon demanded similar quality from the steel industry.

New Products and Processes

Steel production is continually evolving, and new innovative steel products are now in common use. Half the steel grades or types produced today did not exist fifteen years ago. Examples of these new steels include:

• *Corrosion Resistant Steels*: The past decade has witnessed a significant improvement in the manufacture of steels with much higher corrosion resistance, especially through the development of new coating and galvanizing processes, as well as new methods of applying these coatings.

• *High Strength Low Alloy (HSLA) Steels*: These steels are much stronger than traditional steels and can reduce the amount of material required in their production, thus reducing the total weight of the steel.

• *Interstitial Free Steels*: These steels can be formed into intricate shapes without flaws. They are used extensively for exposed applications in the automotive industry.

The development of these new steels was primarily driven by customer demand (Fruehan et al., 1994). However, new processes made it possible to produce new steels with superior properties, and some of those products were developed before market demand existed for them. New processes were generally developed to allow the production of better quality steel or to reduce the cost of production.

The industry has also developed or implemented several major processes in the past decade. These are listed below:

• *Continuous Casting*: Incremental improvements have led to methods that allow all grades of steel to be continuously cast, with fewer surface imperfections and cracks. Today, use of these methods is universal.

• *Secondary Refining*: Improvements in a number of processes, including desulfurization, inclusion removal, and reheating, have significantly improved productivity and steel quality and have given steelmakers much greater control over the composition of their steel output.

- *Vacuum Degassing*: This process, which involves the treatment of steel in a vessel under a vacuum, has enabled the production of interstitial free steels and other special quality steels, which represent over one-fifth of the total U.S. production of steel.
- *Electrogalvanizing*: New processes have been developed to improve the coating and, hence, the corrosion resistance of steels.
- *EAF High Productivity*: A number of process improvements, including ultra high power furnaces, carbon and oxygen injection, and water-cooled panels, have doubled productivity and decreased electrical energy consumption by nearly one third.

Financial Performance

Some researchers have suggested that the competitive decline of the U.S. steel industry in the 1980s has resulted, in large part, from inferior management practices and low labor productivity. Specifically, managers of U.S. steel firms were criticized for promoting an incentive system that rewarded short-term success and failed to encourage capital investment in the new technology needed to compete globally. However, in the last decade, the U.S. steel industry has experienced a steady turnaround in profitability and market share and has invested in additional production capacity. By the late 1980s, the economic performance of the U.S. steel industry, particularly its integrated sector, had improved significantly. Today, U.S. integrated producers have the highest profitability per ton of steel produced in the world. Some key aspects of industry financial performance are discussed in the next section, which is based primarily on a study by Baber and colleagues (Baber et al., 1993) (see Table 1). However, it should be noted that their study extends only to 1993, and industry performance has improved substantially since then.

Using return on assets[3] (ROA) as a measure of profitability, Baber's study (see Figure 7) noted the following:

- The steel industry is less profitable than other U.S. industrial firms. Mean accounting rates of return are 2.95 percent for steel, compared with 9.17 percent for all U.S. industrials.
- The difference in profitability is attributed to the integrated steel firms, which have a mean return of 2.23 percent, far lower than the mean return of 8.09 percent for non-integrated steel firms.
- Non-integrated firms that produce specialty steels are slightly more profitable than non-integrated carbon steel producers.
- The financial performance of the integrated steel firms was worst from 1981-1986.

---

[3]ROA is determined from the product of asset turnover and profit margin.

TABLE 1 Summary of Financial Ratios (1971-1990)

	All Industrials[a]	Integrated	Non-integrated
Return on assets (ROA)	0.09	0.02	0.08
Net income/sales	0.05	−0.01	0.03
Capital expenditure growth[b]	0.10	0.04	0.11

[a]All industrials is defined as the top 30 U.S. industrial firms.
[b]Mean growth rate represents a geometric average over the 1971-1990 time period.
Note: Ratios are presented as mean values.
Source: Baber et al. (1993).

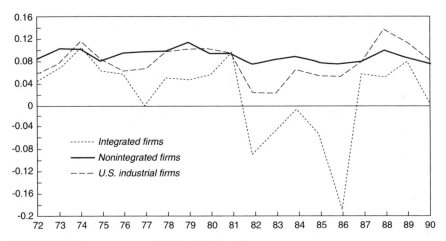

FIGURE 7 Return on assets (ROA): U.S. steel firms.

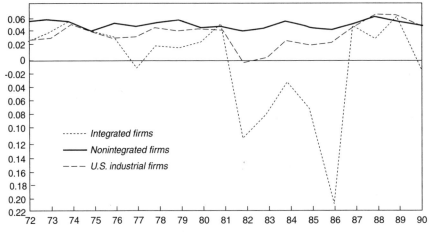

FIGURE 8 Net income/sales: U.S. steel firms.
Source: Baber et al. (1993).

TABLE 2 Profitability Ratio: Return on Assets (%) 1975-1993

	Integrated	Non-integrated	S&P 500 firms
All periods (1975-1993)	–0.52	7.35	6.53
1975-1981	4.33	9.57	7.59
1981-1987	–6.81	5.06	6.59
1987-1993	0.10	7.04	5.27

Source: Fruehan et al. (1997).

Although U.S. integrated steel producers have been generating a persistently lower return on assets than non-integrated producers and other industrial firms, the integrated firms have recently improved in profitability compared with the 1980s (Table 2). This dramatic turnaround has occurred primarily because the industry has reduced its costs, increased production efficiency, and increased overall sales.

However, the production costs of integrated steel producers for some products are higher than the costs for non-integrated steel producers. Thus, integrated firms are unable to produce steel at costs that are less than prevailing prices, which adversely affects profits.

Because the integrated steel industry entails highly capital intensive production, it is difficult to generate the necessary returns to adequately invest in capital improvements and new technology. The capital expenditure growth for U.S. integrated steelmakers is low and has been stagnant for the last three decades (see Figure 9). In contrast, the rapid introduction of the minimill producers in the

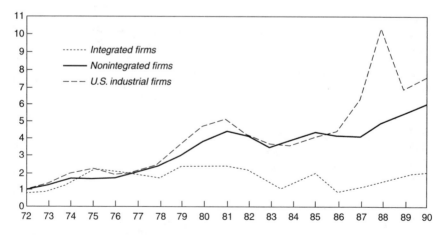

FIGURE 9 Capital expenditure growth.
Source: Baber et al. (1993).

1980s was accompanied by growth in capital spending by non-integrated carbon steel producers, which exceeds the integrated producers.

Overall sales for the U.S. steel industry have risen steadily since the economic decline in the early 1980s. Sales for integrated producers have grown from an average of $3.2 billion in 1986 to $4.8 billion in 1995 (see Figure 10). In comparison, overall sales for the most profitable minimill producer, Nucor, have risen from $0.8 billion in 1986 to $3.5 billion in 1995. These sales trends, combined with lower production costs and increased productivity, suggest a remarkable economic turnaround for the U.S. steel industry. Today, overall sales by all U.S. steel producers are about 3 percent higher than in 1996, continuing an upward trend in recent years, and totaling about $36 billion in 1997 (Pittsburgh Post Gazette, 1997).

CHANGES IN THE STRUCTURE OF INNOVATION PROCESSES

R&D Structure and Operation

The following review of changes in R&D expenditures, personnel, and research effort is based on two recent surveys of R&D activity in the global steel industry (Fruehan, 1994; Fruehan and Uljon, 1995).

R&D Expenditures

Most large North American integrated steel producers spend only about 0.5 percent of sales on R&D activities; a number of other producers spend little or nothing (Fruehan et al., 1995) (see Figure 11). In comparison, the international

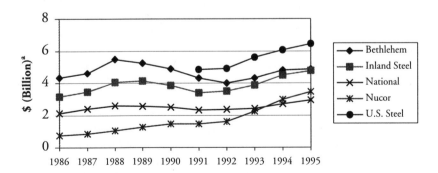

FIGURE 10 Overall sales in billions for selected steel firms.
[a]Current year dollars.
Note: Selected firms surrogates for U.S. integrated steel industry except Nucor, the leading minimill producer. U.S. Steel data begins in 1991.
Source: Compact disclosure (1997).

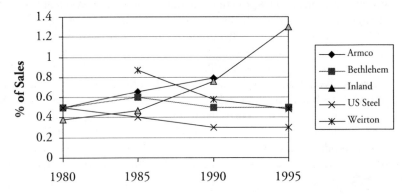

FIGURE 11 R&D expenditures as a percentage of sales for integrated producers.
Note: Selected firms are surrogates for the U.S. integrated steel industry. There were no data avail-
    able for Weirton in 1980 and for Armco in 1995.
Source: Fruehan and Uljon (1995).

firms surveyed indicated that they spend about twice that amount on R&D activities, or 1 percent of sales.

R&D Personnel

As the steel industry grappled with economic decline in the early 1980s, the
average number of R&D personnel in integrated steel companies fell sharply,
from 498 personnel in 1980 to 282 for the five major integrated producers in
1985, a 43 percent decline. A few of the largest integrated firms displaced as
many as 60 to 90 percent of their R&D personnel since 1980 (see Figure 12).
This sharp drop was followed by a steady decline that continued into the 1990s;
between 1985 and 1995, the average number of R&D personnel dropped 46 percent.

Whereas there was a large decrease in R&D personnel for integrated companies since 1980, the percentage of workers in R&D increased from 0.6 percent to
1 percent of the total workforce.[4] Minimill producers have always employed
very few R&D personnel. Nucor had only four employees designated as R&D
personnel in 1991. Although few in number, the R&D personnel in minimills has
grown slightly from the mid-1980s to the present. Even while releasing large
numbers of R&D personnel, U.S. steel firms retained more professional R&D
staff (those with relevant college degrees) than non-degreed staff (see Figure 13).

---

[4]Average percentages determined from the total employees in the four largest integrated producing
firms: Bethlehem, Inland, National, and U.S. Steel.

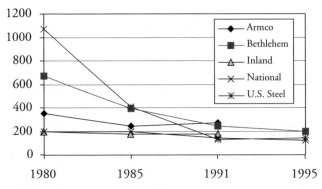

FIGURE 12 R&D personnel in integrated firms.
Note: Selected firms are surrogates for the U.S. integrated steel industry. There were no data available for Armco or National in 1995.
Source: Fruehan et al. (1995).

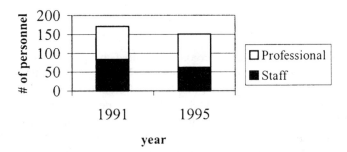

FIGURE 13 Average professional and staff R&D personnel in integrated firms.
Note: Average number represents R&D personnel at Bethlehem, Inland, National, and U.S. Steel. There were no data available for Inland in 1995.
Source: Fruehan et al. (1995).

About 40 percent of the professional R&D personnel at two of the largest integrated companies hold doctorate degrees.

R&D Effort

Even after the economic decline in the early 1980s and until the mid-1990s, R&D organizations at integrated firms concentrated on long-term, applied research, spending about three times more on that effort than on short-term research (see Figure 14). That focus did not extend to fundamental research, however. Only one large integrated firm performed some fundamental research in 1991. Most firms abandoned their fundamental research to focus primarily on applied

R&D and technical assistance projects. In contrast, non-integrated firms divided their effort evenly between long-term and short-term R&D and did not focus on fundamental research at all.

By 1995, R&D effort at U.S. steel firms had shifted dramatically. Integrated producers focused over three times more effort on technical assistance and almost twice the effort on short-term, applied research than on long-term product and process research (see Figure 15). Similarly, non-integrated firms devoted more than twice as much effort to short-term technical assistance as to long-term research.

These changes represent not only a shift in the type of research conducted by R&D organizations in the steel industry but also a shift in the R&D organization's technical objectives and their relationship with the production plants, suppliers, and customers. Because of budget and personnel constraints, the R&D organizations at integrated firms had to focus primarily on the problems, requests, and requirements of the production plants and their suppliers and customers and spend less effort on risky and long-term research.

R&D Structure and Operation

As the steel industry has undergone substantial changes, so has the organization of R&D divisions within the steel industry (Vislosky, 1996; Vislosky, 1998). These changes can be categorized into three different business eras: (1) the decades before the 1980s, when the U.S. steel industry dominated the world marketplace; (2) the early to mid-1980s, when the integrated steel industry experienced financial difficulties and was forced to cut back its R&D operations; and (3) the 1990s, where the integrated steel firms have made a steady financial comeback to

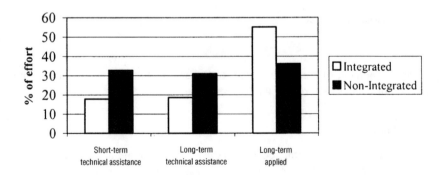

FIGURE 14 Average distribution of R&D effort by project type in 1991.
Note: One integrated firm spent 11% effort on fundamental research.
Note: For integrated firms, 8.3% of the effort is other research.
Source: Fruehan (1994).

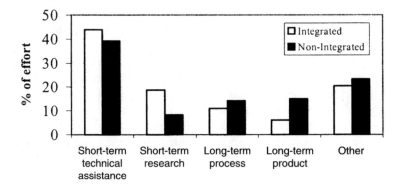

FIGURE 15 Average distribution of R&D effort by project type in 1995.
Note: No firms conduct fundamental research.
Source: Fruehan and Uljon (1995).

become profitable again, and non-integrated firms continue to enjoy seasonally strong profitability.

Integrated Steel Firms

In the 1960s and 1970s, integrated producers dominated the worldwide marketplace for steel. Because of their large market share and control of prices, the integrated firms enjoyed large profit margins. As a result of this economic prosperity, the R&D organizations within the firms enjoyed large budgets and freedom to pursue many different types of relevant research. The R&D budget was funded by the firm, and the number of personnel remained relatively constant. The headquarters for R&D was centrally located, but the R&D organization serviced all business units and holdings of the firm. At the time, many of the integrated producers owned and operated not only the iron and steelmaking operations, but also the raw material's suppliers, and the process equipment producers. The R&D organization retained the skills and capability to service each of these separate business units and its unique needs.

The R&D organizations were typically hierarchical in structure in which a research director, or in some cases a vice president of research, oversaw all R&D operations in the firm. Reporting to the director or vice president were the managers of each division. The divisions were organized by technology area, such as steelmaking, chemicals, and products. Sub-divisions existed to address specific process and product technologies within each area. In addition, there was usually a lab devoted to fundamental research. Since the R&D organizations were staffed by competent and skilled individuals trained in the various disciplines related to the steel firm's operations, most of the firm's research needs were performed in-house. In addition, rather than outsourcing research, U.S. steel firms actually sold some of their technology to other firms in the United States and abroad.

Most R&D projects originated from within the R&D organization as suggestions and proposals by the research staff. The motivation behind new projects was often initiated from a competitor's activities, customer and plant requests, and improvements to products and processes. Most projects at any one time were ongoing from previous years, and new projects were introduced annually. These research projects consisted mainly of applied process and product research with an average life span of three years. In addition, the R&D organization serviced the various business units of the firm by providing technical service and addressing short-term research problems.

After the crisis of the 1980s, the R&D organization was considered an expensive luxury in difficult financial times. The R&D organization was forced to sell its services to the rest of the firm, and research projects were funded by individual business units throughout the firm, such as production plants. In some cases, the research was still funded by the corporation, but the R&D organization was responsible for advocating its worth and the value of each project directly to the production units. Research objectives shifted to an opposite extreme: technical assistance and problem-solving became the primary focus of the R&D organization. Long-term applied research still took place but usually only if such research could directly benefit the customers and the production plants. In addition, the costs, time schedule, and results of applied research were always under scrutiny by upper management, and immediate beneficial outcomes were expected from all research projects.

This environment caused the integrated steel industry to focus on short-term gains and immediate results from research. R&D organizations were more inclined to pursue less risky, incremental research projects that were of direct relevance to their customers and production plants. As a result, the integrated steel industry introduced very few new technological advances in its production processes, and product advances were more often incremental improvements rather than new products or processes.

This cautious and incremental R&D environment continued throughout the 1980s. Only recently has the U.S. steel industry experienced a comeback in the global marketplace. As a result, the remaining R&D organizations in the industry have examined their current operations. Although small in terms of budget and personnel, these organizations are beginning to reexamine their role in the context of the firm by directly incorporating the corporate strategic plan, the firm's marketing plan, and input from the plants, suppliers and end-users into their own technical plan. In addition, these organizations are making efforts to pursue long-term, applied research. They are also entering into partnerships with competing firms and end-users. An example is the ultralight steel auto body partnership between Porsche Engineering Services and 15 steel firms (Porsche Engineering Services, Inc., 1995).

Non-Integrated Steel Firms

Very few non-integrated firms have any formal R&D organization. Most of these firms have small research groups that provide technical assistance to the plants. The non-integrated producers were much less affected by the economic downturn in the U.S. steel industry in the early 1980s. In fact, the non-integrated producers actually contributed to the economic woes of the integrated producers by acquiring some of their market share in high-quality, complex steel products.

Sources of Innovation

The various internal sources of innovation that affect a firm's overall innovative process are examined below. The firm's own R&D laboratories and joint ventures between companies, both domestic and international, are discussed first. Then the discussion shifts to innovations originating with suppliers and turns to university contributions to industrial innovation.

Steel Company's In-house R&D Laboratories

One of the main sources of innovation in the steel industry remains a firm's own internal R&D labs. This has remained the case despite the major cutbacks in in-house R&D activities that most integrated steel firms went through in the mid to late 1980s. These cutbacks have resulted in smaller numbers of available man-hours that can be devoted to general innovative research that has a higher probability of yielding breakthrough innovations. Instead, most of the internal effort has been devoted to research that can result in incremental improvements to existing innovations. In addition, most researchers at firms' central research centers have taken on the role of technical consultants to the firms' various steel-producing plants. For example, researchers may be asked to help the engineers at a plant solve a technical problem that affects the way a machine functions or the quality of its output. Conversely, a plant engineer may contact the company's research center and ask them to perform a research experiment, such as a study of the effect of adding a certain amount of an alloy to a grade of steel.

Joint Ventures with Other Steel Companies

Most U.S. steel firms have joint ventures or general technology agreements (GTA) with other domestic producers. Examples of major joint ventures are listed in Table 3.

The joint ventures between Inland Steel and Nippon Steel involved state-of-the-art facilities. Although they helped reduce the cost and time of production, they do not justify the high capital investment that was required. Also, there has been little innovation spillover to other areas of the firms. The USS-Kobe plant is

TABLE 3 Examples of Joint Ventures in the U.S. Steel Industry

Company	Venture	Partners	Activity
Inland	INTEK	Nippon Steel	Cold rolled sheet
	INKOTE	Nippon Steel	Coated sheet
USX	USS-Kobe Lorain	Kobe	Steel plant
	UPL	POSCO	Finishing plant
LTV	Trico Steel	Sumitomo/ British Steel	Steel plant

Source: Fruehan and Vislosky (1997).

virtually an independent company and in many ways does not perform as well as other USS plants. The USX venture with POSCO has not led to major innovation in USX itself. The Trico plant began operations only in 1997, and its impact is difficult to assess.

The best example of general technology agreements is the GTA between Inland Steel and Nippon Steel. Nippon Steel had as many as 100 engineers teaching Inland Steel engineers how to improve the quality of their automotive steels. Their primary focus was on the Japanese auto transplants. Sumitomo Metals also has long-term agreements with LTV Steel, and other U.S. companies have reasonably successful agreements with Japanese companies. Joint agreements with companies in countries other than Japan have been less productive.

The best example of joint research agreements is the agreement between USS and Bethlehem Steel. About 5 to 10 percent of both companies' research is devoted to selected joint projects. This program has been considered successful and has led to innovations in casting. Other arrangements, such as those on strip casting projects between a number of companies, have been unsuccessful (Fruehan and Vislosky, 1997). To date, joint ventures with foreign producers have had limited innovation spillover to other parts of the company. GTAs have been successful when focused on a specific task, whereas the general exchanges have not led to significant innovation.

Innovations by Suppliers

Suppliers of technology to the steel industry have been a major source of innovation (Fruehan et al., 1994). The best-known example—the SMS thin slab caster—has caused a revolution in steelmaking. Other examples include innovations in EAF steelmaking, continuous casting, and finishing. With the decrease in steel industry research, technology suppliers must continue to take major responsibility for equipment innovations. Joint developments with U.S. firms are extensive and are generally viewed as successful (Dennis, 1991).

University Contribution to Innovation

Universities have not aimed their research to make a major innovation but rather to develop the basic knowledge to aid the steel and steel supply companies in their activities. The two major research consortiums are at the Colorado School of Mines (steel rolling and finishing research) and Carnegie Mellon University (ironmaking, steelmaking and casting). These centers receive nearly all their funding from the steel industry, with each center having over 20 industrial partners. Furthermore, the center at Carnegie Mellon is international with about ten foreign firms participating.

While it is difficult to show that university research alone has produced any major innovations, it is clear that it has provided the fundamental understanding that has supported new innovations. Universities also contribute to innovation through consulting activities between industry engineers and individual professors. This exchange of ideas, although less formal than contacts through the steel centers described above, is nevertheless important. It provides a means for university professors to share results and insights from their research projects that might be of use to industry engineers. Universities also contribute through the transfer of knowledge. When young graduates or more seasoned academics join a steel firm, they bring a fresh perspective and greater creativity.

To help quantify the role of universities and other sources of innovation, a recent Sloan Study project devised a measure—a count of article citations of patents relating to specific innovations. Preliminary results show that university-authored articles accounted for close to 20 percent of article citations in patents issued for interstitial free steel and about 30 percent of article citations in patents relating to direct ironmaking (Cheij, 1997).

Future Directions of Innovation

The gap between the steel industry's technical needs and its R&D resources remains an area of concern. This gap is evident in the study of R&D activity described above. In the study, several potential major new technologies were identified. The companies surveyed were asked which of the new technologies were critical to them and whether they had a related research program. Between one-half and three-quarters of respondents indicated that the technologies were important, but typically less than 35 percent of those indicated that they had a related research program on a given technology (see Figure 16). To address this gap, and to offset project costs, steel producers may be required to participate in collaborative efforts with competitors, customers, and suppliers.

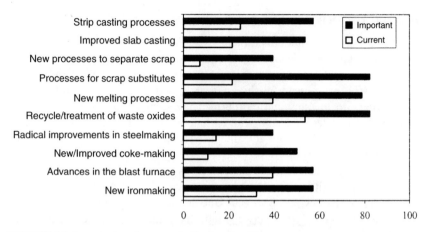

FIGURE 16 Important and current technology areas.

Note: Percentage of respondents from Fruehan and Uljon (1995) survey. Respondents include 28
       domestic and international steel firms.

Source: Fruehan and Uljon (1995).

OTHER FACTORS INFLUENCING COMPETITIVENESS
AND INNOVATION

Technological innovation is only one factor that influences competitiveness.
The impact of the major factors on both the competitiveness and innovation of
U.S. steel firms is summarized in Table 4. Only two factors have had a high

TABLE 4 Relative Impact of Factors other than R&D on Competitiveness and
Innovation

	Competitiveness	Innovation
Minimills	H	H
Customers	H	H
Human resources	H	L
Education and training	M	H
Trade issues	H	L
Foreign investment	M	M
Regulatory policy	H	M
Government support of R&D	L	H[a]
Internationally funded R&D	L	M[b]

[a]Government funded R&D has had a major effect in Japan and Europe, but a medium effect in the
United States.

[b]International funding has had a minor effect in the United States and Japan, but a high one in Europe.

Note: H = high, M = medium, and L = low.

Source: Fruehan and Vislosky (1997).

impact on both competitiveness and innovation—minimills and customers. Most of the other factors that have had a high impact on competitiveness have had a low impact on innovative capacity, or vice versa. This suggests that all the factors mentioned below are important.

Minimills. Minimills have been a tremendous source of innovation, including innovations in technology, management, and human resources. This is due in part to their flexibility in process, management, and labor relations. In general, minimills have not originated technological concepts, but they have implemented, adapted, and optimized processes effectively. The classic example is thin slab casting, which was developed in Germany but successfully commercialized in the United States. Other areas in which minimills are leaders in innovation include scrap substitutes and electric furnace improvements. Minimills have also contributed to improved competitiveness by reducing steel costs and forcing the integrated industry to restructure by closing inefficient plants and concentrating on high-quality steels.

Customers. Customers, particularly in the automotive industry, have been a source of both competitiveness and innovation. Spurred by the Japanese auto transplants, foreign and domestic auto producers placed a significant amount of competitive pressure on steelmakers to improve quality. At the same time, customers also became a source of innovation. Steel producers worked with the auto industry to improve the quality of existing steels and to develop new and improved steels. An example of this collaboration is the optimization of the production of corrosion-resistant steels and their use.

Human Resources. Workers' productivity has increased by nearly 300 percent in the past decade, as discussed earlier. These gains have been achieved not only through new technologies but also through innovative human resources practices, reducing labor costs by over $100 per ton, 25 percent of the total cost of production. Thus, labor considerations have been a driver for technological change, but rarely have they contributed to innovation.

When new technologies are introduced in union facilities, it is usually necessary to negotiate new agreements on working conditions and standards. Of the 20 million tons of new capacity currently being built in the United States from 1990-2000, virtually none is in union plants. Labor represents 10-15 percent of the total costs of steel production in existing plants, and less than 10 percent in new plants. Therefore, there is only room for small improvements in this area.

Education and Training. Education and training of workers in the steel industry is continually evolving to keep up with and respond to new technological innovations that are changing the industry. Approximately three-quarters of the steel industry's on-the-job training is associated with new and emerging technologies. Through such training workers will be better equipped to handle new machinery and to produce higher quality steel.

Trade Issues. Trade, particularly unfair trade, has been a major competitive factor in the United States. In general, the steel industry is profitable when capacity utilization rates are over 90 percent and unprofitable when they are below 80 percent. Imports have averaged about 18-20 percent during the last decade but have exceeded 25 percent in the past. When imports are high, capacity utilization may decrease, resulting in poor financial performance and fewer resources available for innovation.

Imports depend largely on exchange rates and relative production requirements in the United States and abroad. Import sources are shifting from Europe and Japan to developing countries. Imports generally result from the inability of domestic producers to fill all the country's needs or from overproduction in the exporting countries. Currently, the U.S. industry is the low-cost producer for its domestic market.

Foreign Investment. Foreign companies, especially Japanese companies, have invested heavily in the U.S. steel industry. In particular, Nippon Kokan owns much of National Steel, Nippon Steel has invested in Inland Steel, and much of AK Steel (Armco) was at one time largely owned by Kawasaki Steel. Much was expected in terms of technology transfer from Japan. However, these investments proved to be poor and little technological innovation resulted. In fact, these companies have done more poorly than similar integrated companies. Soon after Kawasaki Steel sold its interest in AK Steel, AK became very profitable under U.S. management. Whereas Japanese investment in the U.S. auto industry has been highly successful, its investment in the steel industry has been a relative failure in terms of both profits and innovation.[5]

Regulatory Policy. Regulatory policy to protect the environment has been a major driver of technological innovation, especially in ironmaking, including the elimination of cokemaking and the recycling of waste. In 1997, in response to concerns about global warming, some of the largest U.S. steel firms formed a coalition with the American Institute of Iron and Steelmaking to present a voluntary industry plan to cut emissions of greenhouse gases by 10 percent from 1990 levels by the year 2010 (Pittsburgh Post Gazette, 1997). In return, industry officials requested that the government provide more federal investment in R&D as well as tax incentives for development of new energy-efficient technologies. Thus, environmental concerns strongly influence the types of R&D projects pursued by the steel industry.

---

[5]The reasons for this failure are the subject of a new research project in the Sloan Steel Industry Study.

Government Support of R&D. The federal government, particularly through the Department of Energy, has provided a significant stimulus to technological innovation. DOE committed about $95 million for R&D projects in the steel industry through fiscal year 1994 (Cyert and Fruehan, 1996). In particular, government funding of programs in direct ironmaking and process control have been effective, in part because the government has not attempted to manage the programs. The results of government support are beginning to have some impact on competitiveness, but their full effect may not be fully realized for ten years or more in the U.S. Elsewhere, especially in Europe and Japan, government funding has been much greater and has had more effect on innovation. In Japan, MITI has sponsored many large "National Projects." In Europe, governments have also funded individual projects and institutes devoted to steel.

Internationally Funded R&D. There has been surprisingly little international funding for R&D. Individual companies have engaged in technology exchanges, but there has been little actual joint research or development. Regionally funded R&D has been extensive, particularly within the European Union (EU). For many years, steel companies in countries in the EU have been taxed on each ton of steel produced. The tax has funded a range of R&D activities, from fundamental university and institute research to major commercial demonstration projects, such as coal injection into blast furnaces. The EU program has been reasonably successful and will continue to be so. The American Iron and Steel Institute carries out research sponsored by U.S., Canadian, and Mexican companies, but the program is voluntary and much smaller than the EU program.

One major international program has been launched in response to the "Partnership for a New Generation of Vehicles." More than 20 companies from Japan, Europe, and America are funding work to develop a more fuel-efficient, steel-based automobile, the Ultra Light Steel Body Program (Porsche Engineering Services, 1995).

LINKS BETWEEN THE INNOVATION PROCESS AND INDUSTRY PERFORMANCE

Some analysts argue that an investment in R&D takes five to ten years to begin to yield a substantial return and that the U.S. industry is currently benefiting from previous R&D. This argument is only partially true. Ten years have elapsed since the major R&D restructuring of the 1980s and the industry is doing better than anytime in recent history.

Technological innovation alone does not determine a firm's competitiveness. Other factors including competitors' actions and customer demand, human resources, trade issues, capital availability, market selection, foreign investment, regulatory policies, and funding sources have as great, if not a greater, impact on competitiveness in the U.S. industry.

Nevertheless, there are dramatic examples of technology innovation that clearly affect competitiveness. In particular, the use of thin-slab casting techniques by Nucor and other minimill producers, and other quality improvement innovations that have been implemented by a number of major integrated companies to produce the highest quality steel at the lowest cost, have allowed both types of steel producers to achieve higher levels of productivity and profitability in recent years.

Although new innovations do affect competitiveness in the steel industry, there is no obvious trend between the industry's in-house R&D spending and its economic performance. R&D spending at the major integrated firms decreased drastically in the mid-1980s shortly before these firms began making their greatest increases in productivity, followed by increases in profitability. The minimill producers have had little or no in-house R&D and yet have performed well during this same period. It could be argued that the minimills are living off the research of others. In contrast, it is not clear whether major international firms such as Nippon Steel, Usinor, and POSCO have had good financial performance because of their relatively large investment in R&D, or if they were able to invest heavily in R&D because of good financial performance. Again, the question of how R&D spending is related to economic performance is not obvious in the global steel industry.

The improved economic performance of the U.S. steel industry may be due more to the effective use of R&D resources, capabilities, and the organization and less to the investment in R&D. When the integrated firms restructured their operations and reorganized their in-house R&D to cut costs and improve productivity, they lost a large part of their R&D capability and skills. However, the R&D organization became more efficient and focused more directly on production and issues relevant to customers. The in-house R&D organizations formed tighter relationships with production plants, suppliers, and customers. The acquisition of new technology innovations came more from other sources, including particular suppliers and foreign steel producers. The "not-invented-here" syndrome, which sometimes neglected advances made outside one's own company, that had prevailed prior to the 1980s disappeared almost completely. Today, the R&D organizations of integrated producers remain relatively small and few. However, they are leading the integrated steel industry to sustain a competitive advantage through new process and product innovations that will provide high-quality steel products at the lowest production costs.

In contrast, minimill producers have always effectively utilized innovations developed elsewhere. The U.S. minimills became international leaders in the commercialization of a series of processes that led to the development of continuous steel processing. This process improved the conversion time of raw materials to finished products from several months to ten hours or less. As such, the minimill sector has achieved astounding production efficiency and high profitability in the last two decades. The minimill industry's effective adoption and commercialization of innovations from other sources has been a large determinant of its competitiveness and economic success.

For the U.S. steel industry as a whole, R&D resources have been more effectively utilized, even as R&D resources have decreased dramatically.

SUMMARY AND CONCLUSIONS

The U.S. steel industry has made a remarkable comeback in its competitiveness. By restructuring, massively downsizing operations, closing inefficient plants, and making strategic investments in new plants and technologies, the U.S. steel industry has achieved healthy and growing profitability and productivity. Although a number of different factors discussed in this paper have contributed to the industry's turnaround, the development or acquisition of new innovations, and the efficient implementation of these innovations, played a significant role.

With these innovations, the U.S. steel industry has again become competitive with the best producers in the world. Nevertheless, the industry faces, in some cases, a unique set of economic drivers different from those of its competitors. In the future the industry cannot rely completely on technologies developed elsewhere. In the next decade, the U.S. steel industry may need to rely more on its own innovation or invest more in collaborative developments to continue to improve its competitive position.

REFERENCES

Ahlbrandt, R., R. J. Fruehan, and F. Giarratani. (1996). *The Renaissance of American Steel.* New York: Oxford University Press.

Baber, W.R., Y. Ijiri, and S.H. Kang. (1993). *Financial Analyses of the US, Japanese, and Korean Steel Industries: An Investigation of the Determinants of Global Competitiveness.* Working paper prepared as part of the steel project, 'Competitiveness in the Global Steel Industry', funded by the Alfred. P. Sloan Foundation and the American Iron and Steel Institute, Carnegie Mellon University.

Cheij, D. A. (1997). Case studies conducted as part of the steel project, 'The Economic Impact of University Research in Science and Technology on the Steel Industry,' sponsored by the Alfred P. Sloan Foundation, Carnegie Mellon University.

Cyert, R. M., and R. J. Fruehan. (1996). *The Basic Steel Industry, Meeting the Challenge: U.S. Industry Faces the 21st Century.* Final report for the Office of Technology Policy sponsored by the Sloan Steel Industry Competitiveness Study, Carnegie Mellon University, December 1996.

Dennis, W. E. (1991). *Lessons from a Decade of Collaborative Research.* AISI Ironmaking Conference Proceedings 50:3-10.

Fruehan, R. J. (1994). Survey conducted as part of the steel project, 'Competitiveness in the Global Steel Industry,' sponsored by the Sloan Steel Industry Competitiveness Study, Carnegie Mellon University, January 1994.

Fruehan, R. J., H. W. Paxton, and L. Giarrantani. (1994). *A Vision of the Future Steel Industry.* Prepared for the U.S. Department of Energy, sponsored by the Sloan Steel Industry Competitiveness Study, Carnegie Mellon University, December 1994.

Fruehan, R. J. (1996). *Manufacturing Quarterly.* (April 1-7):11.

Fruehan, R. J., and H. Uljon. (1995). Survey conducted as part of the steel project, 'Competitiveness in the Global Steel Industry,' sponsored by the Sloan Steel Industry Competitiveness Study, Carnegie Mellon University, May 1995.

Fruehan, R. J., R. M. Cyert, and F. Giarratani. (1997). The Steel Industry Study, Final Report to the Sloan Foundation. Carnegie Mellon University, November 1, 1997.

Fruehan, R.J., and D. M. Vislosky. (1997). *R&D Decision-Making in the U.S. Steel Industry*. Case studies conducted as part of the steel project, 'Competitiveness in the Global Steel Industry,' sponsored by the Sloan Steel Industry Competitiveness Study, Carnegie Mellon University.

Hannay, N. B., and L. W. Steele. (1986). *Technology and Trade: A Study of U.S. Competitiveness in Seven Industries*. Research-Technology Management. (Jan-Feb):14-22.

Hoerr, J. P. (1988). *And the Wolf Finally Came: The Decline of the American Steel Industry*. Pittsburgh: University of Pittsburgh Press.

Lieberman, M. B., and D. R. Johnson. (1995). *Comparative Productivity of Japanese & US Steel Producers, 1958-1993*. Working paper prepared as part of the steel project, 'Competitiveness in the Global Steel Industry,' funded by the Alfred. P. Sloan Foundation and the American Iron and Steel Institute, UCLA, May 1995.

Pittsburgh Post Gazette. (1997). (December): A-12.

Porsche Engineering Services, Inc. (1995). Ultra Light Steel Auto Body Consortium Final Report. August 1995.

Vislosky, D. M. (1996). *R&D Decision-Making in the U.S. Steel Industry*. Case studies conducted as part of the steel project, 'Competitiveness in the Global Steel Industry,' sponsored by the Sloan Steel Industry Competitiveness Study, Carnegie Mellon University.

Vislosky, D. M. (1998). *Selecting R&D Projects: Processes and Preferences in the Steel Industry*. Unpublished Ph.D. qualifier paper. Carnegie Mellon University (January).

Powder Metallurgy Parts

DIRAN APELIAN
JOHN J. HEALY
P. ULF GUMMESON
CHICKERY J. KASOUF
Worcester Polytechnic Institute

THE INDUSTRY AND THE TECHNOLOGY

Net shape processing involves metalworking in which the output from the first formation is very close to the final required tolerance and specifications, requiring little additional machining. This manufacturing can be very attractive because of its efficiency, conservative energy use, and relatively minor environmental impact. Net shapes, originating as metal powders, date back thousands of years (e.g., gold, copper, iron), but modern powder metallurgy (P/M) had its beginning in the 1920s with the use of porous self-lubricating bearings in home appliances. This development was followed in the U.S. auto industry by attempts to make structural components from easy-to-handle copper-based powders. These parts are attractive because P/M is a cost-effective metal processing technology with little or no scrap. In Germany, during World War II, iron powder was used to make porous rotating bands for artillery shells to replace scarce guilding metal.

P/M is now the fastest growing net shape metal manufacturing industry in the United States. With the exception of the large captive P/M operations of the auto companies in the early years (1930s-1960s), the balance of the industry has traditionally consisted of relatively small entrepreneurial firms. These firms are squeezed in the supply chain between the large raw material suppliers and big automobile customers, with their heavy pressure on parts prices. Thus, P/M parts manufacturing is a small industry with a history of secrecy, price competition, and margins too narrow to permit any meaningful research. R&D is largely left to raw material and equipment suppliers and to a few universities.

The lack of critical technology mass inhibited R&D among part producers because of the limited resources of any individual firm. It is often the case that the larger firms in the industry can engage in development programs while the smaller firms lag in technology development or rely on their suppliers (e.g., material or equipment producers) for R&D. The dynamics of the external forces affecting the level of innovation and technology commercialization in fragmented manufacturing industries, and in particular the P/M parts industry, is the focus of this chapter.

The basic steps of conventional P/M technology are:

1) manufacturing of powders, predominantly by melting followed by atomization with high-pressure water or gas;

2) mixing and blending the powders and additives (carbon, alloys, lubricants);

3) feeding the mix into a die and consolidating (compacting) the mix, applying pressures of about 50 tons per square inch, resulting in "green" shapes, and

4) sintering the green compacts at about 2100° Fahrenheit, causing solid state diffusion and bonding.

These processing steps result in distinct industry sectors—powder producers, equipment manufacturers, and parts producers. Part producers can be further divided into conventional iron and copper-based P/M and production that uses more specialized materials or production processes such as tungsten or metal injection molding (MIM), a technology similar to injection molding of plastics and ceramics.

P/M has a number of advantages over competing technologies:

• Many metal powders are manufactured from recycled metals or scrap, notably iron/steel and copper, while others are made from virgin ores (tungsten, nickel).

• Net shapes are mass produced to close tolerances over long production runs without scrap residue (machining chips, grinding residue, casting risers, etc.).

• P/M is the lowest energy consumer of all comparable metal working processes and is environmentally benign, producing a minimum of fumes and toxic waste (Bocchini, 1983).

• P/M makes possible otherwise impossible alloys and metal/non-metal combinations of materials or self-lubricating bearings, metal filters, and metal/matrix composites.

• P/M design solutions are often remarkably cost effective.

There are many other applications for metal powders, but this chapter will be restricted to the parts industry, i.e., companies manufacturing structural components, self-lubricating bearings, and friction materials by "compacting and sintering" metal powders, predominantly iron, steel, alloy steel, copper, and copper-based alloys. From a modest beginning of about 2000 tons per year in the mid

1940s this industry has grown to about 12,000 tons in 1950 and 350,000 tons annually today. This evolution is discussed in more detail later in the chapter, and is illustrated in Figure 3 (p. 113).

P/M INDUSTRY STRUCTURE AND STRATEGY

The $1.8 billion North American powder metallurgy parts industry currently includes approximately 213 companies competing at various levels in the manufacture of P/M structural parts, powder forging, bearings, friction materials, and metal injection molded products. More than two-thirds of part sales are automotive applications, the most significant growth segment since 1980 (see Table 1). The industry has responded to several years of real growth that, while currently moderating, is expected to continue. While some managers and analysts have suggested less reliance on automotive parts, these parts continue to exhibit strong growth as auto producers continue to use new P/M applications at the same rate as the industry diversifies into new applications (Roll, 1985). They are attractive for parts producers because of the large volumes that come with a successful contract. Automotive applications have increased from 15 pounds per U.S.-made auto/light truck in 1988 to 29.5 pounds in 1996. Recent forecasts suggest that this volume will increase to 32.5 pounds in 1998 (Winter, 1996, 1997).

Auto company captive P/M plants became the first large-scale P/M operations, but they began to increase their outsourcing during the 1970s, and many of the P/M divisions were divested during the 1980s. This did not change the P/M industry's dependence on the automobile, but it caused major changes in the supply chain and in the industry's pattern of technological innovation and economic performance during the early 1970s. This period saw the auto industry, and thereby the P/M industry, struggle through the energy crisis and the onslaught of

TABLE 1 P/M Markets

| Market | 1980 | | 1996 | | |
	Market short tons	% of Market	Market short tons	% of Market	Market growth (%)
Auto and truck	72,300	44.9	220,400	69.0	205
Recreation and tools	22,200	13.8	36,400	11.4	64
Appliance	13,200	8.2	19,500	6.1	48
Hardware	10,900	6.8	7,300	2.3	(33.0)
Industrial equipment	8,800	5.5	11,800	3.7	34
Business machines	7,800	4.8	3,800	1.2	(51.3)
All other	25,900	16.1	20,200	6.3	(22.0)
TOTAL	161,100	100.0	319,400	100.0	98

Source: White (1996b).

foreign competition, "auto transplants" (domestic production facilities of foreign owned auto producers), and auto imports. Earlier strong demands and tight powder supply were followed during this period by falling P/M part sales and even auto industry restrictions on new P/M parts developments.

Current P/M industry prosperity is based on the success of auto industry restructuring. Longer production runs, lower cost energy and labor, and cost reduction programs initiated by suppliers in response to automotive customers, have made the North American P/M industry the most competitive in the world, with a cost advantage in 1996 of about 20-30 percent over Japanese parts producers. Strengthening of the dollar since then has reduced this advantage, but competition with overseas firms has yet to become a major issue in the North American P/M industry.

Powder metallurgy has thus played a very substantial role in re-engineering powertrain components and has successfully converted other engine parts. This success, in turn, continues to drive P/M growth for automotive and other customers. Much technical innovation in applications originates in the U.S. auto industry with its ongoing acceptance of P/M as a solution in their search for more cost effective net shape manufacturing technologies.

A number of factors have recently contributed to setting the P/M industry apart:

- A healthy economy has led to a surge in demand.
- P/M has been able to meet this demand from latent capacity brought forth with relatively inexpensive productivity improvements.
- P/M still has untapped, latent competitive potential or, expressed differently, has yet to reach that level of commercial maturity, when long-term growth rate levels off and price competition and excess capacity may call for restructuring of individual firms or consolidations within the industry.

The advent of just in time inventory (JIT), the desire for fewer suppliers, and total quality becoming the condition for doing business at any price led to a need for closer relationships, often partnerships, in the supply chain. These requirements frequently span three links in the chain—raw material and equipment suppliers, parts manufacturers, and their customers (i.e., original equipment manufacturers or OEMs). These factors have favored firms with more sophisticated financial, managerial, and technical resources, which sometimes, but not always, translates into size.

P/M part producers face difficult conflicts. Customers are demanding more engineering services, price concessions, and, in some cases, global supply capabilities (Kasouf and George, 1994; Kasouf et al., 1996). Consistent with Lorange (1988), this often leads to a situation in which new business requires investments that part producers may be reluctant to make because of the risk of losing the business downstream after developing it (Kasouf, 1998). Also, the sharing of information to improve quality and reduce cost, desired by the customer, may be at risk, since part producers are sometimes reluctant to share developments when they are unsure about the long-term potential of the relationship with a customer.

Although part producers generally find relationships with their suppliers valuable, few horizontal relationships were observed in the industry, and these are generally limited to areas such as training and trade missions. There is evidence of interfirm cooperation whereby one manufacturer sources some unfavorable product mix (usually a product with a short production run) to a smaller manufacturer who is set up to deal with that type of production profile, thus allowing the larger manufacturer to maximize his capacity. This is not uncommon in western Pennsylvania where 46 of the 213 P/M companies in the United States are located within a thirty-five mile radius of each other and have established mutually beneficial relationships.

Among part producers, the attractiveness of relationships was negatively related to firm size, i.e., smaller firms found alliances more attractive than larger firms (Kasouf and Celuch, 1997). Moreover, firms that found relationships attractive tended to be optimistic about the future growth of the industry and thought that the technology was changing more rapidly than firms that did not find alliances attractive. This may suggest that relationships seem to be attractive for firms willing to share information to deal with growth opportunities that might be difficult to address with limited R&D funds. Thus, the challenge for the supplier in this situation is to develop a satisfied customer yet increase its power vis-à-vis the customer by raising switching costs. This might involve working with its own suppliers to develop a competitive advantage or vertical integration in the case of larger companies.

Like firms in any industry that is greatly dependent on a dominating customer base, such as automotive, P/M part producers have always considered diversifying their risk by developing other markets and applications. While these markets are important and substantial, some of them (e.g., hardware and business machines) have either moved offshore or moved to other materials with more favorable price-performance tradeoffs in those applications (Noted in Table 1). The net result is that these other markets have failed to encroach on the automotive market share. Their comparatively slow growth of only 11.5 percent since 1980, versus 204 percent for the auto and truck market, dramatically illustrates how closely linked the future of the P/M industry is to the fortunes (and misfortunes) of the auto industry. Automotive applications may eventually account for close to 80 percent of the market.

While approximately 70 percent of the P/M part volume is sold to the auto industry, the mean percentage of automotive sales is under 40 percent (Kasouf, 1997). This suggests a "tiering" of the industry, i.e., some firms have little or no automotive sales, while other firms focus mainly on automotive applications. As the auto industry continues its supplier rationalization, many P/M part producers will have to find other markets or means of dealing with other tier one or two suppliers. Strategic redirection will then be critical for some of these firms.

In spite of long-standing predictions of industry consolidation (e.g., Roll, 1987), there are more companies today in P/M parts manufacturing than at any time in the past. In general terms we can identify three different types of companies:

• *Job Shop/Specialty Manufacturers.* These are the small firms who comprise the largest group in the industry. They typically generate less than $10 million per year in revenue and operate with lower press tonnage, up to about 200 tons and under. They produce more complicated, short-run parts, respond quickly to change and have low overheads in terms of organization structure. We estimate that there are about 165 companies in this category (including the metal injection molders), accounting for 77 percent of firms in the industry.
• *Repetitive Process Manufacturers.* An estimated 36 companies (17 percent of the firms in the industry) comprise this group, each generating $10 million to $50 million in annual sales. These firms typically focus on low to medium press tonnage (up to about 500 tons), provide high customer service, and perform many secondary operations. They have the capacity to innovate effectively and generate medium profit margins. However, their limited size makes them vulnerable to supplier rationalization.
• *Large Process Manufacturers.* This group of firms includes the largest producers in the industry. They have large presses from 200 to 1000+ tons and low manufacturing costs due to high volume. They have the most sophisticated quality management systems, a high level of technical support and service, and the lowest prices. We estimate that there are 11 firms in this category, representing 5 percent of the total. They account for approximately one-half of the industry's production (Table 2).

Though the number of parts manufacturing companies has increased from 156 in 1980 to 213 companies in 1997, an increase of 37 percent, the concentration of market share among the largest firms in the industry is increasing, as demonstrated by Table 3. The predicted industry shakeout has not occurred, but the industry may be evolving away from fragmentation. Porter (1980) defines a fragmented industry as one in which no single firm has significant market share

TABLE 2 Number of Firms with Combined 50 Percent Market Share

Year	Number of firms	Combined market share (%)
1982	11	50.5
1987	11	50.5
1991	12	52.1
1996	11	51.8

Source: PMRC.

TABLE 3 Total Market Share of Largest Firms (%)

Year	3 largest firms	4 largest firms	5 largest firms
1982	19	24	29
1987	21	26	31
1991	26	33	38
1996	30	34	38

Source: PMRC.

and can strongly affect industry outcome. He suggests that an industry is fragmented when the four largest competitors have less than 40 percent market share.
Thus, the increasing concentration of market share among the largest competitors
is an important industry trend (Table 3). Given the rationalization of the automotive supply base, and the increasing sophistication required by the auto industry,
we may be observing the emerging importance of size and skill requirements for
P/M part producers. These may evolve into entry barriers, which have historically been very low in this industry. This industry is still fragmented by any
standard, but the most recent mergers and acquisitions among the largest firms
(e.g., the growth of Sinter Metals, recently acquired by the British firm GKN)
may signal the emergence of several relatively large global firms that do have the
capacity to affect the structure of the industry.

A longer history of the market share data is provided in Figure 1, which
illustrates the market share of the eight largest part producers from 1967 to 1996.
The industry exhibited substantial concentration among the four largest firms in

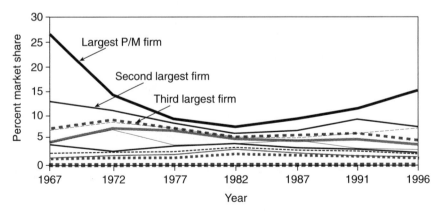

FIGURE 1 Market share of the top eight P/M part producers 1967-1996.
Source: PMRC.

1967. It experienced greater fragmentation in the early 1970s and the 1980s as the auto companies divested their captive P/M units. The 1996 data suggests a trend toward renewed consolidation.

Viewed from a global perspective, P/M part producers as stand-alone metal processing companies are a largely American phenomenon. In Europe, and even more so in Japan, parts fabricators often are captive operations in very large firms that are consumers of P/M parts, without any substantial number of second- and third-tier smaller producers that are independent of the large parts consumers. They therefore remain captives of the existing market, without the ability to influence the growth and acceptance of new P/M applications outside the large established firms, automotive or otherwise, and without the resources or fertile markets needed to test the strengths and opportunities of their own P/M initiatives. There appears to be no "new application pull" to the extent that exists in the North American market.

The U.S. auto companies' struggle to restructure and become competitive set the stage for the recent unprecedented success and growth of the P/M industry, a development that has not been repeated in Europe, Japan, or other foreign markets. For more than fifty years the U.S. P/M industry has remained as large as the rest of the world combined, and the U.S. car and light truck content of 30-35 pounds of P/M parts remains more than twice the weight in any foreign car. One German observer (Huppmann, 1991) comments on the difference between the U.S. and European P/M industries: "While the U.S. industry early on sought cooperation with customers and conscientiously focused on developing parts, e.g. for the auto industry, the European industry all too long fought a battle about properties aimed against wrought steel."

North America has remained the market of P/M acceptance, causing foreign competitors to seek a presence in the U.S. by acquisitions (e.g., European companies) or by establishing transplants here (e.g., Japanese companies, including one powder producer and three parts producers). The P/M industry's ability to respond to the auto industry challenge is in no small measure due to its own success in advancing the technology, stepwise and incrementally, rather than in major breakthroughs. One of the keys to continued inroads against competing technologies is higher physical properties at acceptable cost. This requires bringing together technological advancements in raw materials, process equipment, and parts production for higher density P/M components. Higher density translates directly to higher physical properties. The traditional and relatively costly method to reach higher densities, "double press/double sinter," is giving way to less costly single press warm compaction with much improved properties, notably fatigue life, the key to high stress applications in engines and transmissions. The search for other process innovations continues.

To illustrate accomplishments, the industry has now concluded ten years of manufacturing powder-forged connecting rods for North American vehicles (White, 1996a), and one estimate concluded that over 75 million connecting rods have been produced (White, 1996b). Other innovative product applications include bearing caps and warm formed torque converter turbine hubs.

A useful tool to analyze the competitive success of North American P/M part producers is Porter's (1990) "Diamond of National Advantage." He argues that constantly innovating industries can be explained by four factors characterizing the competitive environment of the home country (Figure 2).

Factor conditions. In addition to the traditional factors of production—land, labor, capital—Porter suggests that specialized resources that support the industry affect the competitive position of a nation or region's firms. In P/M part production, U.S. firms enjoy a skilled work force, especially in the western Pennsylvania area. Moreover, American P/M metallurgists and engineers have become adept at developing conversion applications for the auto industry. This has helped develop an entrepreneurial spirit in the industry; these firms are constantly seeking new applications for the technology.

Demand conditions. Porter suggests that, while many opportunities are global, the characteristics of the home market affect the perception of buyer needs and the development of appropriate responses to those needs. It is a great advan-

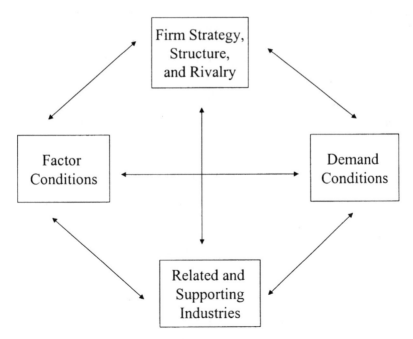

FIGURE 2 The determinants of national competitive advantage.
Source: Porter (1990).

tage when need in the home market mirrors future global demand. As noted above, American P/M fabricators have been responsive to a large customer base pushing them to develop new applications. American-made vehicles contain the largest volume of P/M parts in the world. These applications are a foundation for application development in other parts of the world. If future worldwide demand for P/M parts parallels American part development, then these firms are well positioned for global competition.

Related and supporting industries. Powder and equipment suppliers are largely global competitors. However, the relationships that American part fabricators develop with their suppliers are critical because in many cases upstream R&D or inventory management are essential elements in developing new products. The large, more sophisticated firms in the industry have learned to leverage suppliers effectively to create value with limited resources (Kasouf, 1998).

Firm strategy, structure, and rivalry. The strategies used by firms to respond to customer requirements and the nature of competition among firms also affect the global competitiveness of a nation's industry. As noted, the independent part producers have generated customer-focused, entrepreneurial strategies to generate new developments. This customer focus has served the industry well in conversions. Moreover, the intense rivalry among firms has forced the industry to maintain a cost-effective orientation while adding engineering expertise (Kasouf and George, 1994).

In summary, American part producers feel the dual pressures of providing more expertise at a lower price. This conflict is difficult to resolve, but the demanding U.S. customers in the auto industry have forced these part fabricators to take advantage of factor advantages to develop cost-effective solutions that will serve them well when expanding into global markets. The competitiveness of the 1990s has resulted in an efficient industry ready for global opportunity.

INDUSTRY PERFORMANCE

This section is the first attempt to look at the small but rapidly growing P/M parts industry from the standpoint of economic and competitive performance, and the factors, technical and non-technical, that have caused or forced changes in the industry in the last thirty to forty years. However, few if any statistics are in the public domain. What is available is limited to raw material tonnage, generalized reports, and private sources.

Production and Market Share

Table 4 compares North American and world shipments of metal powder. "Iron & steel," "copper base," and to some extent "nickel" are indicative of the

TABLE 4 Metal Powder Shipments, North America and the World

Metal powders	North America				World	
1000 short tons	1997	1996	1995	1990	1996	1990
Iron and Steel	375	351	347	219	639	600
Copper and Cu base	24	23	23	19		48
Aluminum	40	34	37	36		100
Molybdenum (est.)		3	3	2		
Tungsten, H_2-reduced		1	1	3		30
Tungsten carbide		11	11	5		incl.above
Nickel		12	10	10		22
Tin		1	1	1		
Stainless steel		2	4	3		15
Estimated total		438	434	298		825

Source: MPIF.

U.S. tonnage share of the world market for steel P/M components (about 50+ percent). This share has not changed in any meaningful way the last forty years.

Only modest quantities are imported/exported, either of components or the corresponding raw materials, which if considered, would not change the market share estimate in this report. Figure 3 shows the historical trends of steel powder consumption in North America.

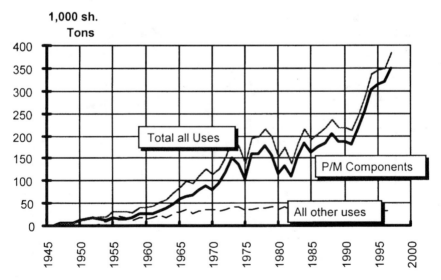

FIGURE 3 North American steel powder shipments.
Source: MPIF.

Historically, U.S. and world steel powder supply has exceeded consumption by a wide margin in all but a few brief periods of shortages. Even in those periods foreign powder capacity never translated into continuous U.S. imports because of high shipping costs, cost of warehousing, JIT, and the difficulty of providing rapid and professional service. The U.S. oversupply has historically made for a competitive powder market with downward pressure on steel powder prices (see Figure 4).

Equipment suppliers have not had import/export constraints and foreign P/M press and furnace manufacturers have long been successful in selling and servicing their products in the United States. Their U.S. counterparts also have substantial exports.

Recent growth of the industry has been substantial but concentrated (Table 5). The increase in iron shipments between 1991 and 1996—a cumulative 75 percent—was primarily shared among the 20 percent of companies that account for 80.6 percent of sales.

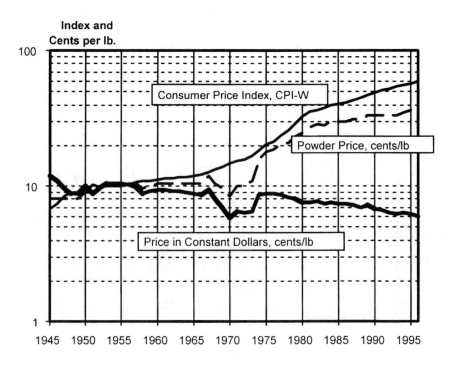

FIGURE 4 North American steel powder prices: primary unalloyed P/M grade; index 1953 = 100.
Source: Private source.

TABLE 5 Steel Powder Shipments in North America

Year	Tons (thousands)	Annual change, (%)	Cumulative Change from 1991, (%)
1992	215		18
1993	255	18.6	40
1994	303	18.9	67
1995	313	3.2	72
1996	319	1.9	75

Source: MPIF.

Total steel powder shipments for all uses for the period 1991-1996 for Japan, Europe, and North America, are shown in Figure 5. One can note that the American industry has grown significantly vis-à-vis Japan and Europe. Moreover, the U.S. P/M industry has succeeded in increasing its share of the structural component market through innovation and new applications over other metal working industries (Table 6).

A desire to be close to customers has often led to clustering of parts manufacturing firms near major markets, notably Michigan. Just as often this clustering has been the result of an entrepreneurial and skill tradition, such as the concentration of parts plants in northwestern Pennsylvania and New England. Recently, some manufacturing has been developed in the south because of lower labor and energy costs.

In the absence of specific industry data by SIC code or a sufficient number of public companies, it is difficult to evaluate the financial health of the industry. Given this void, the Powder Metallurgy Research Center conducted a fi-

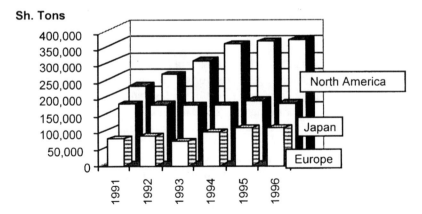

FIGURE 5 Steel powder shipments, all uses.
Source: MPIF.

TABLE 6 Weight of Materials in Pounds in a Typical U.S. Family Vehicle

Material	1997	1996	1994	1990	1980	1978	% change 1978-1997
Reg steel sheet							
Strip, bar, rod	1411.0	1409.0	1389.0	1405.0	1737.0	1913.0	−26
High and medium							
Strength steel	296.0	287.0	263.0	238.0	175.0	133.0	123
Stainless steel	47.5	46.5	45.0	34.0	27.5	26.0	83
Other steels	36.0	38.5	42.5	39.5	54.0	65.9	−45
Iron	387.0	389.0	408.0	454.0	484.0	511.0	−26
Plastic and plastic							
composites	242.0	245.0	246.0	229.0	195.0	180.0	35
Aluminum	206.0	196.0	182.0	159.0	130.0	112.0	83
Copper and brass	46.5	45.0	42.0	48.5	35.0	37.0	26
P/M Parts	**31.0**	**29.5**	**27.0**	**24.0**	**17.0**	**15.6**	**101**
Zinc die casting	14.0	15.5	16.0	18.5	20.0	30.9	−55
Magnesium cast.	6.0	5.5	5.0	3.0	1.5	1.1	445
Fluids/lubricants	198.0	198.0	190.0	182.0	178.0	198.0	−15
Rubber	139.0	139.0	134.0	137.0	131.0	146.0	−5
Glass	96.5	94.0	89.0	86.5	83.5	86.4	12
Other materials	102,0	100.0	94.0	84.0	95.0	120.0	−15
Total	3248.0	3236.0	3171.0	3141.0	3363.0	3576.0	−9

Source: American Metal Market, quoted in the *International Journal of Powder Metallurgy.*

nancial benchmarking study at Worcester Polytechnic Institute (Healy, 1997). This study developed average values for key financial indicators by using a cross section of parts producers. These producers represent firms of all sizes, allowing us to develop a "composite P/M parts producer" representing an estimate on the industry average over a span of ten years. We have also made use of information and statistics culled from published data or made available by private sources (Table 7).

Raw Materials. The raw material share of total manufacturing costs, despite more costly premixes and larger parts, for which raw material costs are a larger share of manufacturing costs, has fallen in the last ten years from about 30 percent to 26 percent. The raw material share of net sales on the other hand continues to be slightly less than 22 percent of net sales for eight of the last ten years.

Direct Labor. While raw materials as a percent of net sales fell modestly over the last ten years, it fared better than direct labor costs, which have remained a constant percentage of net sales, in single batch or average job shop/specialty firms. Barring some immediate manufacturing breakthrough, direct labor costs do appear to continue at slightly less than 10 percent of net sales and seem to remain fixed in relation to volume despite the substantial increase in volume.

TABLE 7 The Composite P/M Parts Firm ($1,000)

	1986	%	1991	%	1996	%	% change from 1986 1991	1996
Net sales	$8284	100	$10,514	100	$13,402	100	27	62
Raw materials	2,013	24	2,239	21	2,882	22	11	43
Direct labor	944	11	1,041	10	1,327	10	10	41
Other mfg costs	3819	46	5,405	51	6,827	51	41	79
Total mfg costs	6776	82	8,685	83	11,035	82	28	63
Gross margin	$1508	18	$1,829	17	$2,367	18	21	57

Note: Values are in constant 1986 dollars, as reported by the U.S. Bureau of Labor statistics for "Bars, Cold formed, Carbon."

Productivity. We earlier pointed to the apparent lack of variability of labor costs with volume. This also related to the age of existing production equipment. Despite the design advances in new equipment, most of the existing equipment in the P/M parts industry, though often rebuilt, is very old, which has a direct effect on productivity (Table 8).

Approximately 70 percent of the press equipment currently in use is over 10 years old, and over 30 percent is over 20 years old. Maintenance costs in P/M at 5.6 percent of net sales are twice the National Association of Manufacturers average of 2.3 percent for standard industry cost. In 1997, the total P/M parts industry equipment purchases are approaching $100 million for the first time, or approximately 5 percent of the parts industry's sales.

The P/M industry is very close to the National Association of Manufacturers' net sales per employee, which reflects the minimal spending on activities outside the production area by P/M companies (Table 9). As can be seen, the productivity reflects a 7 percent drop in real dollars adjusted for the Bureau of Labor Statistics producer price index for corresponding metal products (e.g., cold finished bars, carbon) during this period.

TABLE 8 Age of P/M Equipment

Equipment Years old	Age Distribution (%) <5	6-10	11-20	>20	% >10+
Presses					
Mechanical	15	16	36	33	69
Hydraulic	12	16	44	28	72
Sizing	13	18	39	30	69
Furnaces	23	24	30	23	53

Source: MPIF/PMPA.

TABLE 9 Productivity Indices (Constant 1986 Dollars)

Year	Pounds per employee	Net sales per employee	Price per pound
1986	30,900	$73,900	$2.39
1989	27,500	$79,062	$2.88
1991	28,400	$83,183	$2.93
1995	41,200	$91,913	$2.23
(%) Change	+33	+24	−7

Research, Intellectual Property, and Technology Diffusion

The industry trade association, Metal Powder Industries Federation (MPIF), and the P/M professional organization, the American Powder Metallurgy Institute (APMI International) have been major catalysts in fostering professional development and technology advancements in the P/M industry. The annual International P/M Technical Conference and Exhibits has been a good forum for cross-fertilization and development of ideas, a precursor for technology development. This is clearly manifested in the growing number of technical presentations made at the conferences (Table 10). In the recent past, MPIF has also been active in arranging trade missions to Japan and China.

As reported earlier, R&D have traditionally been nominal at best in the P/M parts industry, and we have not seen any significant increase in the last five to ten years. The industry has concentrated on gradual process refinements relying heavily on raw material and equipment manufacturers to carry out whatever material and equipment improvements would be most conducive to increasing the market for P/M parts. Cognizance of the importance of intellectual property protection is noticeable in the number of patents issued. Table 11 illustrates this growth.

Research and development in the P/M industry, essentially only material and process development, was until very recently directly focused on perfecting existing processes and products and metal powders and P/M parts. Little fundamental research has been performed in or by the industry over the years. Fundamental research on new atomization techniques, alloying during atomization

TABLE 10 Growth of APMI Conference Presentations

Year	No. of technical presentations	No. of pages in proceedings
1982	39	571
1987	55	895
1991	146	1888
1995	192	2443

TABLE 11 Patent Activity

Period	No. patents issued	% change 1981-1985	% change from prior year
1981-1985	159		
1986-1990	180	+13	+13
1991-1995	219	+37	+21
1996-1997[a]	260	+64	+19

[a]Annualize for a five-year period based on the number of patents issued through October 1997.

process via gaseous reactions, new composite P/M materials, and novel compaction processing methodologies, among other areas, has not been pursued. As a result, the basic science of P/M technology has taken a back seat to developmental projects.

Globally and in the United States, the academic community has not been engaged in R&D and teaching in the fields relevant to P/M parts manufacturing. There is clearly a need for more research. The recent increase in industry consortia and scattered university-based cooperative research in the United States is an encouraging sign. Government funding for research, with one exception, has been nonexistent. At CTC in Johnstown, Pennsylvania, the Navy Manufacturing Program has financed a multi-year program to develop P/M industry standards, an effort unlikely to have been supported by the industry on its own. On the other hand, in Europe, Japan and Russia, government investments in P/M research have been consistent and substantial. Yet the return on these investments has been poor compared to the much greater success of the U.S. industry. This raises the question of focus and timing of government support of technology.

CONCLUSIONS AND RECOMMENDATIONS

Although this chapter has focused on the P/M industry, within the spectrum of metals processing industries there are several industry clusters with similar characteristics. The gap between firms with R&D capabilities and smaller firms with limited resources can result in a tiered industry, in which the smaller firms are unable to compete effectively for contracts with demanding OEMs. In P/M, most of the new applications and increases in market share have occurred because of innovations and technology commercialization driven by the customer, principally the automotive industry, and not due to major investments by P/M parts producers. Suppliers, or powder producers, have invested in R&D to assist the parts producers. There is a symbiotic relationship between suppliers and parts producers.

P/M industry investments in R&D, taken as a whole, are minimal and are not evenly distributed across the industry. Facilitating the development of R&D capabilities in smaller firms is important for the growth of the P/M parts industry.

The Metal Powder Industries Federation and the trade associations have played an important role in establishing forums for "cross-fertilization," information dissemination, and professional development programs; but these efforts are not sufficient. Innovation and technology commercialization require investments in knowledge of workers and in research and development

The metals processing industry in general, and the powder metallurgy industry in particular, are in transition. To strengthen their supply drain, the OEMs have been seeking a smaller base of more sophisticated suppliers (e.g., Sage et al., 1991; Bertodo, 1991; Helper, 1991; Lyon et al., 1990). To avoid being squeezed out, smaller firms need to recognize the value of investment in technology development and commercialization. One reason for optimism is the considerable enthusiasm for solving the technical problems of the industry within alliances of multiple firms centered in universities (Table 12). Interestingly, there is

TABLE 12 Attractiveness of R&D Alliance Options for P/M Part Producers

	FF	FU	FFU
General Research and Development			
Effects of trace impurities on properties	5	12	22
Detection of green cracks	10	6	23
Effects on side wall lamellar sheer	8	13	19
Advanced process automation capabilities	12	11	14
New joining techniques	7	15	16
High temperature sintering	10	13	19
Improved powder delivery systems	20	9	11
Corrosion resistance	6	13	23
Further process developments	12	11	18
Improved material properties	5	14	24
Computer Software Applications			
Cost/Investment alternative analysis	12	9	16
Applications database to assist designers	11	6	20
Automatic tool design generation	13	15	12
Process plan generation	10	18	8
On line standards database	12	2	21
Expert systems to help design parts	6	8	21
Process models and analyses	4	13	19
Process Models and Analyses			
Sintering simulation	3	11	22
Compaction simulation	5	14	18
Tool and press deflection analysis	11	14	14
Part dimensional analysis	13	12	10
Process cost analysis	14	11	9
Powder flow analysis	4	10	20

Note: FF = interfirm cooperation; FU = single firm and university; FFU = multiple firm and university.
Source: Kasouf et al. (1994).

an inverse relationship between size and the relationship orientation, with smaller firms seeing more value in developing alliances (Kasouf and Celuch, 1997).

REFERENCES

Bertodo, R. (1991). "Alignment of Automotive Suppliers to a Strategic Vision," *International Journal of Vehicle Design* 12(3):255-267.

Bocchini, G.F. (1983). "Energy Requirements of Structural Components: Powder Metallurgy vs. Other Production Processes," *Powder Metallurgy* 26(2):101-113.

Drucker, P. (1998). *Managing in a Time of Great Change.* New York: Truman Talley/Plume.

Frame, P. (1995). "GM Exec seeks Suppliers with Global Savvy," *Automotive News* (May 1).

Healy, J. (1997). "Financial Benchmarking of the P/M Parts Industry," work in progress, Worcester, MA: Powder Metallurgy Research Center.

Helper, S. (1991). "How Much Has Really Changed Between U.S. Automakers and Their Suppliers?" *Sloan Management Review* 32(Summer):15-28

Huppman, W. J. (1991). "Wettbewerbschancen der Pulvenmetallurale," *P/M International* 24(2):124

Kasouf, C. (1997). "Interfirm Relationships in the Powder Metallurgy Parts Industry," testimony to the Federal Trade Commission Joint Venture Project, June 24, 1997.

Kasouf, C. (1998). "Interfirm Relationships in the P/M Value Stream: Case Studies," working paper, Worcester, MA: Powder Metallurgy Research Institute.

Kasouf, C., and K. George (1994). "Interfirm Relationships in the P/M Industry," research report, Worcester, MA: Powder Metallurgy Research Institute.

Kasouf, C. and K. Celuch. (1997). "Interfirm Relationships in the Supply Chain: the Small Supplier's View," *Industrial Marketing Management,* 26(6):475-486.

Kasouf, C., D. Zenger, P. Ulf Gummeson, and D. Apelian. (1994). *The P/M Industry Study: A Final Report.* Princeton, NJ: Metal Powder Industries Federation.

Kasouf, C., S. Nigam, and K. Celuch. (1996). "Globalization in the P/M Parts Industry," working paper, Worcester, MA: Powder Metallurgy Research Center.

Lorange, P. (1988). "New Strategic Challenges for the Materials Oriented Firm: Requirements of Management to Steer Towards the Year 2000," *Materials Research Society:*139-146.

Lyon, T., A. Krachenberg, and J. Henke, Jr. (1990). "Mixed Motive Marriages: What's Next for Buyer-Supplier Relations?" *Sloan Management Review* (Spring):29-36.

Monts, R. (1995). "Ford Wants Suppliers as World Partners," *Automotive News* (April 17).

National Association of Manufacturers. (1996). "Benchmarks for U.S. Manufacturing Productivity," Schaumburg, IL: McGladrey & Pullen.

Porter, M. (1980). "Competitive Strategy," *New York Free Press.*

Porter, M. (1990). "The Competitive Advantage of Nations," *Harvard Business Review* (March-April):73-93.

Price, B. (1995). "Extended Enterprises," *presented at Creating and Managing Corporate Technology Supply Chains,* Cambridge, MA: Massachusetts Institute of Technology, May 10, 1995.

Roll, K. (1985). "The State of the P/M Industry," *1985 Annual Powder Metallurgy Conference Proceedings,* Princeton, NJ: Metal Powder Industries Federation.

Roll, K. (1987). "Powder Metallurgy at the Turn of the New Century," *1987 Annual Powder Metallurgy Conference Proceedings,* C. Freeby and H. Hjort, eds. Princeton, NJ: Metal Powder Industries Federation.

Sage, L., T. Ozan, D. Cole, and M. Flynn (1991). *The Car Company of the Future: A Study of People and Change, A Joint Research Project of Ernst & Young and The University of Michigan.* Ernst & Young and The University of Michigan.

White, D. (1996a). "Review of P/M in North America," *Advances in P/M and Particulate Materials: Proceedings of the 1996 PM2TEC Meeting,* Princeton, NJ: Metal Powder Industries Federation, A-3 to A-13.

White, D. (1996b). "P/M Technology Trends," *International Journal of Powder Metallurgy* 32(3):225-228.

Winter, D. (1996). "Materials '96 - Powder Metal: '96 Best Yet for Powder Metal," *Wards Auto World*, Detroit, MI, Sept. 95, 31(9):56-57.

Winter, D. (1997). "Brisk Growth for Powder Metals," *Wards Auto World*, Detroit, MI, Sept. 97, 39(9):78-81.

Winter, D. (1998). "Powder Metals Take a Breather," *Wards Auto World*, Detroit, MI, Sept. 96, 32(9):63-64.

Trucking[1]

ANURADHA NAGARAJAN
JAMES L. BANDER
CHELSEA C. WHITE III
University of Michigan

INTRODUCTION

This chapter examines the technological and non-technological factors that have influenced recent service and process innovations in the trucking industry. Intense competition, low margins, and relative ease of entry in the trucking industry motivate firms to develop or adopt many innovations. Unstructured and semi-structured interviews with trucking industry stakeholders indicate the following:

- Technological factors have enabled many *process* innovations.
- Non-technological factors have motivated many *service* innovations.
- As is typical of other service industries, several service innovations have been generated from within the industry, by attempting to emulate competition or through assessment of customer needs.
- Innovations adopted by the trucking industry have extended the business of moving freight into the realm of managing information.
- Innovations, developed outside the trucking industry, particularly in electronic commerce and navigation, tracking, and sensing, have been adopted by specific segments within the industry to enhance customer satisfaction and improve business processes.

Deregulation, globalization, the availability of novel, modern technologies, and new demands by customers for advanced logistics and other services have

[1]The authors gratefully acknowledge the generous grant from the Alfred P. Sloan Foundation program on Centers for Study of Industry. We thank Pete Swan for his insightful suggestions and Harish Krishnan for helping with the data collection.

changed the competitive landscape for trucking firms. The movement of freight is no longer the single strategic focus of the trucking industry. Trucking firms are becoming involved in the generation and movement of timely, accurate information. Customers and trucking firms can use information relating to the exact location of shipments to enhance operational efficiency. This paper identifies several significant forces that are driving the development and adoption of innovations in the trucking industry, and discusses their influence on industry performance. The significant conclusion of the paper is that innovations in the trucking industry have addressed two basic issues: the enhancement of value to customers at an affordable price and the utilization of information to improve business practices through the application of technology. In general, trucking firms have invested in technology that is particularly relevant to the key success factors in their segment in an attempt to enhance productivity and increase competitive advantage.[2]

Freight activity is increasing worldwide with road transportation and air freight becoming the dominant modes. OECD countries generally increased their freight activity at an annual rate of between 1 percent (e.g., France, the United Kingdom, and the Netherlands) and 4 percent (e.g., Italy, Japan, and Spain) between 1970 and 1994. In the United States, freight activity increased annually by about 2 percent. The United States dominates the world in domestic freight activity. In 1994, U.S. domestic freight activity was estimated at 5.13 trillion metric ton kilometers (mtk). In comparison, domestic freight activity in Western Europe was 1430 billion mtk and in Japan was 557 billion mtk (BTS 1997a). Trucking's modal share of the freight activity has been growing fast at the expense of other modes. Table 1 provides an international comparison of domestic freight activity for selected countries and regions with particular attention to freight moved by road.

The trucking industry moved an estimated 27 percent of U.S. freight traffic in 1996 (measured in ton miles) and accounted for 81 percent of the nation's freight bill, valued at about $367 billion (Bank of America, 1997). Competitive pressures and technological advances have combined to make innovation critical to the growth and sustainability of the trucking industry. Many of the innovations created and adopted by the trucking industry extend beyond new products to include broader processes and activities, as emphasized by Kline and Rosenberg (1986). In their view, innovation may be thought of not only as a new product, but also as:

---

[2]Many of the innovations discussed in the paper have been implemented only recently. We therefore do not draw any inferences relating to innovation adoption and firm performance based on empirical data. In our judgment, sufficient time has not elapsed to realistically measure the impact of innovation on the survival and profitability of individual firms. For example, Swan (1997) used a three year time frame to study the impacts of change on trucking firm survival and performance. Our conclusions about innovation in the industry are based on our observations, stakeholder interviews, and popular press articles.

TABLE 1 Domestic Freight Activity for Selected Countries and Regions (billions of mtk)

Country/ region	Year	Road	Total domestic freight	Average annual growth rate in domestic freight activity (%)	Real average annual GDP rate (%)
United States	1970	602.0	3216.5	2.0	2.8
	1994	1326.0	5130.3		
Canada	1984	43.6	296.6	0.3	2.5
	1994	60.1	305.5		
Mexico	1980	82.2	141.8	2.5	1.6
	1993	139.7	194.8		
Japan	1970	135.9	350.5	2.2	4.2
	1991	281.6	557.0	(1970-1991)	(1970-1992)
Western Europe	1970	420.6	839.3	2.3	Unavailable
	1994	1010.2	1430.0		
China	1970	13.8	414.6	7.5	7.5

Source: Transportation Annual Statistics (1997).

- a new process of production;
- the substitution of a cheaper material, newly developed for a given task, in an essentially unaltered product;
- the reorganization of production, internal functions, or distribution, arrangements leading to increased efficiency, better support for a given product, or lower costs; or
- an improvement in instruments or methods of doing innovation.

For the purposes of this chapter, we embrace this broad definition and apply it to the study of innovation in the trucking industry.

Figure 1 illustrates the innovation process in the trucking industry showing how technological and non-technological factors motivate and enable service and process innovations. Service and process innovations, in turn, may be expected to improve firm performance. We consider that a service or a process innovation is motivated by a factor when it is intended to fulfill a need created by the factor.[3] An innovation may be enabled by a factor when the knowledge embodied in its software and hardware is instrumental to its effective utilization. For example, the just-in-time manufacturing environment demands that the location of parts be known at all times. Real time tracking is an innovation that is motivated by the just-in-time manufacturing environment. Satellite communication has been core to the development of real time tracking systems and can be considered as an enabler of the innovation.

Our study reveals that customers are the primary sources of innovation in the trucking industry, as can be expected in a service industry. Changes in manufacturing and retailing practices recognize that significant value is created when

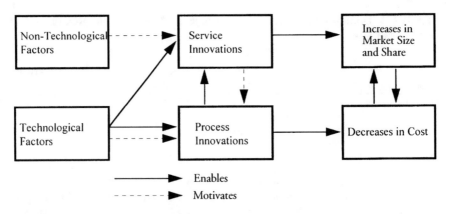

FIGURE 1 Interactive model of factors, innovations, and outcomes.

inventory levels are reduced and when goods are produced closer in time to the point when the goods are consumed. Such modern time-sensitive management practices have drastically altered the role of the trucking service provider in the economy by altering the size, distance, and frequency of shipments and by increasing the importance of transportation reliability, timeliness, and speed. The emphasis on customers and their important role in the innovation process is consistent with the current industry dynamic. Loyal customers become critical in a competitive environment where there are a large number of trucking firms for the customer to choose from and changes in customer preference entail no significant switching costs.

Fierce competition compels firms in the trucking industry to develop and adopt appropriate innovations. Large firms in the trucking industry are often lead users of new products intended for the industry. Many of the technology-driven innovations that have been adopted by the trucking industry have been developed outside the industry. The communications and computer industry have had a significant stake in the innovations adopted by the trucking industry. The widespread adoption of some of these innovations has been enabled by the close coordination between the user (the trucking industry) and the developer (communications industry, for example). This is consistent with Von Hippel (1976) who found that lead users in the scientific instrument industry often play a major role in the innovation-development process. Smaller trucking firms adopt these innovations in order to achieve competitive parity with their larger rivals in the competition for customers.

In this chapter section 2 presents a brief overview of the trucking industry and discusses the transformation of the trucking environment through the birth of

---

[3]Rogers (1995) defines a need as a state of dissatisfaction or frustration that occurs when one's desires outweighs one's actualities.

the logistics industry and the transportation services industry. Section 3 discusses some of the non-technological factors that influence innovation in the trucking industry, such as the globalization of markets and resources, intermodal transportation, changing business practices, competitive pressure on price and service, labor productivity and skill, and environmental and safety considerations. Section 4 presents some of the important technological factors that have influenced innovations in the industry, including telecommunications, computer hardware and software, navigation and positioning systems, surveillance, sensing and tagging technologies, and data exchange and fusion capabilities (BTS, 1997a). Section 5 discusses the relative contribution of technological and non-technological factors to innovation and firm performance. Section 6 concludes the chapter with a look at the future of freight and expectations for the trucking industry.

AN OVERVIEW OF THE TRUCKING INDUSTRY

Traditionally, the industry has been segmented into three categories, depending on the size of the shipments that are carried by each firm: truck load, less-than-truckload, and package express. Trucking appears to be expanding into a fourth segment—logistics. Recent industry trends indicate significant industry consolidation, with large firms participating in multiple segments of the industry through subsidiary relationships. These firms offer a "one stop shop" for a variety of transportation services.

The Traditional Trucking Industry Segments

Truckload (TL) carriers specialize in hauling large shipments (often weighing over 10,000 pounds). The average TL shipment weighs about 27,000 pounds. An owner-operator[4] or a driver for a TL firm will pick up the load from the shipper and carry it directly to the consignee, without transferring the freight from one trailer to another. Thus, TL carriers do not need a network of terminals. The TL segment of the industry is highly competitive because there are very low barriers to entry. Key issues for managers of TL firms are the management of backhaul routes[5] and driver turnover.

Less-Than-Truck Load (LTL) carriers haul shipments that usually weigh between 150 and 10,000 pounds. The average LTL shipment weighs slightly over 1000 pounds. The key economies of scale and density for an LTL carrier come from consolidating many shipments going to the same area. Such consolidation requires a terminal network. Thus, an LTL shipment will typically be picked up

---

[4]An owner-operator is a sole proprietorship or other small company whose primary purpose is to operate one or more trucks for hire.

[5]Backhaul routes: after a load goes from point A to point B, the firm must either find a shipment originating near point B or incur the costs of operating an empty truck (deadheading).

at the shipper's dock by a pickup and delivery truck and hauled to the trucking firm's local terminal, where it will be unloaded and placed with other shipments going to similar destinations. The process of moving groups of shipments from one city terminal to another is known as line-haul operations. It is usually accomplished by large trucks, often with tandem trailers, or by rail or some other mode, depending on price and service considerations. Once the shipment arrives at its destination city terminal, it is moved to a pickup and delivery truck at a terminal and hauled to the consignee. Key issues for managers of LTL firms are increasing density and linehaul network optimization (Swan, 1997).

Package Express (PE) carriers usually haul shipments that weigh less than 150 pounds. A typical package weighs less than 50 pounds. PE carriers offer at premium prices time-sensitive or other specialized services such as "air express" shipments, many of which are carried by truck and not by airplane. The firms in the PE segment have been experiencing tremendous growth; some of the large firms have been maintaining higher levels of operating income than firms in every other segment of the trucking industry. Cost and competitive pressures are moderating this success. Key issues for managers of PE firms are increasing customer density and delivering freight on time.

The Modern Outlook

Globalization, technology, and specialization have combined to bring a new dimension to the trucking industry: logistics. Logistics can be defined as a concept to guide economic processes and as a tool of rationalization to optimize purchasing, transport, reshipment, and warehousing (Danckwerts, 1991; Plehwe, 1997). Logistics uses the right information to move materials to the right place, at the right time, for the right cost. While logistics once belonged in the realm of the manufacturing firm, today trucking firms are seizing the initiative and absorbing the logistics function into their value chains.[6]

As customers focus on cutting costs and developing core competencies, trucking firms are restructuring to offer the total transportation solution by including logistics and a variety of other transportation options in their corporate portfolio. The logistics business, almost nonexistent ten years ago, is now approximately a $20-30 billion industry segment and is projected to grow at about 20 percent a year (Industry Week, 1997).

Logistics may not only provide functionality and lower costs to the customer; it may also improve service and increase the customer's perception of value. This is especially true because many customers are focusing on ways to reduce costs and improve quality in response to international competition. Consequently, many U.S. businesses are steadily reducing their investment in inventory. Manu-

---

[6]A company's value chain identifies the primary activities that create value for customers and the related support activities (Thompson and Strickland, 1996).

facturers are also faced with the need to reduce cycle time. In the past, many manufacturing firms included inbound and outbound logistics among their management activities. Competitive pressures are forcing firms to focus all their energies on their core competencies and primary activities. Support functions to the manufacturing activity, such as the logistics function, are outsourced to specialists in transportation.

These specialists might be the firm's transportation provider or a third-party logistics provider. Cost reduction seems to be the primary motivator for outsourcing logistics, followed closely by customer service, according to the Exel Annual Third Party Logistics Key Market/ Key Customer survey (Industry Week, 1997). Some 60 percent of the nation's largest manufacturers use third-party logistics providers, according to a study by Mercer Management, Inc., and Northeastern University (Purchasing, 1996).

The availability of appropriate technology has facilitated the growth of logistics. Logistics providers are using large databases, complex software and algorithms, supporting hardware, and the latest tracking and communication technologies to track fleets, organize customers and loads, and provide the most efficient way to satisfy the customer.

Logistics providers in the trucking industry have unique industry-specific knowledge that can be applied to enhance the logistics function. Through specialization, firms achieve learning curve advantages in leveraging technological and transportation planning knowledge. Typically logistics firms contract with shippers or consignees to assume responsibilities ranging from transportation and material handling to warehouse management and the management of inventory levels and distribution throughout large portions of the value chain.

Further, logistics providers may be in a position to leverage freight volume to achieve the gains that accrue to network densities. The logistics firm coordinates information relating to many shippers and consignees in the same geographical area. The firm is then in a position to optimize freight movement by increasing the volume carried by each truck carrying loads to or from a certain location.

Firms have used different organizational arrangements to incorporate logistics in their arsenal. Schneider, the nation's largest TL firm, is associated with logistics provider, Schneider Logistics. The logistics arm of Schneider innovates and develops products to enable Schneider to compete effectively and efficiently. In contrast, J.B. Hunt, another TL firm and a close competitor of Schneider, has a logistics arm, a wholly owned subsidiary called Hunt Logistics, which provides independent logistics services. J.B. Hunt Transport is but one of the transport companies that are a part of the portfolio of trucking firms used by Hunt Logistics. Hunt Logistics, which has been in operation for three years, has customers in the retail segment, consumer and industrial goods, paper, and automotive industries.

Smaller firms specializing in logistics are usually organizing in one of two ways: either as dedicated contract carriers or as non-asset based supply chain management companies. Dedicated Contract Carriers (DCC) are strongly asset based. Tractors, trailers, and drivers are their focus, along with the management of information. Their core objective is better truck transportation, and they tend to have a trucking perspective. Ryder Integrated Logistics, with a revenue of nearly $6 billion, is the largest DCC firm.

Supply chain management companies focus on information technology management through software applications. They emphasize shipment control and visibility. Ideally, these companies manage every part of the host company's inventory as tightly as possible to reduce cost and cycle times. Multiple segment transportation, warehousing, and inventory are their target areas.

Logistics and supply chain management have brought about some restructuring of the trucking industry. Firms are now offering a variety of transportation services including TL, LTL, logistics, package express, and intermodal as a one-stop transportation solution. They are accomplishing the feat of "one call, one carrier" primarily through acquisitions, mergers, and alliances. For the purposes of this paper we consider the newly restructured firms to be providers of transportation services (TS).

TS firms are new organizational forms that are emerging in the trucking industry. They cross traditional boundaries and integrate across segments and modes of transport. These hybrid forms have emerged in response to changing competitive conditions. Firms such as CNF Transportation and Caliber Systems, through mergers and acquisitions, have developed a portfolio of transportation services. CNF transportation has, among its operating units, a package express firm (Emery Worldwide), an LTL firm (Con-Way Transportation Services), and a logistics provider (Menlo Logistics). Caliber Systems[7] has in its portfolio of firms a Package Express firm (RPS), an LTL firm (Viking Freight), and a logistics provider (Caliber Logistics). CRST International has recently restructured itself into a single transportation services company by combining its six units into one operating unit. In the past, each unit served customers separately in their niche markets. Through the restructuring, CRST International combines CRST in TL, Malone Freight lines and Three 1 Truck line in flat beds, CRST logistics, and an express LTL service. According to company President John Smith, "It didn't take a genius to figure out it was better approaching this as one team of professionals totally focused on the customer and making transportation as easy as possible as our customers" (Traffic World, 1997a).

Competitively, these organizations have to contend with the challenges posed in each of the segments in which they participate. Many of the TS firms have yet to find the synergies they were looking for through restructuring and consolidation. It is expected that the next few months will produce some of the biggest mergers ever. As more firms present themselves as providing total transportation services, the formidable task that lies ahead of them is to achieve the close coordination that is required to capture the benefits of being a single entity.

---

[7]Federal Express acquired Caliber Systems in 3Q 1997 in a $2.4 billion bid.

TABLE 2 Non-technological Factors Influencing the Trucking Industry

Non-technological factor	Area of focus
Globalization	Markets and resources spread throughout the world
Intermodalism	Coordination between different modes of freight transport
Changing business practices	Just-in-time, Quick Response, inventory reduction
Standard weight limits	Standardization of load limits across state and national boundaries
Competitive pressures on price and service	Lower operating costs and relationship-specific assets
Labor productivity and workforce skill	Training and technology
Environmental and safety considerations	Sustainable trucking

NON-TECHNOLOGICAL FACTORS INFLUENCING THE TRUCKING INDUSTRY[8]

Several non-technological factors have induced innovation and changed the competitive landscape. These factors include globalization of markets and resources; intermodal transportation; changing business practices; and competitive pressure on price and service, labor productivity and skill, and environmental and safety considerations. Table 2 provides the area of focus for the non-technological factors.

Globalization

The fundamental nature of the overall business environment is changing. With the lowering of trade barriers and advances in technology and communication, the competitive landscape has been transformed into a global economy where goods, services, people, skills, and ideas move freely across geographic borders (Hitt et al., 1996). Since 1969, the number of multinational corporations in the world's 14 richest countries has more than tripled, from 7000 to 24,000. These companies control one-third of all private sector assets and enjoy worldwide sales of $6 trillion (Alden, 1997). International trade now accounts for 24.7 percent of the U.S. GDP, up from only 11.3 percent in 1970 (BTS, 1997a). Global competition has increased performance standards in many dimensions, including cost, quality, new product development, and service.

Globalization increases the range of opportunities for firms in many industries. The implication for the trucking industry is that there is an advantage for a trucking firm to be a single source provider in order to meet a global firm's trans-

---

[8]Sections 3 & 4 draw upon our findings in our case studies and field work. Archival sources on the trucking industry are also drawn upon to illustrate innovations and the innovation process. See Appendix A for details of the study method.

portation needs. The emergence of the TS segment of the industry is the industry's response to the global challenge.

Firms in the trucking industry are providing freight movement through alliances and international operations. Package express firms that have air freight as a critical component of their transportation portfolio, such as UPS and FedEx, have extensive international operations. UPS World Wide Logistics was formed in 1993 and serves North America, Latin America, Asia, and Europe. Logistics firms are now beginning to venture abroad as their domestic customers expand their requirements. In 1996, Schneider Logistics, Inc., entered the European logistics market after being selected to provide inland transportation management services for Case Corporation's European manufacturing and service parts operations. Schneider Logistics is one of the first companies to engineer and implement a European freight management program to manage across multiple shippers, carriers, transportation modes, and countries, according to the company's press release. The company uses engineered solutions and systems technology to perform shipment optimization, electronic date interchange (EDI) with carriers, freight payment, and reporting. Schneider Logistics provides logistics management services for Case's eight European manufacturing locations. Its European services mirror the service Schneider Logistics currently provides Case within North America. These services include management of all ground transportation, optimization of Case freight, inbound and outbound transportation from manufacturing facilities, interplant movements, transportation required for Case's service parts operation, and engineered solutions.

Within North America, there has been significant growth in international trade since the North American Free Trade Agreement (NAFTA) went into effect in January 1994. In 1995, nearly $274 billion worth of goods moved by land between Canada and the United States—up 10.5 percent from 1994, according to information from the BTS Transborder Surface Freight Dataset, collected by the census bureau. By value, 68 percent of this trade moved by truck. Over $97 billion worth of goods moved by land between the United States and Mexico in 1995, up 7.8 percent from 1994. By value, 81 percent of this trade moved by truck (BTS, 1997b).

To adapt to this new opportunity and to adjust to evolving cabotage[9] rules, trucking firms are now engaged in a new process innovation called "sweeping." Analogous to a milkrun,[10] the trucking firm sweeps a region for exports and moves

---

[9]Cabotage rules are laws prohibiting motor carriers from hauling freight between two points outside of the carrier's home region. For example, Canadian truckers may haul freight from Ontario to Florida, and may return with freight destined for Ontario. It may be illegal to make deliveries from Florida to Michigan along the way.

[10]Milkrun is a type of less-than-truckload service in which a truck visits several origins in sequence to pick up freight with a common destination. For example, a truck might visit several auto parts suppliers to collect parts destined for an assembly plant.

the swept freight to a single terminal or focal point for line haul movement. The sweep allows a firm to build exports to a critical freight volume and optimize the mode of transport subsequently.

Intermodalism

Intermodal has been railroads' successful answer to change in the transportation environment. By offering low cost solutions and innovations such as double stack service, intermodal has become dominant in certain corridors of freight movement. In 1993, intermodal shipments exceeded 200 million tons of goods valued at about $660 billion. The classic intermodal combination of truck and rail accounted for 41 million tons and $83 billion. In addition to the $660 billion, about 3 million tons, valued at about $134 billion, is estimated to have moved by truck and air combination (BTS, 1997a).

Intermodal shipments are higher in value per pound on average than typical single-mode shipments. The average value of goods shipped by air (including truck and air) was $22.15 per pound, followed by parcel, postal, and courier services ($14.91 per pound) and by truck and rail combination ($1.02 per pound). Goods shipped only by truck averaged about 34 cents per pound and goods shipped by rail, water, and pipeline averaged less than 10 cents per pound (BTS 1997a).

Transportation providers have become more capable of substituting one mode of transportation for another when such a substitution creates an economic advantage. Some of this effect can be explained by recent innovations in containerization. JB Hunt has been a pioneer in the use of intermodal double stack containers. Traditionally, a loaded trailer was put on a flat car, and there was one trailer for each flat car. By separating the chassis from the container, JB Hunt can stack two containers on one special rail car. The cost of moving freight decreases dramatically. The use of double stacked containers and other container innovations allows firms to offer as many modal choices as possible to lower total transportation cost without compromising time sensitivity.

Changing Business Practices

Changes in manufacturing and retailing practices (such as "just-in-time," JIT, and "quick response," or QR) recognize that significant value is created when inventory levels are reduced and when goods are produced closer in time to the point when the goods are consumed. Such modern time-sensitive management practices have drastically altered the role of the trucking service provider in the economy by altering the size, distance, and frequency of shipments and by increasing the importance of transportation reliability, timeliness, and speed. In general, shipping rates (prices) are falling, while trucking firms are providing an increased level of service to their customers. Trucking firms are becoming more sensitive to the importance of time and the critical nature of pickup and delivery appointments.

For example, one major shipper, an automobile assembler, has reduced the number of trucking firms with which it contracts by over a factor of ten in order to encourage the remaining carriers to coordinate their operations with the assembler and its trading partners. That assembler has reduced in-plant parts inventories to four hours or less. When an inbound truck is delayed sufficiently to stop an assembly line, this assembler could incur a loss of $8000 to $12,000 per minute for the delay. If the carrier is the cause of the delay, some of this loss is charged to the carrier in the form of a penalty. In general, by reducing the number of carriers they deal with, shippers sacrifice the bargaining advantages that could be obtained from a large set of carriers. However, the shipper minimizes transactions costs, in terms of control and coordination, and builds closer partnerships with the few chosen carriers.

An increasing number of manufacturers are also engaging in a new process— disintermediation—in which they send products directly to retailers or consumers. Giant retailers deal directly with manufacturers, eliminating the need for shipments to and from wholesalers. All such disintermediation creates an increase in the frequency of shipping activity, while reducing the average size of shipments. For example, the advent of the "world's largest bookstore," Amazon.com, on the Internet has provided an alternate method by which books can be purchased. Customers place their orders through the Internet, and publishers directly send them the books via PE firms. The wholesale and retail distributor are no longer a part of this value chain, and consequently neither are the TL and LTL firms. The number of shipments that move directly from plant to customer with no warehousing or middleman will go from 31 percent in 1994 to 36 percent in 2000, according to a survey conducted by an industry expert (Traffic World, 1997b).

Trucking companies must therefore be prepared to haul smaller shipments and make more frequent deliveries. Besides expecting their carriers to be efficient and cost-effective, shippers now demand on-time deliveries and consistent cycle times, with little damage or loss to valuable freight. The trucking industry has responded to the critical role caused by changing business practices by offering new products, such as time-sensitive delivery, enabled by process innovations, such as sleeper teams and enhanced quality standards, including QS 9000 certification.

The emphasis on quality and process is changing the business environment. As large shippers such as the Big-3 automotive companies cope with globalization, they are adopting quality standards, such as ISO 9000 and QS 9000, throughout their organizations. As these large shippers reduce the number of companies with which they will do business, they increasingly require their transportation companies to comply with these standards. As we found among many of the firms we surveyed—including Mark VII and TNT—QS 9000 certification has become a competitive necessity for firms hoping to do business with large firms, especially in the automotive sector.

Global competition and technological advances are changing business practices including the way goods are manufactured and distributed. Innovations such as time-sensitive delivery, sleeper teams, and QS 9000 are enabling the trucking industry to respond effectively to these changes. Truck manufacturers, in turn, are enabling the execution of these innovations.

Competitive Pressures on Price and Service

Because switching costs are low for shippers and entry barriers to the trucking industry are also low—especially in the TL segment—price competition has been fierce. Shippers, consignees, and third parties seem increasingly sensitive to the price of transportation. Consequently, the operating costs for the trucking firms have often exceeded operating revenues. The squeeze on revenue has led trucking firms to focus on costs as their means to greater profits. Asset productivity is critical to operational efficiency. The source of competitive advantage lies in the ability of the firm to cut costs, to increase productivity, and to adapt to the customer's specific environment.

CF Motor Freight (CFMF), now Consolidated Freightways, redesigned its hub-and-spoke[11] network in late 1995 for a directional loading system in order to maintain a focus on costs. The new system uses fewer break bulk terminals with regional hubs that cover greater areas. Results of this change include a reduction in the firm's 24 former hubs to 14 flow centers and a reduction in its 21,533-person work force by 670 drivers and 440 dock workers. According to the claims, the new load planning program, called the Business Accelerator System, has sliced a day off average transit time, reducing it from four to three days, and dramatically improved on-time delivery. The Business Accelerator System also has eliminated excessive freight handling and cut claims, according to company reports. CFMF, using the new system, can deliver to 70 percent of the United States in two days or less, compared with 42 percent before this initiative. Only half the freight requires intermediate handling, compared with three-quarters before the system was implemented. The current firm goal is to reduce the percentage of intermediate handling to one-third.

To implement the new network, CFMF invested about $25 million in new facilities and another $25 million in 370 sleeper tractors equipped with Qualcomm

---

[11]Hub-and-spoke is a network structure in which an LTL carrier operates one or more major terminals (known as hubs) and a larger number of smaller terminals (known as spokes). Freight is taken from its origin terminal to a spoke terminal, where it is consolidated with other freight going toward similar destinations. It is then moved (by line-haul truck or another mode of transportation) to a nearby hub, possibly for forwarding to a hub closer to the destination. Finally the freight is moved from a hub to a spoke near the destination terminal, and ultimately to its destination.

satellite communications units. It also persuaded Teamsters union officials to approve what the company said was the largest change of operations in trucking industry history, which resulted in the transfer of 850 employees in a single week. Under the hub-and-spoke system, freight heading from Detroit to Boston took 77 hours to travel 1041 miles. Four intermediate stops added 53.5 hours of non-driving time to 23.5 hours of driving time. Under the Business Accelerator System, the same load now takes about 43 hours to travel 716 miles with only one stop in Buffalo. Nondriving time was reduced to about 26 hours and driving time to 16.5 hours.

An organizational innovation to control costs and enhance service—the introduction of cross-functional service teams—was implemented at TNT Express. The company has five service teams. Each team is assigned a geographic area. Each service team typically is comprised of a dispatcher, a representative from safety, finance, and operations, and a recruiter. The service teams are assigned key performance indicators, and the teams are responsible for control and continuous improvement of those key performance indicators. These teams foster teamwork and cooperation while keeping focused on achieving best business practices.

Trucking firms are also calling for greater productivity through the use of longer, heavier trucks traveling longer routes. Between 1982 and 1992, the number of trucks operating weights above 80,000 pounds increased by 180 percent from 18,000 to 50,000 (BTS, 1997b). The total number of vehicle miles traveled by this weight class rose by 193 percent.

Trucking firms are finding innovative ways to deal with the pressures of cost containment and service enhancement. The use of new network systems and service teams indicates an attention to process improvements to contain costs. Increasing asset productivity and investing in relationship-specific assets have also addressed the need to create competitive advantage.

Labor Productivity and Workforce Skills

Measured by output per worker, labor productivity for trucking rose 2.8 percent annually from 1954 to 1989—considerably higher than the annual average rate of 2 percent for the overall business sector from 1954 to 1994 (BTS 1997a). Enhancements in truck technology and process improvements in the manner in which freight is moved indicate that labor productivity in the trucking segment of the transportation sector is likely to increase.

Important segments of the industry such as TL have long experienced a driver shortage because of high employee turnover at individual firms. However, a recent study by the Gallup organization, commissioned by the ATA, suggests that the problem may lie elsewhere. The study found that roughly 320,000 drivers switch jobs within the industry every year. This phenomenon, also known as "churning," presents a different challenge than one posed by driver shortage.

Firms are working hard to retain their drivers so as to minimize recruiting and training costs. JB Hunt has recently boosted average driver pay by 34 percent, to about $0.38 cents per mile on average, and early results indicate that this move has cut driver turnover in half and has reduced recruiting, insurance, and training costs almost enough to offset the high wages (Traffic World, 1997c).

The increased availability of information and enhancements attributable to advances in truck engineering are changing the job descriptions of drivers and dispatchers in the trucking industry. Drivers must become more skilled in the use of technology, especially information technology. In response to the increase in skill required of drivers, firms such as Roadway Express have moved from hiring at multiple locations to a single recruiting and training center. The resulting coordination and monitoring benefits enable the firm to attract and train quality employees and develop essential human capital.

Dispatchers are increasingly able to rely on computer systems for routine parts of their work. Each suitably equipped dispatcher can now track more vehicles through the use of AVL systems. Innovations have transformed the job from routine activities to planning and anomaly management. New systems are designed to automate routine inquiries. For example, if a load is picked up on time, the information goes directly into the database. The dispatcher is informed only if expected events do not occur. Innovation and new technology adoption offer an opportunity for employees to enhance their skills and earn higher wages while increasing job mobility. However, the firms are unable to either find or attract appropriately qualified individuals. According to a study by the Information Technology Association of America, U.S. industry in general is short about 190,000 knowledgeable workers (Traffic World, 1997d). The challenge for the trucking industry is to find and retain a workforce that can interpret information and react rapidly.

Environmental and Safety Factors

Environmental impacts, particularly air pollution, has been a concern for the trucking industry. Federal regulation of heavy-duty trucks has proved to be quite effective, and the Environmental Protection Agency (EPA) is currently proposing more stringent emissions standards. EPA estimates that these standards could initially increase vehicle retail prices by $200 to $500, but this cost is expected to decrease rapidly.

EPA has also required onboard diagnostic systems on all new light-duty vehicles and trucks since 1994. These systems monitor emissions control components for any malfunction or deterioration that would cause certain emission thresholds to be exceeded and alert vehicle operators to the need for repair.

The issue of safety has always been paramount in the trucking industry. Safety becomes a greater concern as the industry moves toward using longer trailers and carrying heavier loads. While drivers of trucks account for only 3.3

percent of drowsy driver crashes, they are generally involved in accidents with more severe injuries and property damage. Among workers as a whole, the rate of occupational fatalities in the United States remained fairly constant from 1992 to 1995. Truck drivers alone made up 12.1 percent of all occupational fatalities, with 749 killed. Of the 150,000 truck drivers involved in traffic accidents in 1994, an estimated 22,000 sustained injuries (USDOL, 1996).

To provide safety training to its drivers, Ryder Integrated Logistics is using an innovative instruction package. Developed by Vortex Interactive, it contains generic programs on hazardous materials, hours of service, and drug and alcohol awareness. Ryder's customized programs include "Ryder Backing Techniques," Ryder "Policies and Procedures," and "Your Back at Work." The entire program is on CD-interactive equipment and does not need a computer; it can be hooked to a regular television monitor. According to the company, drivers significantly prefer this new system to the old video and lecture format.

The need for safety enhancements in trucks is also being met through aggressive innovation among the truck manufacturers. For example, major improvements in safety are incorporated in the recently announced Volvo VN Series. The hood, sloped at an angle of 18 degrees, and a large, one-piece windshield, in combination with a new mirror system, provide better visibility around the vehicle than do current Volvo tractors. The cab material is High Strength Steel (HSS), found to have the best strength-to-weight ratio of any material evaluated by Volvo GM Heavy Truck. While the overall weight of the Volvo VN Series is less than comparable current Volvo models, the new tractors have passed the Swedish Impact Test and the Volvo Barrier Test. The Swedish test, recommended as a standard by the Maintenance Council of the American Trucking Associations, has been determined by industry researchers to be the most severe impact test in the world (SAE, 1993).

TECHNOLOGICAL FACTORS INFLUENCING THE TRUCKING INDUSTRY[12]

At the foundation of much of the technological innovation in the trucking industry lie five basic building blocks: telecommunications; computer hardware and software; navigation and positioning systems; surveillance, sensing, and tagging technologies; and data exchange and fusion capabilities. The rapid pace of technological change in these building blocks presents a quandary to firms. They may procrastinate to see how technologies evolve and may be less likely to adopt technologies that might become quickly obsolete. However, the new technologies present opportunities for new products, processes, and services that could not have been conceived, developed, or commercialized a few years ago.

---

[12]The categorization and descriptions of the technologies are drawn in large part from the Bureau of Transportation Statistics: *Transportation Statistics Annual Report 1997—Mobility and Access.*

TABLE 3 Technological Factors Influencing the Trucking Industry

Technological factor	Area of focus
Telecommunications	Wireless and wireline communications technologies
Computer hardware/software	PC and mainframe computer systems
Navigation and positioning	Computer-aided dispatch, automatic vehicle location, and route guidance
Sensing and tagging	Electronic toll tags and marking
Data exchange	Data interchange between the trucking firm and outside customers or partners
Data fusion	Integration of data from multiple documents, databases
Onboard diagnostic system	Computer system built into power unit that helps monitor vehicle performance
Internet	Provides interconnections between firms' computers and/or customers' computers
Bar codes	Support tracking of freight
Sixth axle	Mechanical design that allow higher weight loads
Sleeper cabs	Power units with specialized living facilities
Larger trailers	Longer, wider, or taller
Soft-sided trailers	Vinyl sides simplify loading and unloading of trailer and increase variety of freight

Table 3 lists some relevant technological factors, along with the area of focus for each factor.

Telecommunications

Advances in telecommunications facilitate many innovations that enhance value to the customer and increase the control and flexibility of the trucking firm. An especially powerful tool is the ability to make real-time operational information available to a dispersed transportation network. Trucking firms are improving traditional wireline communications with advances in satellite, cellular, and fiber optic technologies.

Satellite systems have been popular because they provide coverage of a wide area. However, satellite bandwidth is quite expensive. As a result, freight mobility systems based on satellite communications typically rely on short, infrequent bursts of data transmitted between a data center and the vehicle. Wireline communications such as telephone or leased-line service are used to transmit data from a dispatch center to the data center. Cellular telephone is relatively inexpensive, and bandwidth costs are falling quickly. Cellular is considered ideal for applications with infrequent position reporting rates and modest need for voice communications, such as applications where a trucker is required to report position data only a small number of times per day and only occasionally needs to speak directly with dispatcher or to call home.

Increasingly, communication in support of trade between firms uses some means of Electronic Data Interchange (EDI), the paperless exchange of business documents via computer. EDI systems support electronic commerce. Trading partners send computer messages, rather than paperwork, for their routine transactions, such as quotes, orders, and shipping notices.

Trucking companies have moved into the role of technology integrator by coordinating their computer systems with the computer systems of their customers (shippers and consignees). Wireless, vehicle-to-roadside EDI communications can also occur over a very short range: for example, between trucks and wayside inspection systems and toll collection systems.

Network linkages such as the Internet are being combined with electronic data interchange to offer innovative new service enhancements. For example, CFMF exchanges EDI shipping information through the Internet with one of its customers, Westinghouse Hanford. Typically, EDI information is exchanged between CFMF and the customer either directly or through a third-party Value-Added Network (VAN). Through EDI, customers have access to financial data, bills of lading, shipment status, and freight invoices. Use of the Internet to exchange EDI information is less costly because the electronic documentation is sent through a public network, as opposed to a private phone line; it is also a more efficient means of transferring documentation, according to company press releases. In most cases, the response time is shorter through the Internet, giving users opportunity to interact with one another in a more timely fashion, compared to time-delayed, batch-style interactions. The process vastly improves the reliability of the data and reduces the time needed to transfer EDI files.

Computer Hardware and Software

The price of computer processing power has dropped by approximately 30 percent per year for the last two decades. Such dramatic cost decreases permit highly sophisticated sensors and control systems to be built into transportation vehicles. Digital electronics and computer chips have become commonplace in truck braking, engine control, security systems, and climate control.

Fleet management systems use combinations of microprocessors and other computing and location devises that allow data to be gathered and processed for strategic and operational decision making. These systems may be grouped in three categories (Hubbard, 1997):

• Trip recorders are data-gathering devices aboard the vehicle that gather information about the truck's operation. Data are recorded about such events as whether the engine is on, when the truck is idling, how it is shifted, and how quickly it accelerates and decelerates. Such information can be collected when the truck returns to its base, and managers can use the data to determine how the truck was driven. Such information aids management with many activities, such as determining fuel economy, monitoring tire wear, identifying good and bad driving practices, assessing responsibility after an accident, and performing preventive maintenance.

- Electronic vehicle management systems are onboard computers that combine the functions of trip recorders with automatic vehicle location and two-way communications. Information about the vehicle's activities is communicated between the vehicle and a central dispatch center. The Omni-Tracs system from Qualcomm is an example of such a system. Approximately 200,000 Class 8 trucks in the United States—approximately one in every six such trucks—are equipped with electronic vehicle management systems.
- Computer-aided dispatch and freight management systems support the operations of the trucking company office. Such systems are more traditional office-based data processing applications, rather than vehicle-based systems. They support the assignment of freight to vehicles, the assignment of drivers to vehicles and routes, the tracking of shipments, as well as such traditional data processing functions such as accounting, bill-of-lading processing, payroll, billing, and claims processing. Activity-based costing systems are increasingly common. Such systems help trucking firms determine which customers and what locations are profitable and align their freight rates with their costs.

The availability of low cost, high-powered computer hardware and software is providing the trucking industry with an information-rich environment. The new focus therefore turns to the management of information in all facets of the trucking business in order to effectively and efficiently move freight (Bander et al., 1997a).

Navigation and Positioning Systems

For the trucking industry, near real-time tracking of dispersed assets is a critical function. Technology has moved the trucking industry from the use of pay phones to relay location information to automatic vehicle location (AVL) systems that permit the central operations to be aware of the location of the entire fleet automatically at all times in real time.

Navigation systems use location aids and computerized maps to provide route guidance. When combined with real time traffic, weather, and construction information, these systems are becoming critical assets in the increasingly time-sensitive trucking environment. Location and navigation systems are also used along with computer-aided dispatch systems for route optimization.

One satellite-based vehicle location technology, the global positioning system (GPS), has the potential to make vehicle location and navigation a universal practice. GPS was designed as a dual-use system with the primary purpose of enhancing the effectiveness of U.S. and allied military forces. The growing demand from military, civil, commercial, and scientific users has generated a U.S. commercial GPS equipment and service industry that is a world leader. GPS receivers cost only a few hundred dollars, with prices falling rapidly. GPS works anywhere in the world where there is a direct line of sight from the receiver to the satellite. This makes it an appropriate service for commercial freight applications except while the truck is traveling in an urban canyon or tree-covered area. During those periods, GPS can be combined with other navigation techniques.

GPS is not a communications system; it is a satellite-based navigation (positioning) system. Many current satellite-based fleet management systems use a constellation of satellites and other communication technologies—GPS for positioning with a different medium (such as cellular or low-earth orbit satellite) for communications in order to provide full positioning, navigation, and communication capabilities.

Surveillance, Sensing, and Tagging Technologies

These technologies identify a vehicle's location, characterize its environment, or permit the exchange of information. Tagging freight-carrying vehicles and cargo is revolutionizing logistics. Tags are capable of storing situation-specific data, such as container contents and destination, and of being read over long distances by satellite in some cases.

These technologies are also being applied in an innovative manner to enhance processes and improve truck safety. Bar code technology (and, to a lesser degree, radio frequency ID tags) have advanced the practice of item-level and shipment-level tracking. RPS is using two-dimensional bar codes that can represent significantly more information than traditional linear bar codes. UPS is using three bar codes for package tracking: two linear bar codes plus a circular "bulls-eye" code that can be read at any angle. UPS has also recently introduced ring scanners, which are worn in the hands of sorters and allow workers to capture more precise information about packages.

Imaging systems are used to store paperwork electronically. Such systems consist of databases that store images of documents that have been scanned or faxed. Imaging systems are being used to help reduce the tremendous numbers of freight bills and other shipping documents that are handled by shippers.

Sensor technologies are also being used to improve truck safety. The Eaton Vorad system is an example of a radar-based collision warning device that has been used on some trucks and buses. In general, the sensors are coupled to computers and displays that provide warnings and sometimes suggested responses. Surveillance, sensing, and tagging technologies are providing intelligence about the trucks' environment while also enabling process improvements.

Data Exchange and Fusion Capabilities

The broad impact of information technology (IT) on transportation is to provide a more knowledge-rich system that is interoperable and can be used for decision making. The combination of disparate information from diverse sources is central to many IT applications. Beyond technical interoperability and the establishment of institutional and organizational relationships for effective data exchange, the new discipline of "data fusion" has emerged. Data fusion is the design of computer systems that can facilitate the convenient, flexible, secure, and adaptable blending of information from a wide range of independent sources.

A significant innovation enabled through data exchange and fusion capabilities is real time Internet package tracking. Package express firms such as Federal Express and UPS pioneered Web-based shipment tracking. Now several express freight companies, including Airborne Express, DHL Worldwide, Emery Worldwide, and Roadway Express, offer on-line tracking via the World Wide Web. The ability to locate a package or load in real time appears to be as important to the shippers as it is to the trucking firm.

The availability of computing power and data exchange and fusion capabilities is also spurring innovation among truck manufacturers. For example, Freightliner Corporation recently introduced a new milestone software package, CustomerLink, an electronic inventory replenishment system connecting Freightliner dealers and their customers. According to company reports, the Customer-Link package is the first phase of supply chain integration at Freightliner, a fully automated and integrated parts distribution network designed to facilitate inventory planning and replenishment along the entire line of supply: suppliers, parts distribution centers, dealers, and customers. CustomerLink allows Freightliner dealers to relieve customers of the burden of inventory management by providing the tools to automatically poll customers' parts usage data to maintain their inventories at predetermined levels.

CustomerLink incorporates bar code technology and telecommunications links to capture the usage of parts at a customer's site. Each time a part is taken from inventory, its bar code is scanned and other critical information, such as vehicle ID, repair order number and quantity, is entered into the CustomerLink system. At non-peak times, the Freightliner dealer can electronically retrieve this information to generate a restocking order and pull the needed parts for delivery by the next morning.

Sleeper Cabs

New, advanced technology is being used to improve driver comfort, especially in sleeper cabs. Sleeper teams use sleeper cabs, which allow one driver to sleep while the other drives, so that the vehicle can keep moving over long distances in spite of hours-of-service regulations and human limitations. Sleeper teams are not new to the industry. They are reemerging, since the Master Freight Agreement of 1994,[13] as a solution to the need for speed, timeliness, and long distance hauling. One of the early proponents in the use of sleeper teams is Roadway Express. When Roadway began implementing sleeper teams, management played an active role in ensuring driver satisfaction. The sleeper cabs are

---

[13]The Master Freight Agreement governs major unionized LTL carriers and their employees.

built for luxury and comfort, and on their return to the terminal, the cabs are cleaned thoroughly. Management regularly conducts on-site inspections, often solicits the driving team's opinion on improvements that could enhance the effectiveness of the process, and takes any needed action. In one case, an on-site inspection resulted in a negligent supervisor being fired, and this action by management reinforced the drivers' view of the firm's commitment to the drivers and the new sleeper team process.

Truck manufacturers are catering to this new process by making comfort a top priority. Freightliner Corporation recently introduced the Penthouse sleeper option, which features a single 50-inch-wide upper bunk, or loft, that can be raised by an electric motor to the ceiling of the sleeper or lowered to its standard sleeping position. It features a flat panel display, which has several locations available for use as either a monitor for the built-in computer or as a television screen. The mounting options allow the screen to be seen from either the bunk or the seating area. A kitchenette is provided on the driver's side, featuring a sink, refrigerator, microwave oven, coffee maker, countertop, cutting board, and fluorescent lighting. The water system is equipped with a fresh water reservoir, a sink, a water heater, and a "gray water" tank, so a driver can not only cook a meal, but also clean up afterwards. Closets, storage drawers, and even a pantry for food storage are included. Packaged into a new sidewall cabinet, is a novel integrated computer and entertainment package. This system interconnects the stereo, the television, and a computer together into a single unit. A modem connection through a Highway Master communications module allows the driver to make computer contact with home base or even the Internet while the truck is in operation. A separate headphone system is provided to allow the user of this system to operate without interfering with the driver's radio choice or to allow silent operation when the other partner is sleeping. Products such as the Freightliner Penthouse improve the work environment for innovations such as sleeper teams.

Larger Trailers

Productivity gains are also being made through the use of longer trailers, such as 48 foot and 53 foot trailers. Some states allow 57 foot trailers; however, some firms are reluctant to use them since their operation is restricted by region. Trucking firms have also begun to use wider 112-inch trailers. These enhancements in length and width increase the volume of freight that can be carried within the same weight and height limits, increasing productivity without increasing direct cost significantly.

The use of doubles and triples—combinations of two or three 28-foot trailers—has also been on the increase. These combinations allow the firm to avoid the expense of labor and time to load and unload freight since smaller trailers increase the probability that a trailer can be filled with freight for a single destination. The volume per driver is also increased by this process. Drivers are using terminals or meet-and-turn points[14] to exchange trailer loads. Reengineering the freight movement process allows the balance of equipment and trailers while allowing the drivers the opportunity to get home and improve their quality of life. According to Don Schneider, president of Schneider National, "As we pay our drivers more, one of the ways (to cope) is to make them more productive. If the driver does more on the highway, that benefits the customer and the industry benefits from those kind of productivity adjustments" (Traffic World, 1997e).

Sixth Axle

To increase productivity, the president of the National Private Truck Council, John McQuaid, recommends raising the standard weight from 80,000 to 97,000 pounds per semi-trailer via a sixth axle (Traffic World, 1997f). This action could save shippers $2.6 billion annually, according to a 1990 report by the Transportation Research Board. Adding a sixth axle and increasing the payload would also better align U.S. truckers with those from Canada and Mexico. Canada currently has a 107,000-pound weight limit while Mexican truckers are allowed 119,000 pounds.[15]

Soft-sided Trailers

The introduction of soft-sided trailers at Trans-National Freight Systems is an example of an innovation that came about in response to specific customer needs. The soft-sided trailers are being built to carry vinyl siding and cost nearly twice as much as regular trailers. However, Trans-National Freight Systems sees this as a business opportunity. The company is also providing reverse logistics for its customers by contacting the consignees and picking up empty pallets as they make deliveries. This provides a win-win situation for the two companies. The customer does not need to buy more pallets to keep supplying the market, and Trans-National has the opportunity to capture backhaul without waiting for uncertain loads. By investing in transaction specific assets and providing value-added services through pallet management, the company describes itself as not just a trucking company but an asset to their customer.

In summary, advancements in a variety of technologies have made a plethora of data available to managers in trucking firms. However, trucking firms need to

---

[14]Meet-and-turn points are another means of keeping the freight moving quickly. At companies like CCX, two drivers (in two different trucks) will start at origins 500 miles away from one another. They will meet at a point roughly halfway in the middle, swap trucks, and each return to their origin. That way, the truck moves 500 miles at a time, but the drivers wind up back at home after a full day's work.

[15]Heavier trailers could impact the highways adversely. However, according to the Federal Highway Cost Allocation study (Department of Transportation, 1997), the more axles a vehicle has, the lower its cost responsibility at any given weight—and the more nearly it comes to paying its share of highway costs.

have the financial and human capital to convert these data into information. The corporate culture has to become conversant in the strategic use of information technology. The management of information may be a critical success factor in the new age of trucking (Bander et al., 1997b). The vast capabilities of computer and communications technology too often cause companies to lose sight of the main purpose of innovation and new technology adoption—the fulfillment of business needs. In the successful cases, the innovation push created by emerging technologies is matched by the innovation pull created by non-technological factors in the business environment and technological advancements in truck and trailer production.

RELATIVE CONTRIBUTIONS OF TECHNOLOGICAL AND NON-TECHNOLOGICAL FACTORS TO FIRM PERFORMANCE

The trucking industry appears to be on the rebound, and the future appears to be bright. General freight rates were up 3.5 percent in 1997, compared with 2.2 percent in 1996. Table 4 presents some other encouraging data about recent trends in the industry.

The improved performance of the trucking industry can be attributed, in part, to the agility and innovative behavior of the firms in the industry. The growing economy and the rationalization of capacity have also contributed to the trucking industry gains. Table 5 lists several services and process innovations already described above. Many of these are responses to recent competitive and environmental challenges. Accordingly, the impact of these innovations will be difficult to measure with any accuracy until more time has elapsed.

Innovations can create new markets (for example, the soft-sided trailer), capture additional market share (UPS second day guaranteed delivery), increase customer loyalty (QS 9000), increase revenue, and/or reduce costs (the Business Accelerator System). These innovations can have industry-wide impact, such as the growth of third-party logistics companies. They may also have impact outside the industry, such as the growth of trucking industry-focused information

TABLE 4 Trucking Trends

	1997	1996	1995
Increase in LTL rate	5.7%	4%	2.5%
Increase in TL rate	1.0%	0.2%	2.4%
Average operating ratio	93.6%	97.1%	96.9%
Collective net profits	3.65%[a]	1.18%	1.54%

[a]Second Quarter 1997.
Source: Traffic World (1997a).

TABLE 5 Some Innovations in Trucking

Innovation	Innovation Category
Next day delivery/time-sensitive delivery	Service
Supply chain management	Service
Customer package tracking	Service
Electronic data interchange (EDI)	Process and service
Package tracking (within firm)	Process
Service teams (interdisciplinary teams empowered to service important customers)	Process
Sleeper teams	Process
Redesigned consolidation terminals	Process
Meet-and-turn points	Process
Flexible backhauls (designed to reduce the number of trips a truck makes empty)	Process

technology systems integrators and software houses and an increase in labor pool skills and wages.

There are complex interactions between the factors that have been presented in sections 3 and 4 and a large number of innovations. Many factors, both technological and non-technological, *motivate* innovations, while other factors *enable* innovations that are motivated for other reasons. Figure 2 presents an illustration of the interactive influence that technological and non-technological factors have on the innovation process.

Non-technological factors, such as the changing customer and competitive environment, organizational restructuring, and globalization, are playing an important role in motivating new service innovations, thereby allowing firms to be

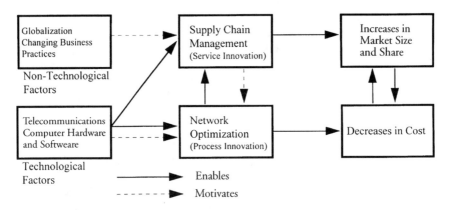

FIGURE 2 An illustration of the influence of some technological and non-technological factors on innovation and firm performance.

more responsive to customer needs and specifications. Service innovations, often motivated by non-technological factors, are enabled through the use of technology.

Technology plays a dual role in innovation. As an enabler of service innovations that interface with the customer, technology plays a role in enhancing the
revenue stream of the firm. Technology also provides the means to introduce
process improvements that reduce the cost and/or cut the time of providing services. Thus technological factors can also enhance firm profitability through
strategic cost management.

Sometimes innovations motivate other innovations. Many of these innovations improve processes within the firm. The robust management of the business
processes within the firm is critical to providing effective and efficient service.
Innovations motivated by other innovations provide a means to this end. Moreover, these innovations are often key to the eventual success of the innovation
that motivated them. For example, in order to provide time-sensitive delivery,
firms have invested in technology to track trucks and packages. The ring scanner
adopted by UPS, and described earlier, is an example of an innovation that was
motivated by the service innovation in time-specific delivery.

To summarize, many factors can motivate and/or enable innovation. Non-
technological factors appear to motivate innovations that improve market share,
while technological factors enable innovations that reduce costs and improve processes within the firm. Trucking firms look to technology to differentiate themselves. Already many advanced technologies, such as AVL, satellite and cellular
communication, on-board computers, and EDI are becoming commonplace in the
industry. Computer-aided routing and dispatch, automated hubs, and load and
container tracking seem to be on the verge of widespread adoption. While some
of these technologies offer competitive parity at best, they nevertheless offer a
means of survival in the tightly contested industry.

Technology has brought excitement to an industry that acknowledges that
the old way of doing business is no longer feasible. With their low margins and
fickle and demanding customers, firms in the industry innovate so as to provide
value-added services that attract and retain customers. While the influence of
non-technological factors on firm performance appears to be significant in motivating the need for action and response, it is technology that is providing immediate solutions for some competitive challenges.

Consistent with the findings of Cooper and Merrill (1997), the innovations
enabled by technological factors have come from outside the trucking industry,
but the innovations have been significantly influenced by the trucking industry's
idiosyncratic requirements. For example, Schneider was one of the early adopters of the Qualcomm communication system and the system's design was greatly
enhanced through the active interaction between user (Schneider) and developer
(Qualcomm). Similarly, the implementation of EDI within a trucking firm is
often driven by the trucking firm's customer specifications. The trucking firm

has to ensure compatibility between its internal management information systems and the customer's EDI interface. Often systems integrators are involved in the development and coordination of the process of implementing EDI.

The development and adoption of innovation has been motivated by the business needs of specific segments within the industry. Satellite communication systems have enabled many adopters to know the location and status of the fleet at all times. Information about the whereabouts of a specific truck are especially critical to the TL segment and, to a lesser extent, logistics providers. Many of the early adopters of the system have come from these two segments of the industry. In contrast, the LTL segment of the industry emphasizes precise information relating to the contents of the truck and to network optimization. Accordingly, package tracking and route optimization have been more frequently adopted within that segment.

The management of information is becoming as important as the management of freight for the trucking industry. The ability to provide end-to-end information services becomes a critical capability for firms in the trucking industry. Recent trends indicate that the Internet and Web-enabled software may provide financial, operational, and structural assistance for shippers and carriers. The acquisition and processing of operational, financial, and related data should directly support higher system capacity, greater labor and capital productivity, improved efficiency, more effective resource allocation, and better integration of processes and activities (BTS, 1997a).

THE FUTURE OF FREIGHT AND EXPECTATIONS FOR THE TRUCKING INDUSTRY

The trucking industry has seen tremendous growth, and that growth is expected to continue. Table 6 shows some of the predictions of a study commissioned by the American Trucking Association (ATA). It is expected that trucking will rise from its 1996 share of 64 percent of the transportation sector to over 66 percent in 2006. The truckload sector is expected to have the largest gains in

TABLE 6 Market Share Forecasts

Mode	1996 share (%)	2006 forecast (%)
All trucking	64	66.3
Truckload	30	34.3
LTL	1.1	1.3
Private	32.9	30.7
All rail	14	13.5
Intermodal	1.2	1.4

Note: Totals do not sum to 100%. Remainder is air freight, maritime, and pipeline.
Source: Martin Labbe Associates, ATA.

market share, rising from 30 percent in 1996 to 34.3 percent in 2006 (Traffic World, 1997a).

To sustain growth and improve profitability, firms in the trucking industry must continue to innovate and adopt innovations. This chapter has presented an overview of innovation and the innovation process in the current trucking environment. As is typical of other service industries, the emphasis on the customer is a clear imperative for competitive success. Some preliminary evidence suggests that innovation is playing a major role in the rejuvenation of the industry. The next few years will indicate whether firms' investments in creating new services and adopting new technologies will have commensurate payoff. In general, as new technologies emerge, as the rate of technological change accelerates, and as the degree of international competition increases, firms must enhance their ability to develop and introduce new products, services, and processes (Penrose, 1959). Firms in the trucking industry have had to respond to the challenge of intense competition and demanding customers with new innovations in order to survive.

There are several barriers to innovation in trucking. These barriers include the lack of standards in many new technologies, an inadequately skilled labor force, and the lack of adequate capital.

When there are multiple technological paths to achieve the same functionality, manufacturing firms have to decide on a particular technology path in order to commercialize a new product. When firms choose incompatible technology trajectories, customers are forced to evaluate the relative advantages of each technology as part of their purchase decision. This can inhibit widespread adoption of the product since the customers are often not technological experts and have limited information about the comparative merit of each technology. In an environment where network externalities exist, customers benefit from large scale adoption of a product based on a single technology. Compatibility standards assure the user that an intermediate product or component can be successfully incorporated in a larger system of closely specified inputs and outputs (David and Greenstein, 1990). The existence of standards enables product development and innovation diffusion (Nagarajan, 1996). For example, adoption of electronic toll collection tags might save time for trucks. However, the lack of a common standard and problems relating to cost allocation have hindered widespread adoption of this product. For many firms, the competitive circumstances of the trucking industry has created an environment in which profit margins are small. The decision to invest in innovation rather than, for example, trucks is often difficult for trucking firms to make.

The introduction of new technology-based innovations in the trucking industry has increased the need for individuals trained in computer and other technology use. The skills required to work in the trucking industry have changed dramatically across all ranks of employees. All employees, including drivers, dispatchers, administrative, and managerial personnel, are exposed to varying levels of new technologies. For example, the driver sometimes has an onboard computer system that tells him his next destination. He is expected to interact with the system in order to make the entire network optimization process more efficient. The dispatcher deals with a computer that monitors the status of the fleet and updates routes in response to changing customer demands. The trucking industry must look to educational institutions to provide it with a workforce for the new competitive environment.

According to managers, the key success factors for firms in the industry are customer satisfaction, cost management, employee retention, and safety. Technology, such as EDI, GPS, information technology developments, and advanced communications, have enabled many innovations that directly address the key success factors. The role of these innovations in firm survival and profitability requires attention. Our initial conclusion—based on the limited information available—is that innovations enabled by technological factors often appear to enhance profitability by reducing costs through process reengineering, strategic information, and cost management. Innovations attributable to non-technological factors appear to be the competitive weapons motivating new ways of doing business that provide strategic advantage.

The growing importance of logistics, as it progresses from being a value-added service to a critical product in the trucking industry, is evident. Industry consolidation has become the strategic response to the demands of the competitive environment. Indeed, these are interesting times for firms in the trucking industry.

REFERENCES

Alden, J. (1997). "Delivering the Global Village." Speech before Friday Group of Dallas, TX, June 20, 1997.

Bander, J. L., A. Nagarajan, and C. C. White. (1997a). "Strategic Management of Technology: The Case of Freight Mobility Systems in the Trucking Industry." working paper.

Bander, J. L., A. Nagarajan, and C. C. White. (1997b). "Information and the Trucking Industry." Paper presented at the IEEE-SMC Conference in Orlando, FL. Oct 1997.

Bank of America. (1997). "Economic and Business Outlook—Transportation," August.

BTS. (1997a). *Transportation Statistics Annual Report 1997: Mobility and Access*, U.S. Department of Transportation Bureau of Transportation Statistics.

BTS (1997b). *Transportation in the United States: a review*, U.S. Department of Transportation Bureau of Transportation Statistics.

Cooper, R.S., and S. Merrill. (1997). *Industrial Research and Innovation Indicators. Report of a Workshop*, Washington, DC: National Academy Press.

Danckwerts, Dankwart (Hg). (1991). *Logistik und Arbeit im Gutertransportsystem*. Rahmenbedingungen, Verlaufsformen und soziale Folgen der Rationalisierung, Opladen.

David, P.A., and S.M. Greenstein. (1990). *The Economics of Compatibility Standards: An Introduction to Recent Research*. Economics of Innovation and New Technology Vol. 1.

Department of Transportation. (1997). 1997 Federal Highway Cost Allocation Study.

Hitt, M. A., R. D. Ireland, and R. E. Hoskisson. (1996). *Strategic Management: Competitiveness and Globalization*. Minneapolis/St. Paul: West Publishing.

Hubbard, T. N. (1997). *Why are process monitoring technologies valuable? The Use of Trip Recorders and Electronic Vehicle Management Systems in the Trucking Industry.* Unpublished report.

Industry Week. (1997). 246(19):136.

Kline, S., and N. Rosenberg. (1986). "An overview of innovation" in *The Positive Sum Strategy— Harnessing Technology For Economic Growth.* R. Landau and N. Rosenberg, eds. Washington, DC: National Academy Press.

Nagarajan, A. (1996). *Acquisition of Technology in an Emerging Industry—A Study of the Intelligent Vehicle Highway System Industry.* Ph.D. dissertation, University of Michigan School of Business Administration.

Penrose, E. T. (1959). *The Theory of the Growth of the Firm.* New York: Wiley.

Plehwe, D. (1997). "Deregulation and Integration of Transport Industries: The emergence of transnational transportation systems." Presented at the Council of Logistics Management annual conference in Chicago. Oct. 1997.

Purchasing. (1996). "Hype vs. Reality" 120(3):48S2-48S12.

Rogers, E.M. (1995). *Diffusion of Innovations.* New York: Free Press.

Society of Automotive Engineers (1993) Technical Paper 932975, "Tomorrow's Trucks—A Progress Review and Reappraisal of Future Needs."

Swan, P. F. (1997). "The Effect of Changes in Operations on LTL Motor Carrier Efficiency and Survival." Ph.D. dissertation, University of Michigan School of Business Administration.

Thompson, A. A., and A. J. Strickland. (1996). *Strategic Management: Concepts and Cases.* Chicago: Irwin.

Traffic World. (1997a). "Blue Skies" 252(5):22.

Traffic World. (1997b). "LaLonde: Ante Up" 250(1):21.

Traffic World. (1997c). "Churning a Problem" 252(4):27.

Traffic World. (1997d). "Viking Connected" 252(3):21.

Traffic World. (1997e). "The 'R' Word" 252(4):29.

Traffic World. (1997f). "Changing Direction" 252(5):23.

U.S. Department of Labor. (1996). Occupational Safety and Health Administration, 1996 Census of Fatal Occupational Injuries, Washington, DC.

Von Hippel, E. (1976). "The Dominant Role of Users in the Scientific Instrument Innovation Process" *Research Policy* 5(3).

APPENDIX A: THE STUDY METHOD

This study used a multi-pronged approach to studying the innovation process in the trucking industry. The study began with unstructured interviews with 25 industry stakeholders. Of our 25 initial respondents, 9 were trucking and logistics firms from the different segments of the trucking industry, while 16 were experts and regulators from outside the industry. We focused on understanding the changing competitive dimensions of the industry as a context for studying the innovation processes within the firms. Using the discussions as the basis, we developed semi-structured questionnaires for trucking firms and for technology vendors. The questionnaire for the trucking firms consists of three sections. Two qualitative sections focus on the innovative process and the factors enabling or hindering the process. The third section seeks to obtain quantitative information about financial performance and the firm's operational characteristics. Seven firms drawn from a cross-section of industry segments, were interviewed.

A second semi-structured questionnaire was developed to administer to technology vendors. This instrument focused on competitive issues in the vendors' industry and the barriers to proliferation of new, advanced technologies in the trucking industry. Seven technology vendors were interviewed.

Grocery Retailing[1]

JAY COGGINS
BEN SENAUER
University of Minnesota

INTRODUCTION

The essential output of the U.S. grocery industry is services. A supermarket is in the business of selling products, but few are produced by the store itself. Rather groceries are brought into the store through a distribution system and are then sold to consumers who have ever-increasing expectations of service quality. Consumers want a wide selection, high quality, shopping convenience, and competitive prices.

The fact that services *are* the product presents certain difficulties in the study of innovation in the supermarket industry. A manufacturing enterprise, for example, can usually measure its expenditures on new-product development. It can relate these expenditures to some measure of the output of new products and their profitability. In grocery stores, as in retailing generally, innovation appears to be a more elusive concept. Something as subtle as redesigning a store's checkout counters or a display case may constitute an innovation. However, the idea and the budget to carry out such changes are unlikely to be assigned to a particular unit in the corporate structure. The line between marketing and research and development blurs as different units seek to improve the consumer's experience in the store.

Although the study of innovation for this industry is difficult, we shall nevertheless attempt it. Even though this chapter will perhaps contain less in the way of hard data than many of the other contributions, our goal is to describe a variety

[1]This research was funded by the Retail Food Industry Center at the University of Minnesota.

of innovations, their sources, and their effects on industry performance. The forces driving innovation are also important, and these are also addressed.

It is useful to divide the forces driving change in the supermarket industry into three categories—competition, consumer preferences, and technology. The first of these, competition, comes both from within and from outside the industry. As recently as the 1960s, the supermarket industry was highly efficient by the standards of the day. In the 1970s, as prices soared and holding goods was preferable to holding cash, the entire distribution network accumulated large inventories. Although this strategy made sense while inflation was high, as inflation fell the expense of owning inventory caused a decrease in efficiency, creating in turn a decline in profitability and industry performance. Inflation itself, while an important force in the history of the industry, had an indirect effect on innovation. It helped cause inventories to increase, reducing efficiency and opening the industry to a competitive threat from outside.

During the 1980s, the growth and superior efficiency of large general retail outlets such as Wal-Mart exerted competitive pressure on supermarkets. These retailers demonstrated that U.S. supermarkets and the distribution system behind them were not so efficient after all. When these new retailers began expanding their food offerings, pressure on supermarkets grew even further. Innovation and change became necessary for survival, and the industry entered a period of upheaval that continues to the present.

The second source of innovation is the preferences of consumers, both for types of food and for certain attributes of a shopping experience. It could be said of almost any industry that consumers drive the innovation process. But seldom is this as true as in retailing, where contact with consumers is immediate and direct. Some of the consumer changes that have forced supermarkets to change and improve their service offerings will be discussed in this chapter.

The third source of innovation is the technology employed in managing inventories, transmitting information up and down the supply chain, and recording information about consumers' store-level purchasing behavior. Development of this type of technology, mostly electronic, is almost entirely external to the industry. For the most part, supermarkets adopt equipment and software developed by third parties. Computer-related technologies are crucial to many of the efficiency-enhancing advances, but the innovative role of the supermarket industry is in using the technology, not in producing it. The role of technology is itself somewhat ambiguous. In some cases the availability of a new technology drives innovation. In other cases, technology is the means by which a new innovation is implemented.

Innovation can take many forms in a retail industry. Many of the most important innovations are related to processes: of inventory management, information control and transmission, and in-store product flow. These and a variety of related changes are addressed by a major industry-wide initiative that began in 1993, efficient consumer response (ECR). A section of this chapter is devoted to

ECR. Other innovations are directly related to consumer service, such as store layout and meal replacement. Included in this category is the wide variety of new products that cater to consumers' wish to purchase meal-ready foods for home consumption.

The following section provides a brief historical overview of the industry and the events that led to the aggressive change currently experienced. The third section contains a brief discussion of the role of external sources of innovation. A variety of performance measures are presented in section four. The fifth section describes the basic approach of ECR and its effects, and section six addresses service-related innovations. The seventh section presents the linkages between the various innovations and changes in industry performance.

MAJOR FORCES DRIVING INNOVATION

For the past decade or so, the supermarket industry has been constantly racing to keep up with its competitors in reducing distribution costs and improving efficiency. It was not always this way. As recently as the 1960s, the industry was regarded by many as the benchmark against which to compare the performance of distribution systems. When Toyota, the Japanese automobile company, developed early versions of "lean inventory management" or "just-in-time delivery," Toyota management gave substantial credit for the innovation to the efficiency they observed in the U.S. supermarket industry at that time (Kurt Salmon Associates, 1993).

Then inflation rose during the 1970s. Firms throughout the food distribution system changed their strategies dramatically. Now excess inventories could generate a return by increasing in value as prices rose (Kahn and McAlister, 1997). The age of "the deal" came into being. Product promotions became more prominent and had a major impact on distribution in the food industry. Manufacturers produced the same item in large production runs and then pushed large batches of product out to wholesalers on special discounted deals, causing inventory to build up in warehouses. Likewise, wholesalers would push the product out to retailers with promotional discounts. The excess inventory would then be promoted by retailers and sold to consumers by discounting the price or providing coupons.

In this environment, the goal was to maximize sales volume. The former emphasis on reducing costs faded along with concern for distributional efficiency. Industry-wide returns on net worth, which have seldom been below 15 percent in the past decade, were just over 6 percent in 1972-1973, the first year for which such data are available. The economy as a whole did not perform well during this period, but the supermarket industry was particularly ripe for change.

Competition as a Driving Force for Innovation

Change did indeed occur. Several factors combined during the late 1980s to deliver a wake-up call to the industry. Inflation plummeted and the economy entered a recession. Holding inventories, especially large inventories, became very expensive. Most important, perhaps, was a dramatic increase in competition caused by the spread of mass merchandisers such as Wal-Mart and the entrance of alternative formats such as warehouse clubs and supercenters owned by Wal-Mart and others (Kurt Salmon Associates, 1996). Wal-Mart's initial impact was not due to its presence in the grocery market. Rather, its electronically driven distribution system and the way in which it dealt with suppliers shook other retailers to the core. It quickly developed sufficient market power to alter the way the product was purchased, warehoused, and distributed.

Wal-Mart demonstrated to suppliers that it was not interested in promotional deals but wanted the lowest price possible on all its purchases and would determine the quantities and delivery schedule. Its retail prices were based on an "everyday-low-price" (EDLP) approach. Wal-Mart became the leader in distributional efficiency. Inventory turnover was high and excess supply in the system was all but eliminated (Senauer and Kinsey, 1997). Today, Wal-Mart's reduced operating costs are 17.5 percent of sales, compared with around 22 percent typical for supermarkets (Blattberg, 1996).

Wal-Mart's effect grew along with its market penetration. The number of Wal-Mart stores increased from 859 in 1985 to 2314 Wal-Mart stores and 439 Sam's Clubs in 1997 (Larson, 1997). Their success with the EDLP approach caused retailers generally, and the grocery store industry in particular, to rethink their reliance on volume purchases, volatile prices, and maximum dollar sales. It became clear that survival required reducing costs and improving efficiency of the distribution system.

If this message was not clear before, it certainly became clear in the early 1990s when Wal-Mart and other discount merchandisers, including Target, K-Mart, and Meijer, began developing "supercenters," which combine general-purpose retail with full-scale grocery stores. Supercenter sales are expected to exceed $52 billion in 1998 (Food Institute Report, 1998), of which about 40 percent, or nearly $21 billion, is grocery sales. A 1996 report of the Food Institute estimated that the supercenter presence in groceries would increase from 2 percent of sales in 1994 to 7.4 percent in 1999 (Food Institute Report, 1996). Another study, by the Food Marketing Institute, predicted that sales of grocery products by nontraditional food retailers will reach about $70 billion by the year 2000 (Food Marketing Institute, 1997). This would be around 14 percent of total grocery sales. Clearly the threat to the supermarket industry posed by Wal-Mart and other discount retailers is and continues to be serious.

In a recent case study, Capps (1997) examined the effect of Wal-Mart on sales at traditional grocery outlets in rural Texas. He used data from 30 stores in the David's Supermarket chain, covering the period from 1987 to 1994. Of the 30 stores, Wal-Mart competed directly (that is, had a store nearby for at least part of the study period) with 22. Linking monthly sales at each store to local demographic information, sales in the previous period, the number of competing supermarkets, and the proximity of Wal-Mart, Capps found that, on average, Wal-Mart led to a reduction of 17 percent in sales.

Technology as a Driving Force for Innovation

In 1992, President George Bush made headlines with his reaction of surprise to the novelty of an electronic scanner at a grocery store he visited during his re-election campaign. This device, familiar to most of us, is now used in more than 97 percent of supermarkets. Supermarkets were the first retailers to use electronic scanning at the point of sale (POS). In 1972, supermarkets worked with the Uniform Code Council to develop Uniform Product Codes, setting industry-wide standards for POS information (scanners). Supermarkets adopted the standard early as a way to speed checkout and eliminate the need to put a price tag on every item, thus reducing labor costs. This provided some gains in efficiency for grocery stores. In addition, supermarket POS scanner data were sold to food manufacturers and market research companies (Senauer and Kinsey, 1997).

However, traditional supermarkets fell behind the leading general retailers, including Wal-Mart, in the use of POS scanner information and electronic data interchange (EDI) in supply chain management. "Quick response" is the term applied to the restructuring of inventory replenishment in general merchandising in which the logistics system is driven by POS data. Quick response was itself based on lean-inventory management or just-in-time delivery, which has been widely adopted by U.S. automobile manufacturers and other industrial companies in restructuring to become more efficient. Recent attempts to catch up to retail competitors in exploiting this fundamental technology mark some of the key innovations in the supermarket industry.

In the area of labor, the biggest challenge that food retailers currently face is simply finding enough workers in areas with very low unemployment levels and tight labor markets. Most supermarkets are staffed by a core of stable long-term employees and a substantial group that has a high turnover rate. Hiring new employees is, therefore, a continuous process. Most entry-level positions require few skills and do not even require a high school diploma. The industry has a high demand for part-time workers because most of its business is in concentrated periods, between 4 and 6 p.m. Monday to Thursday and between Friday afternoon and Sunday night, when half of all sales now occur. Many of the industry's jobs are not only part-time, but also low-skill and relatively low-paying positions.

Thus supermarkets have an incentive to find labor-saving capital that can substitute for workers. One example is self-checkout, in which the cashier is eliminated and shoppers scan the items themselves. Self-checkout systems have been the subject of experiments for some time, but the systems are not yet effective enough to be widely adopted. An obvious problem is how to prevent shoppers from cheating by not scanning all their items. Although the need for technologically sophisticated workers is not as great as in many industries, the technological demand placed upon workers is growing and will probably continue to do so. As the number of jobs requiring the use of electronic equipment increases along with the sophistication of the ordering and in-store management systems, the need for more skilled workers will increase.

Innovation Is Consumer-Driven

Innovation in the grocery industry, as in all retailing, is driven not only by competition and technology but also by consumers. The most successful retailers respond to the wants and needs of customers. An especially important customer demand is convenience in food purchasing and consumption.

The trend behind much of the rising demand for convenience is the increased participation of women in the labor force. Between 1970 and 1994, women's labor force participation rate grew from 43.3 percent to 58.8 percent. For women ages 35-44, the participation rate in 1994 was 77.1 percent (U.S. Department of Commerce, 1996). Many consumers do not have the time to prepare traditional meals and increasingly even lack knowledge of how to cook. After work, they want a meal to eat or, at most, to assemble at home, but not ingredients to cook. They also want to relax in the comfort of their own home and not spend time and money at a full-service, sit-down restaurant (Kinsey and Senauer, 1996). Grocery retailers are increasingly providing meals that can be taken home and eaten, referred to as meal solutions or home meal replacement in the industry. Meal solutions are a challenge for the industry because it has more in common with food service than traditional grocery retailing.

THE ROLE OF EXTERNAL SOURCES OF INNOVATION

The major trade associations and several key consulting firms play an important intermediary role in facilitating innovation by food retailers. The Food Marketing Institute (FMI) is the single most important trade association for the industry. FMI holds the industry's major annual trade show in Chicago each May, as well as other conferences and workshops throughout the year. The topics of these conferences frequently relate to innovation in the industry. Recent conferences, for example, have focused on category management and home meal replacement. FMI also does a substantial amount of research for the industry. Some is done by in-house staff, but much of it is contracted out to consulting firms. The results are then reviewed, published, and distributed by FMI. FMI receives substantial funding through its membership dues from grocery stores and supermarkets. Dues are in proportion to sales. They vary from as little as about $100 for a small single grocery store operation to well over $100,000 for large supermarket chains.

Three of the major consulting firms used by the industry are Willard Bishop Consulting, Kurt Salmon Associates, and Arthur Andersen. Each is known for its work in particular areas. Kurt Salmon, for example, did the original ECR study.

Willard Bishop, whose offices are located outside of Chicago, focuses on helping retailers develop loyalty or frequent shopper programs, pricing strategies, and category management programs specifically for perishables. Bishop has developed computer software that can be purchased and used to implement programs in several areas. Willard (Bill) Bishop, the firm's leader, is widely respected and regarded as one of the industry's "thought leaders," a source of innovative ideas for the industry.

MEASURES OF CHANGE IN INDUSTRY PERFORMANCE

Consumer Food Expenditures

The performance of the retail food industry is tied to consumers' spending patterns, and a crucial fact is that the share of income spent on food has long been in decline. Real median family income has generally increased since 1970, although during parts of this period it flattened or even fell. The share of income spent on food, on the other hand, fell from 13.9 percent in 1970 to 10.9 percent in 1996. Expenditures on food eaten at home fell even more precipitously, from 10.3 percent in 1970 to 6.8 percent in 1996. Expenditures on food eaten away from home rose slightly, from 3.7 percent in 1970 to 4.5 percent in 1996 (see Figure 1). Although real expenditures on food eaten at home rose, the increase in industry sales did not keep pace with population and income growth.

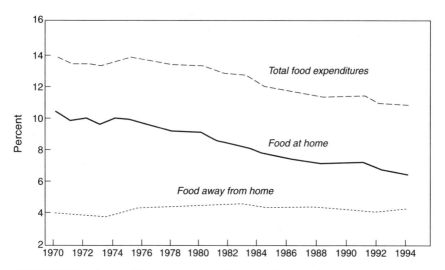

FIGURE 1 Food expenditure as a percent of income (average family income).
Source: "Food Consumption, Prices, and Expenditures, 1996," U.S. Department of Agriculture, Economic Research Service Bulletin Number 928, 1996.

Financial Measures of Productivity

The financial performance of the industry, however, has been quite good, especially in recent years. Nevertheless, profit margins have always been low. This is perhaps not surprising given the highly capitalized nature of the industry and its high sales volume. Indeed, industry members and observers have often thought of the low profit margins as an indicator of the efficiency of the industry. Defined as after-tax net profit as a percent of sales, profit margins have recently hovered around 1 percent. The industry-wide figure stood at 0.50 percent in the 1972-1973 fiscal year, the earliest year for which the data are reported. The figure trended upward through the 1970s and 1980s, reaching 1.19 percent in 1985-1986. Net profitability declined in the late 1980s and early 1990s and re-covered in 1993-1994. It was 1.20 percent in 1995-1996 before dropping back to 1.08 percent in 1996-1997 (Figure 2).

Profit margins may not be the best indicator either of efficiency or of financial performance, however. Investors in publicly held supermarket companies have done very well recently. Return on net worth, measured as a percent of net worth, has been strong, reaching 16.33 percent in 1996-1997 (Figure 3). In early 1997, *Supermarket News* reported that its index of 48 stocks climbed 30.3 percent between 1994 and 1995 and 45.4 percent between 1995 and 1996. The index

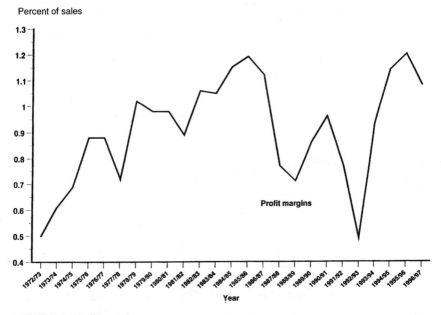

FIGURE 2 Profit margins.
Source: Food Marketing Institute's *Annual Financial Review,* various years. Compiled by Gerne and Associates.

Percent of net worth

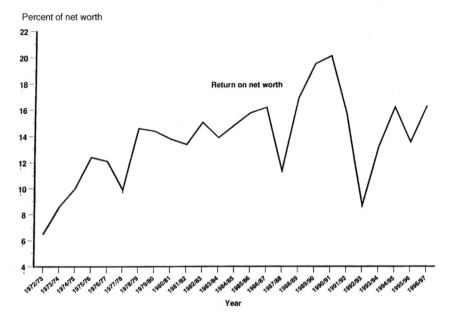

FIGURE 3 Return on net worth as a percent of net worth.
Source: Food Marketing Institute's *Annual Financial Review,* various years. Compiled by Gerne and Associates.

outperformed both the Dow Jones Industrial Average and the S&P 500 (Food Marketing Institute, 1997).

In-Store Measures of Productivity

Traditional measures of industry performance include measures such as weekly sales per square foot, weekly sales per labor hour, and sales per transaction. Median real weekly sales per square foot fell from $11.71 in 1960 to $6.34 in 1996 (Figure 4). This figure is affected by many factors, including the size of stores. Median store size has long been increasing, growing from 31,000 square feet in 1987 to 38,600 in 1996. Also, aisles have grown wider and checkout areas have become larger. Both of these changes are meant to make shopping more pleasant for consumers, and both tend to reduce sales per square foot.

During the past four decades, median real weekly sales per labor hour first rose, from $78.86 in 1960 to a high of $98.72 in 1980, before retreating to a new low of $74.45 in 1996 (Figure 5). The drop in recent years is primarily due to an increase in labor-intensive supermarket services such as deli counters.

Real median sales per transaction, another indicator of a supermarket's productivity, have also fallen in recent years (Figure 6). After several decades dur-

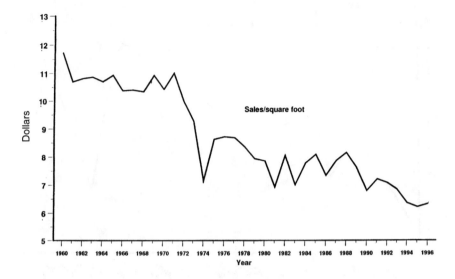

FIGURE 4 Real median sales per square foot.
Source: Food Marketing Institute, *FMI Speaks,* various issues.

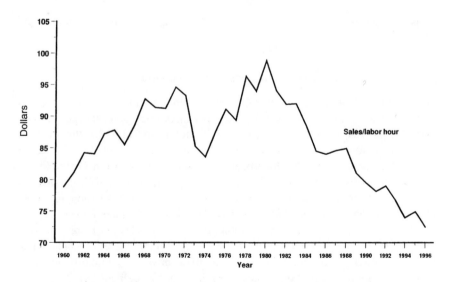

FIGURE 5 Real median sales per labor hour.
Source: Food Marketing Institute, *FMI Speaks,* various issues.

FIGURE 6 Real median sales per transaction.
Source: Food Marketing Institute, *FMI Speaks,* various issues.

ing which this measure stayed at or above $14, it fell in the 1990s all the way to $12.44 in 1996. This reflects changes in consumers' shopping habits. They tend to make more shopping trips than previously and to buy less on each trip.

These measures of industry performance paint a mixed picture of an industry that is facing intense competitive pressure but has provided a healthy return to its investors. How has the retail food industry managed to maintain a relatively healthy financial situation in the face of such pressure? Much of the answer has to do with its ability to adopt new and innovative ways of operating. We now turn to a description of several of the industry's central innovative strategies.

PROCESS INNOVATION: ECR

ECR is U.S. supermarkets' answer to their more competitive environment. The major goals are to produce and ship products in response to consumer demand, eliminate costs that do not add to value, reduce inventories, spoilage, and paperwork, and simplify transactions between companies.

The ECR movement was launched after Wal-Mart and other discount mass merchandisers entered food retailing with supercenters. ECR is akin to "lean-inventory management" or "just-in-time delivery" in manufacturing. The purpose is to reduce costs by increasing the efficiency of distribution. The strategy calls for grocery retailers, wholesale distributors, and manufacturer suppliers to be linked together electronically and to cooperate closely.

ECR focuses on improving the efficiency of the entire system rather than just certain activities. According to the key report on ECR by Kurt Salmon Associations (1993):

> The ultimate goal of ECR is a responsive, consumer-driven system in which distributors and suppliers work together as business allies to maximize consumer satisfaction and minimize cost. Accurate information and high-quality products flow through a paperless system between manufacturing line and check-out counter with minimum degradation or interruption both within and between trading partners.

Widespread adoption of ECR could lead to a 41 percent cut in dry grocery inventories and a $30 billion decline in total distribution costs—along with an increase in product quality and freshness, the report predicts.

The vision of ECR, as shown in Figure 7, is that a timely, accurate, paperless flow of information starts at the checkout counter and facilitates a smooth, continuous flow of product that matches consumer purchases. Computers and software programs allow data to be transmitted directly to distributors and/or manufacturers in real time. This flow of information allows fast-moving items to be replenished automatically and makes it possible for manufacturers to adjust production lines in response to consumer demand. In contrast, in the past information circulated much more slowly and only in closed circles—between consumers and retailers, between retailers and wholesale distributors, and between wholesalers and food manufacturers and other suppliers (Senauer and Kinsey, 1997).

The Development of ECR

To respond to the increased competition and need to improve efficiency, industry leaders formed the ECR working group in mid-1992. ECR was developed through the main trade associations to ensure that its benefits would be widely available. The working group commissioned a major study by the con-

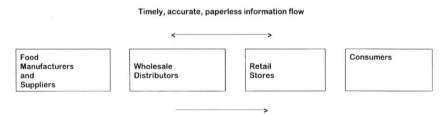

FIGURE 7 The ECR system.

sulting firm Kurt Salmon Associates. Kurt Salmon had helped develop quick response in the general merchandise industry. Their major report, *Efficient Consumer Response: Enhancing Consumer Value in the Grocery Industry*, was published in 1993 by the major trade association, the Food Marketing Institute (Kurt Salmon Associates, 1993).

The fundamental stimulus for the ECR initiative was the intensified competition from nonfood retailers, such as Wal-Mart. Moreover, it was known that Wal-Mart had plans to enter food retailing, which it has since done with its supercenters, combining discount general merchandise and food. Food retailing is a relatively low-tech, fragmented industry. The success of the ECR initiative depended on the backing and financial support of trade associations, especially FMI and the Grocery Manufacturers Association (GMA). In addition, the large manufacturers, such as Proctor and Gamble, were behind the ECR initiative and provided much of the necessary funding. Major food product manufacturers, such as Proctor and Gamble and Coca-Cola, have substantial research and development budgets, in contrast to the retailers. The manufacturers saw ECR as a way to increase their own efficiency and profitability by streamlining distribution in partnership with the retailers.

In 1993, the working group was replaced by the Joint Industry Project on ECR with representatives from ten trade associations and the Uniform Code Council. More than 200 people representing over 100 companies participated. The Joint Industry Project analyzed distribution in detail to find ways to cut costs and improve efficiency. By 1995, most of the analysis was done and the task forces were disbanded. Two groups now manage the project: an executive committee on ECR to set overall direction; and an ECR operating committee to carry out additional research. Actual implementation is up to individual companies and their suppliers (Kurt Salmon Associates, 1996).

Objectives and Major Components of ECR

In the new system, consumers will directly elicit supply. Product is pulled through the system by consumers, rather than pushed by suppliers. Data on consumer purchase behavior is paramount in this system since it drives the decisions up and down the supply chain (King and Phumpiu, 1996).

ECR has several major objectives (Kurt Salmon, 1993):

- *Efficient store assortment* optimizes the use of space within the grocery store to reduce costs and increase profitability.
- *Efficient replenishment* creates a smooth synchronized flow of product based on consumer purchases, using EDI linkages between trading partners.
- *Efficient promotion* directs promotional activities away from trade promotions that pushed product through the system and toward increasing and responding to consumer demand.

- *Efficient product introduction* uses consumer POS information and cooperation among trading partners to develop new products with a better chance of
success and at lower cost.

To achieve these objectives, ECR emphasizes several major components:

- *Category management.* Product categories are managed as strategic businesses to maximize profits.
- *Continuous replenishment.* Products are delivered based on actual consumer sales, reducing time in shipment and storage, using computer-assisted ordering (CAO), directly via POS scanner data.
- *Electronic data interchange.* Retailers and vendors (suppliers) are linked
through computer-to-computer ordering, billing, and payment.
- *Direct store delivery.* Products are delivered directly from the manufacturer to the supermarket without the use of warehouses or intermediate distributors.

To succeed, ECR requires an effective and flexible management and good use of
information technology. These and other neceesary conditions of innovation
adoption are discussed below.

Innovation Adoption

A characteristic shared by many of the most innovative food retailers that
have succeessfully adapted ECR practices is strong leadership at the top level of
management that clearly supports innovation. The importance of management
leadership is true for the single owner-operated store or for the large supermarket
chain. H. E. Butt, a regional supermarket chain, operates in south Texas, which is
one of the first areas in which Wal-Mart opened supercenters and began to compete in the food business. H. E. Butt is considered one of the most well-managed
and innovative food retailers in terms of supply chain management. Strong, effective leadership is provided by Charles Butt, the grandson of the company's
founder and the current chief executive. The company has been a leader in developing "activity-based costing," whose aim is to attribute to each product all of the
costs—inventory, transportation, and the like for which it is responsible in the
entire supply chain. In another example of H. E. Butt's innovative spirit, the
company's major distribution warehouse for San Antonio, where summer tem-
peratures can exceed 100° Fahrenheit, was air conditioned a few years ago. This
made working conditions much more pleasant for the warehouse employees and
substantially reduced labor turnover, raising productivity.

The "activity-based" idea can take several forms. Fleming Company, a
wholesaler based in Oklahoma City, Oklahoma, recently put in place an activity-
based pay scheme for its drivers at three distribution centers in Texas and Tennes-
see. Rather than hourly pay, the drivers receive a fixed amount per mile driven,
deliveries and pickups made, and pallets moved. Fleming reports a reduction of

40 percent in time spent at each stop by the affected drivers. Workers responded to the new pay scheme by working faster and smarter. A survey of stores served by the three centers revealed that 38 percent of respondents felt that delivery services had improved, while only 2 percent felt they had declined.

Supervalu, a Minneapolis-based wholesale distributor, has been fairly aggressive in adopting certain components of the ECR program. In a 1997 activity-based selling pilot program in Denver, the company separated the cost of providing retailer services from the cost of the products themselves. Retailers pay for each product based on the base price of the item, plus all deals and pick-up charges. Store managers quickly learned where and how they could cut costs by ordering more products in pallet quantities, reducing the frequency of deliveries, combining stops on more efficient routes, and reducing the time spent on each delivery. As a result of the pilot program, Supervalu's Denver facility was able to reduce its total miles driven by all trucks by 338,000 per year, with no reduction in sales volume.

Supervalu's category-management program is used to align product assortments as closely as possible with each store's consumer demand. Some manufacturers worry that category management, by reducing the number of brands carried by a store in a given category, will harm certain brands disproportionately—especially private-label or store brands. In Supervalu's experience, this has not been the case in most stores. Those stores whose consumers demand the more economical store brands should maintain a place for them on the shelves.

Another type of innovation adoption employs information technology to implement frequent-shopper or customer-loyalty programs. Consumers are somewhat fickle, changing stores regularly. Woolf (1996) found that the top 20 percent of a store's customers, measured by total purchases, account for as much as 64 percent of store sales. Supermarkets are increasingly working to foster loyalty among their most valuable customers. Shopping cards and other programs are designed to reward consumers for attaining various levels of total purchases.

The programs rely upon POS scanner data, which are recorded at checkout time and linked to individual consumers by use of identification numbers and cards. These programs attract customers voluntarily with various promotional benefits for participation. When enrolling, the consumer provides basic household demographic data. The linkage of POS scanner data to consumer purchases and demographic information opens the door to sophisticated database marketing programs, promotions, and incentives that reflect the individual customer's demographics and purchases. The goal is to increase sales and decrease defectors to other stores.

SERVICE INNOVATIONS

Supermarkets are facing competition not only from the supercenters and mass merchandisers for the price-conscious market, but also from fast food outlets and other food retailers for the convenience-oriented market for food. Although grocery sales in real terms are not growing, the demand for food that is ready-to-eat that can be purchased and taken-out to eat at home or elsewhere is growing. The terms "home meal replacement" and "meal solutions" have been coined to reflect this demand. Supermarkets have responded to this competition by providing more ready-to-eat or easy-to-prepare foods.

Convenience is about saving time. Reducing the preparation required before actually consuming a product is one form of convenience. However, there are at least two other dimensions to convenience from the food shopper's perspective. One relates to the number of tasks that can be accomplished during a single shopping trip or in a single store; the other relates to the ease of shopping and time required to shop. To improve their one-stop shopping appeal, supermarkets have been adding new services, such as banks, florists, video rental, and pharmacies (Kinsey and Senauer, 1996).

To make shopping easier, supermarkets are changing their interior designs or floor layouts. A model of a typical traditional grocery store layout is shown in Figure 8. Grocery products (in boxes, bottles, and cans) occupy the center of the store. The perishable products are around the perimeter. The layout is designed to produce a specific circulation pattern for customers. In a traditional store it is inconvenient to shop for just a few items or just part of the store. Dairy products, for example, are frequently in the back of the store. The customer who wants simply to purchase some milk might be induced, during his or her long walk to the dairy case, to buy additional products. The traditional layout frustrates some customers, however, who find it much easier to shop at convenience stores for

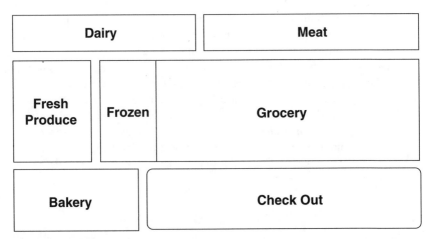

FIGURE 8 Traditional grocery store design.
Source: Kinsey and Senauer, 1966.

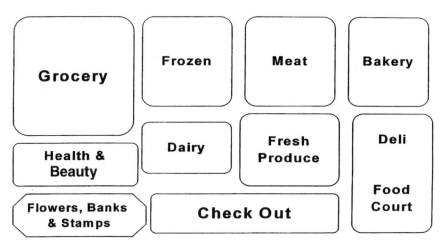

FIGURE 9 New grocery store design.
Source: Kinsey and Senauer, 1966.

such items. Supermarkets have discovered that their traditional layout can reduce sales compared to a more user-friendly layout.

A study for the FMI classified consumer food shopping trips into three categories—stock-up, representing purchase of more than 60 percent of weekly groceries, routine (20-60 percent); and fill-in trips (less than 20 percent) (Willard Bishop, 1991). Stock-up shopping accounted for less than one-third of the trips. In the stores studied, only 32 percent of shoppers were on stock-up trips, whereas 41 percent were on a routine trip, and 27 percent were on a fill-in shopping trip. The number of stock-up trips continues to decline, whereas the routine and fill-in trips increase. A traditional store reflected in Figure 8 is built for stock-up shopping.

Some supermarkets have responded by dramatically redesigning their stores so that the very process of shopping is altered. Figure 9 shows a typical layout for a new or remodeled store. The perishable departments, which consumers shop frequently, are all grouped to one side. A shopper on the way home from work can conveniently buy some fresh fruit, milk, and a ready-to-eat meal for dinner. Grocery products, which many shoppers buy only once or twice a month, are now in back. The deli may have been expanded into a "food court" with several meal solution stations and perhaps even a sit-down area to eat. Frozen products are close to the checkout counters so those purchases can be made last and the items gotten home without defrosting.

The question of shopping convenience was tackled by Cub Foods, a chain of supermarket stores in the Midwest owned by SuperValu. The design of many of Cub's new stores is similar to the plan in Figure 9. Cub saw that consumer demands were changing. The convenience factor was becoming crucial to many consumers. They wanted to be able to shop more quickly. A typical customer could take more than an hour to shop a traditional Cub store. Cub needed to find a way to build convenience into their supermarkets.

An internal Cub team was in charge of the redesign project. Consultants were employed to collect and analyze consumer information. Consumer panels were used and the participants asked, if they were to build a supermarket, how would they improve it. Cub used Plan Mark, an in-house service of SuperValu, for technical store design work. The first prototype store opened in 1991 in Apple Valley, Minnesota. Most Cub stores that have been built recently incorporate the new design.

The new design improves financial performance, according to Cub management. There is a substantial increase in sales of perishables and usually a small decrease in grocery sales. Perishables have much larger gross margins than groceries: typically 30 percent or more, versus around 20 percent.[2] Perishables also have higher labor costs. Previously Cub was oriented toward the stock-up shopper. With the new design, Cub gets more routine and fill-in shoppers. Although the average sales per customer has declined, customers shop more often and the number of customers per week has increased. Therefore, total sales have increasd. On average, customers are purchasing items with a larger profit margin.

Yet another innovation that can have a significant effect on sales is rearranging items on store shelves or improving displays generally. In a recent study, Dreze et al. (1994) estimated that, by rearranging products both along shelves and between shelves (from a lower to a higher shelf, for example), profits could rise by up to 15 percent. Optimizing the design of displays and product placement in this way relies crucially upon the POS information gathered at the checkout counter.

EFFECTS OF INNOVATIONS ON THE CHANGE IN PERFORMANCE

It is not easy to draw clear and definitive connections between the various innovations we have described and the performance of the supermarket industry. A research study by The Retail Food Industry Center analyzed the adoption of ECR practices by grocery stores in Minnesota and the impact on performance (Phumpiu and King, 1997). Data were gathered by interviewing managers at 40 stores varying in store size, location (metropolitan and out-state), and organizational form (chains and independents). An ECR "readiness index" was developed, which indicated the level of adoption by a store of key business practices and technologies that support ECR. Table 1 gives the indicators included in the ECR "readiness index" and the average adoption rate by stores. The index would equal 100 percent for a store that had adopted all 17 practices in Table 1 and zero

---

[2]Gross margin is the difference between the retail price and the cost of the goods sold.

TABLE 1 ECR Readiness Indicators

ECR readiness indicator	Adoption rate in stores interviewed (%)
Scan merchandise	88
Scan coupons	33
Manager has access to a personal computer	15
EDI transmission of order	98
EDI transmission of movement data	60
Scanning of incoming shipments	40
Shelf tags have movement and/or reorder information	20
Weekly sales forecasts based on POS data	65
POS coordinator has formal training on scan data quality	60
Resets based on formal planograms	20
Non-DSD resets coordinated with outside parties	60
Non-DSD product sssortment decisions coordinated with outside parties	60
DSD reset and product assortment decisions coordinated with outside parties	40
Manager has attended training on category management	43
Promotion and pricing decisions are coordinated with outside parties	53
Telxon units are used for price verification	80
Store uses competitor price information	68
ECR readiness score	53

Source: Phumpiu and King (1997).

for a store that had implemented none. The index ranged between 15 percent and 100 percent for the stores interviewed.

ECR adoption was closely associated with better productivity. Stores with a higher ECR "readiness index" had substantially higher sales per labor hour, sales per square foot, and annual inventory turnover, as show in Table 2. High-readiness stores were defined as those with an index of more than 75 percent, moderate as 75-40 percent, and low as stores that had adopted less than 40 percent of the 17 practices listed in Table 1. However, what this study could not conclude was whether ECR readiness increases productivity or whether better performance simply makes it easier to adopt ECR practices.

Category management, a part of ECR that seeks to reduce the number of

TABLE 2 ECR Readiness and Store Productivity Measures

	ECR readiness		
Productivity measures	High	Moderate	Low
Weekly sales per labor hour	$124.01	$104.61	$78.07
Weekly sales per square foot of selling area	$13.65	$10.70	$6.06
Annual inventory turnover	37	26	16

Source: Phumpiu and King (1997).

products in a store, has been shown to improve the bottom line. Although it might seem that consumers would prefer more choice rather than less, the evidence shows that this may not be so. Too many choices can lead to shelf clutter, making it hard for a shopper to decide what to buy. In a study conducted in 15 Chicago stores, Stickel (1996) reports that when 25 percent of individual items were removed in one category, sales in that category rose by 2 percent compared with control stores.

The annual U.S. supermarket industry data for 1995-1996 compiled by the FMI were used to suggest that ECR adoption was having a positive impact on performance (Zwiebach, 1997). Although sales fell for the third year in a row in 1996 after adjusting for inflation, average operating income rose to 3.05 percent, compared with 2.68 percent a year earlier. The average net profit margin reached 1.2 percent, as shown in Figure 2, the highest level since FMI began tracking it in 1972, compared with 1.14 percent the previous year. However, net profits fell in the most recent year, 1996-1997, and many factors affect net profits. Companies using ECR practices had the best performance. Earnings were strongest among those using category management and EDI. According to the 1996 FMI data, 36.7 percent of companies are using EDI, 35.9 percent have implemented category management, and 13.4 percent are using continuous replenishment.

FMI also found that 45.3 percent are operating frequent-shopper programs, with higher average sales and margins as a result. Average transaction size was $36 for frequent shoppers last year, compared with $22 for others. The average gross margin was 25 percent on frequent-shopper purchases, versus 23 percent for others. Only 18 percent of frequent-shopper participants defected to other stores, compared with 21 percent of regular shoppers (Zwiebach, 1997).

CONCLUSIONS

The U.S. supermarket is currently undergoing significant changes in the way it operates. Competition, especially from retailers outside the supermarket industry, is increasing. Pressures to reduce prices likewise have hit the industry in recent years. Consumers are driving much of the change. They have become accustomed to buying high-quality products at low prices from general-purpose merchandisers. Americans devote less time to food preparation than they ever had in the past, so they demand—and are willing to pay for—food products that require little preparation at home. The share of food expenditures spent away from home continues to rise, squeezing the supermarket industry yet again.

The industry's responses to these external changes have come in two broad categories: process innovations and service innovations. ECR is a managerial initiative aimed at improving the use of information to control the supply chain. ECR also involves significant changes within a store. Many supermarkets are currently trying to reduce the number of items they carry. By replenishing their stocks continuously, in a way not possible without electronic information-management systems, grocery stores are trying to reduce inventories and associated costs.

Many of the innovations witnessed in the industry are purely service oriented. Changes in store layout, for example, designed to make shopping easier for the hurried shopper, do not require significant technological inputs. However, consumers respond to these changes. Supermarket companies have expanded the number of non-food departments in their stores. These innovations attract consumers who face many options for purchasing the items once thought to be the supermarket's special domain, such as soft drinks or household paper products, but who can now shop for them at discount general merchandisers and other retail outlets.

Consumers shop differently than they did a decade ago, and they also buy different types of foods. Families who have little time for food preparation increasingly choose to buy prepared foods. These include whole-meal replacement and ready-to-eat foods. Supermarkets have responded to consumer changes and to the new pressures from innovative competitors by increasing their offerings of ready-to-eat and ready-to-cook foods. These items have led to many changes in store operations. Convenience requires space, and increasing store size has caused a reduction in sales per square foot of selling space. Sales per labor hour, likewise, have fallen over the past several years as labor-intensive departments such as the deli have become larger and more important. Savings due to ECR may offset these additional costs.

The necessary data on a national level are simply not available to definitely assess the impact of service and process innovations in grocery retailing, and in particular ECR practices, on productivity and performance. With this in mind, The Retail Food Industry Center (TRFIC) intends to initiate a major data-gathering effort using supermarket panels. A panel will be composed of a sample of similar stores across the country that will complete an annual questionnaire on a confidential basis. The panel data will allow TRFIC researchers to compare innovations and practices across comparable stores and track the impact over time. An ultimate goal is to be able to evaluate their impact on productivity and performance.

The source of innovation in the grocery store industry is as eclectic as the innovations themselves. The majority of electronic technologies employed by supermarkets are developed by third parties, which market them to the industry. Many service innovations are developed by consultants and trade associations as well as by supermarket operators. These innovations are so diffuse that it is virtually impossible to capture their costs and to get a clear picture of what would be called "research and development" in many product-oriented industries.

Food retailing is in a period of consolidation with a substantial number of mergers and acquisitions. Several European food retailers have purchased U.S. supermarket operations. Royal Ahold of the Netherlands, for example, recently bought Giant Food. The consolidations may stimulate innovation as the firms seek to cut costs and improve performance. More capital may also be available to large operations to finance innovation.

The innovations outlined in this chapter require changes in the supermarket labor force. The industry has long been known as a low-tech industry. It draws much of its labor from the younger, less-educated portion of the population. As the use of technology increases, this traditional view has changed. A major constraint to the implementation of ECR practices has been the education and skill level of people in the industry, especially in the area of information technology at the store level (Kurt Salmon, 1996). The grocery industry, like so many others, is increasingly in need of employees with technical abilities, particularly computer-oriented skills. Firms are making increasing use of technology, such as computer-based courses and videos, in training and also screening their employees.

The ultimate measure of productivity in a service industry like the supermarket industry is consumer welfare. This is very difficult to measure, especially when one considers the quality component of the industry's output. Consumers spend less of their income on food than their parents did, but this is natural given that real incomes have risen over the past quarter century. Today's consumers pay for the convenience engineered into their food items. Evidently they find this to be a good buy, or more people would be cooking their meals from scratch. Clearly changes in prices are not a sufficient measure and consumer satisfaction is what really ought to be measured in order to determine how well this service industry, with its many changes, is performing.

REFERENCES

Blattberg, R. C. (1996). "Changing Store Image Through Category Revitalization." Paper presented at Food Marketing Institute Conference. Chicago, IL. May 7. Photocopy.

Capps, O. Jr. (1997). "New Competition for Supermarkets: A Case Study." Working Paper 97-05, The Retail Food Industry Center, University of Minnesota, St. Paul, MN. February.

Dreze, X., S. J. Hoch, and M. E. Purk. (1994). "Shelf Management and Space Elasticity." *Journal of Retailing* 70:301-326.

Food Institute. (1996). *Food Retailing Review, 1996.* Fair Lawn, NJ.

Food Institute. (1998). *Food Institute Report, No. 24.* Fair Lawn, NJ.

Food Marketing Institute. (1997). *The Food Marketing Industry Speaks, 1997.* Washington, DC.

Kahn, B. E., and L. McAlister. (1997). *Grocery Revolution: The New Focus on the Consumer.* New York: Addison-Wesley.

King, R. P., and P. F. Phumpiu. (1996). "Reengineering the Grocery Supply Chain, in Changes in Retail Food Delivery: Signals for Producers, Processors and Distributors." Working Paper 96-03, The Retail Food Industry Center, University of Minnesota, St. Paul, MN.

Kinsey, J., and B. Senauer. (1996). "Consumer Trends and Changing Food Retailing Formats, in Changes in Retail Food Delivery: Signals for Producers, Processors and Distributors." Working Paper 96-03, The Retail Food Industry Center, University of Minnesota, St. Paul, MN.

Kurt Salmon Associates. (1993). *Efficient Consumer Response, 1993: Enhancing Consumer Value in the Grocery Industry.* Washington, DC: Food Marketing Institute.

Kurt Salmon Associates. (1996). *ERC 1995: Progress Report*. Joint Industry Project on Efficient Consumer Response, Grocery Manufacturers of America. Washington, DC.

Larson, R. B. (1997). "Key Developments in the Food Distribution System." Working Paper 97-08, The Retail Food Industry Center, University of Minnesota, St. Paul, MN.

Phumpiu, P. F., and R. P. King. (1997). "Adoption of ECR Practices in Minnesota Grocery Stores." Working Paper 97-01, The Retail Food Industry Center, University of Minnesota, St. Paul, MN.

Senauer, B., and J. Kinsey. (1997). "The Efficient Consumer Response Initiative: Implications for Vertical Relationships Throughout the U.S. Food System." Paper presented at the International Conference on Vertical Relationships and Coordination in the Food System, Piacenza, Italy, June 12-13, 1997.

Stickel, A. I. (1996). "Dominick's Changes Mix to Cut Out-of-Stocks." *Supermarket News* 46(June 3):20.

U.S. Department of Agriculture. (1996). "Food Consumption, Prices, and Expenditures," Research Service Bulletin Number 928.

U.S. Department of Commerce, Bureau of the Census. (1996). U.S. Statistical Abstract. Washington, DC.

Willard Bishop Consulting. (1991). *How Consumers Are Shopping the Supermarket 1991*. Food Marketing Institute, Washington, DC.

Woolf, B. P. (1996). *Customer Specific Marketing*. Greenville, SC: Teal Books.

Zwiebach, E. (1997). "Filling Out the Bottom Line." *Supermarket News* 47(May 12):14,19.

Retail Banking[1]

FRANCES X. FREI
Harvard University
PATRICK T. HARKER
LARRY W. HUNTER
University of Pennsylvania

INTRODUCTION

How does a retail bank innovate? According to the view in traditional innovation literature, organizations innovate by getting new and/or improved products to market. However, in a service industry like retail banking, the product is the process of serving customers. Thus, innovation in retail banking lies more in process and organizational changes than in new product development in a traditional sense. This chapter reviews a multiyear research effort on innovation and efficiency in retail banking and discusses the means by which innovation occurs in retail banks. It also examines factors that make one institution better at innovating than another. The chapter draws implications to the broader service sector.

We conclude that there is simply no "silver bullet"—no single set of management practices, capital investments, and strategies that lead to success. Rather, it appears that the "devil" is truly in the details. The key to efficiency in this industry appears to be the alignment of technology, human resources management, and capital investments with an appropriate production "technology." To achieve this alignment, banks need to invest in a cadre of "organizational architects" that are capable of integrating these varied pieces together to form a coherent structure.

The biggest challenge facing retail banks with respect to efficient and effective innovation lies in the management of the "New Age Industrial Engineers"

[1]This research was supported by the Wharton Financial Institutions Center through a grant from the Sloan Foundation and the National Science Foundation's Transformation to Quality Organizations, Grant SBR-9514886.

that must combine technological knowledge with process design to create the delivery systems of the future.

THE INNOVATION CHALLENGE IN FINANCIAL SERVICES

Financial services are a huge and critical sector of the U.S. economy, comprising over 4 percent of the U.S. Gross Domestic Product (GDP) and employing over 5.4 million people—more than double the combined number of people employed in the manufacture of apparel, automobiles, computers, pharmaceuticals, and steel.[2] While impressive, these numbers belie the much larger role that this industry plays in the economy (Herring and Santomero, 1991). Financial services firms provide the payment services and financial products that enable households and firms to participate in the broader economy. By offering vehicles for investment of savings, extension of credit, and risk management, they fuel modern capitalistic society.

While the essential functions performed by the industry—the provision of payment services, and facilitation of the allocation of economic resources over time and space—have remained relatively constant over the past several decades, the structure of the industry has altered dramatically. Liberalized domestic regulation, intensified international competition, rapid innovations in new financial instruments, and the explosive growth in information technology are fueling this change. Against this backdrop, managers and workers face intensified pressure to improve productivity and financial performance. Competition has created a fast-paced industry where firms must adapt and innovate to survive.

Given the increasing competition in the financial services industry and rapid technological evolution, how do retail consumer banks innovate to meet these challenges? This chapter attempts to answer this question by considering general trends in retail banking and by describing a detailed field study at a major U.S. retail consumer bank. We discuss the forces that are driving retail banks' need to innovate and describe the means by which banks innovate. In the process we discuss what constitutes efficient and effective innovation in banking. After all, not all innovation is good, and even if the innovation is a good idea, the costs of execution can substantially exceed the benefits.

The Changing World of Retail Banking

Nowhere are the changes sweeping the financial services industry more strongly felt than in retail consumer financial services. Once the sole domain of

---

[2]Comparison based on average 1991 data reported by the U.S. Bureau of Labor Statistics, *Employment and Earnings Report*, March 1992. Data for the financial services industry includes SIC codes 60-64 and 67. Data for the apparel, automobile, computer, pharmaceutical, and steel industries include SIC codes 239 (less 23), 371, 357, 283, 331, and 332.

TABLE 1 Changes in the U.S. Banking Industry 1979-1994

Item	1979	1994
Total number of banking organizations	12,463	7,926
No. of small banks	10,014	5,636
Real industry gross total assets (trillions of 1994 dollars)	3.26	4.02
Industry assets in megabanks (percent of total)	9.4%	18.8%
Industry assets in small banks (percent of total)	13.9%	7.0%
Total loans and leases (trillions of 1994 dollars)	1.50	2.36
Loans made to consumers (percent of total)	19.9%	20.6%
Total number of employees	1,396,970	1,489,171
Number of automated teller machines	13,800	109,080
Real cost (1994 dollars) of processing a paper check	0.0199	0.0253
Real cost (1994 dollars) of an electronic deposit	0.0910	0.0138

Note: A "megabank" in this table is a bank with over $100 billion in assets in real 1994 dollars. A
     "small" bank is one with assets under $100 million in 1994 real dollars.
Source: Berger et al. (1995).

the retail bank, various non-bank competitors—including mutual funds and brokerage firms—are increasingly offering competing services, eroding the market
share of the traditional retail banking sector. Consider the changes depicted in
Table 1. It is also investing heavily in new information technology—primarily to
facilitate new electronic means of transactions, which hold a cost advantage over
traditional paper-based banking.

The major forces for these changes will be described in detail in the next
section, but a quick glance at Figure 1 confirms that increased competition from
other players in the financial services industry continues to erode the market-
share of banks. This competition, along with the explosive changes in information technology (IT) and changes in consumer demand, fuels the need for banks
to innovate in products, services, and delivery channels.

THE FORCES OF CHANGE IN RETAIL BANKING

Various forces are driving change in retail banking, but the principal ones are
regulatory changes, technological innovation, and changing consumer demand.

Regulatory Change and Consolidation

As shown in Table 1, the retail banking industry is undergoing a period of
rapid consolidation as well as expansion into non-traditional banking products
and services. Between 1979 and 1994, approximately 5000 banking organizations were taken over by other depository institutions. Why?

First, regulations restricting interstate banking and the broadening of product
lines of the banks continue to weaken. Changes regarding reserve limits, bank

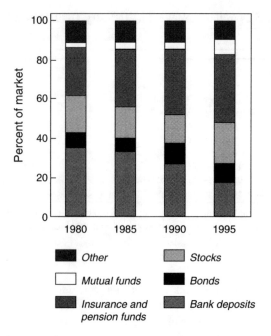

FIGURE 1 Share of U.S. consumer financial assets, 1980-1995.
Source: Federal Reserve data, reproduced in Council on Financial Competition (1996).

powers, geographic restrictions, and the Glass-Steagall Act restrictions on product offerings have all fueled merger activity.[3] Banks are also responding quickly
to the removal of limits on interstate banking activities, as shown in Table 2.

Similarly, the relaxation of the Glass-Steagall restrictions on bank holding
companies has permitted banks to merge across product lines. Bank holding

TABLE 2 Changes in the Geographic Focus of the U.S. Banking Industry
1979-1994

Item	1979	1989	1994
Total national banking assets (%) legally accessible from a typical U.S. state	6.5%	29.0%	69.4%
Typical state's banking assets controlled by out-of-state multibank holding companies	2.1%	18.9%	27.9%

Source: Berger et al. (1995).

---

[3]See Berger et al. (1995) for a detailed discussion of these regulatory changes.

companies are increasingly purchasing mutual fund companies, brokerage houses, and insurance firms in order to offer a full spectrum of financial products to their customers. These cross-industry acquisitions are aimed at stemming the continued erosion of market share depicted in Figure 1. The driving force in every bank is "share of wallet"—the desire to attract and retain more and more of a consumer's financial business.

Do these mergers work? At present, the evidence is quite mixed in terms of both cost reduction and profit efficiency.[4] In terms of shareholder value, recent research suggests that these mergers have tended to destroy, not enhance, value as shown in Figure 2.

One major explanation for retail banking's consolidation is the desire to have sufficient size to exploit scale economies in transaction processing and scope economies in cross-selling multiple financial products to a household. However, numerous studies of efficiency in the banking industry show that neither scale nor scope efficiency is the main cause of inefficiency. Summarizing this research, Berger et al. (1993) focus on a measure—X-efficiency—that isolates all technical and allocative efficiencies of individual firms that are not dependent on scope or scale. That is, X-efficiency captures how well management is aligning technology, human resources, and other assets to produce a given level of output. They note, "The one result upon which there is virtual consensus is that X-efficiency differences across banks are relatively large and dominate scale and scope efficiencies."

Other results, such as those reported by Fried et al. (1993) in the context of credit unions, add additional weight to the importance of X-efficiency by providing evidence that it is a dominant factor in both large and small institutions. Based on this evidence, it is clear that scale and scope economies are not the driving factor in explaining firm-level efficiency and the driving force behind mergers.

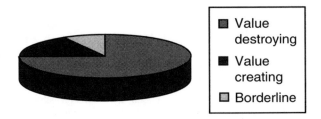

FIGURE 2 Shareholder value analysis of bank mergers and acquisitions 1983-1988. Source: D.C. Cates (1991).

---

[4]Some studies, such as Shaffer (1993) and Akhavein et al. (1997), show that banks can obtain lower costs and increased profits, while others (Rhoades, 1993; Peristiani, 1997) show little to no post-merger gains.

Summarizing the problems of inefficiency in this industry, Berger et al. (1993) state:

> Our results suggest that inefficiencies in U.S. banking are quite large—the industry appears to lose about half of its potential variable profits to inefficiency. Not surprisingly, technical inefficiencies dominate allocative inefficiencies, suggesting that banks are not particularly poor at choosing input and output plans, but rather are poor at carrying out these plans.

What then drives the consolidation of the industry? When questioned on their strategic response to increased competition, bank directors stated that acquisitions were the most important method for overcoming competitive threats and positioning themselves for the future (see Figure 3). Thus, much of the consolidation can be viewed as a strategic response to an acceleration of change in the industry. Many bankers are worried about firms like Microsoft entering the banking business. To face this competition, they feel that they must extend both scale and scope in order to compete in the future.

Obviously, not all banks that merge or acquire other institutions are achieving negative results. Just like the inefficiencies described above, there is a distribution of talent when it comes to consolidation. In a recent paper, Singh and Zollo (1997) discuss the role of organizational experience and learning in the bank acquisition process. Summarizing their results, the authors state: "The probability of a high level of integration [of banks] is strongly determined by the degree to which the acquirer has codified its understanding of how to accomplish this extremely complex and relatively infrequent task." Thus, the acquisition process itself can be viewed as a major source of innovation in banking.

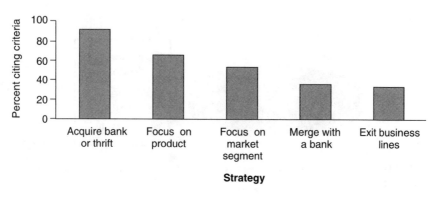

FIGURE 3 Bank director's response to the following question: What will you most likely do to overcome competitive threats and better position yourself for the future?
Note: Over 200 bankers were surveyed.
Source: Towers Perrin's 1994 survey (1996).

Mergers and acquisitions, therefore, are a powerful force of change in the banking industry, impacting not only the geographic scope and product variety of the organization, but also affecting the underlying technological and managerial infrastructures of the banks. For the foreseeable future, consolidation will continue to position the organizations against current and future players in the marketplace.

Technological Innovation

Technology plays a key role in the performance of banks. Large U.S. banks spend approximately 20 percent of non-interest expense on information technology, and this investment shows no signs of abating. Even with these large investments, it is still difficult to ascertain the payoffs associated with these projects. In manufacturing, recent studies have found large payoffs in IT investments, in terms of both equipment and personnel (Brynjolfsson and Hitt, 1993; Lichtenberg, 1995). For example, Lichtenberg (1995) states that "...the estimated marginal rate of substitution between IT and non-IT employees, evaluated at the sample mean, is six: one IS employee can substitute for six non-IT employees without affecting output."

Unfortunately, similar results for financial services are not available. For example, in the recent study by the National Research Council (1994) on IT in services, the problem in the context of banking is summarized as follows:

> Neither approach [for productivity measurement] is able to account for improvements in the quality of service offered to customers or for the availability of a much wider array of banking services. For example, the speed with which the processing of a loan application is completed is an indicator of service that is important to the applicant, as is the 24-hour availability through automated teller machines (ATMs) of many deposit and withdrawal services previously accessible only during bank hours. Neither of these services is captured as higher banking output at the macroeconomic level.

While hard-and-fast data are not yet available, many believe that financial services are at the brink of major performance improvements due to technology. However, this will not occur in the traditional "back-office" functions such as check processing. Rather, the performance improvements will result from the integration of front- and back-office functions—that is, in integrating business processes. Roach (1993) points out that the consolidation of back-office operations is due in large part to scale economies resulting from to IT investments but that these investments are becoming increasingly difficult to find. However, he states that "...new productivity opportunities are now spreading rapidly across the sales function of the service sector...." It is precisely in these front-office functions that major investments will occur. Philip Kotler (as cited in Pine, 1993) states this trend clearly:

Instead of viewing the bank as an assembly line provider of standardized services, the bank can be viewed as a job shop with flexible production capabilities. At the heart of the bank would be a comprehensive customer database and a product profit database. The bank would be able to identify all the services used by any customer, the profit (or loss) on these services and the potentially profitable services which may be proposed to that customer.... This movement away from mass marketing, mass production, and mass distribution is widespread throughout the financial services industry.

Technological innovation in the retail banking industry has been spurred on by the forces described by Kotler, particularly in terms of new distribution channel systems, such as PC banking. As the industry has provided more ways for consumers to access their accounts, they have added significant costs to each institution. A need to combat these costs resulted in a major effort to reduce cots in back-office operations through automation contributing to productivity improvements in functions and the processing of loan applications. Now, after adding significant costs through added distribution channels and cutting as much as possible in the back-office, banks have realized that the key to profitability is through revenue enhancement.

Banks are now forced to consider new ways to drive revenue through their distribution system. The most common way to do so is to try to increase the share of the customer's wallet. As explained above, the share of wallet is the portion of a customer's entire financial relationship that any particular bank has with the customer. The prevailing hypothesis is that the more products that a customer has with the bank, the cheaper it is to serve them per product and the more difficult it would be for the customer to switch to another bank.

The primary revenue-enhancing innovations occurring today are in platform automation, i.e., the automation of the functions performed by front-line employees, for branch and phone center employees, and in the newest distribution channel, PC banking. While these innovations have aspects in common, they serve different needs in the distribution strategy of retail banks.

Platform automation is the retail banking industry's first major attempt at giving employees a single view of the customer. Prior to this innovation, it was not possible for an employee to view the entire customer relationship at one time. Why is this important? First, a single view lets the employees understand how important a customer is based on their portfolio of products rather than on their current checking account balance. If hidden behind that low checking balance is a series of CDs and a home equity loan, for example, then the employee may want to think twice before refusing to waive a small fee associated with the checking account. However, although the concept of bringing together all of a customer's relationships with the bank is quite simple, in reality it has proven to be an extremely difficult task.

Retail banks collect and process information by product and transaction, not by customer. While it is quite easy to access all the information on checking account customers or on credit card customers, for example, taking a slice of the data per customer is technologically difficult. Virtually every bank has been faced with this same problem. Legacy systems (i.e., 1970s-style data processing software) were built with transaction processing per product in mind. Banks have excellent check processing systems, but such systems do not easily interact with the mortgage processing system, for example. Moreover, the data are scattered among a variety of systems and locales. It is quite common to have credit card processing in a different state from the rest of the retail bank, so that bringing this data together is a massive undertaking.

PC banking represents a new distribution channel and an area for significant technological innovation. With this new channel, there are many alternatives available to each bank, and with these alternatives come managerial decisions regarding alliances, outsourcing, new product development, and a host of other critical factors that will influence future profitability. At the surface, one could consider the PC channel similar to the phone center in that a customer is simply contacting the bank remotely, in one case over the phone and in the other by the PC. The major difference between the channels comes in the variety of ways that a bank can offer PC banking and in the implications resulting in each model. We describe the four most common PC banking models below in order to demonstrate the variety of alliances and outsourcing practices as well as to discuss the implications of each in terms of potential loyalty and increased share of wallet.

Coincident with the retail banking industry moving from cost-savings innovation to revenue-enhancing innovation is the move from in-house development to outsourcing and alliances. While there are many arguments favoring this shift, including the most common view that banks are not software companies and should not be developing these systems in house, it remains to be seen if this shift will loosen the bank's stronghold as the predominant financial intermediary. As payment systems in the United States catch up to the rest of the world in terms of the ability to have end-to-end electronic processing, it is not clear where the profits will be made. By making choices today in terms of platform automation and PC banking models, banks are making explicit choices about where they see themselves in the future.

The Changing Consumer

The final, and perhaps the most important, force of change in the banking industry is the rapid evolution of consumer wants and desires. Consumers are demanding anytime-anywhere delivery of financial services, along with an increased variety in deposit and investment products.

Consider first the desire for greater product diversity. Whereas Fidelity Investment and Merrill Lynch both offer over 100 different choices for mutual funds, the typical bank offers 17.[5] As a result, banks continue to lose market

---

[5]"Mutual Fund Review," *Wall Street Journal*, April 1996.

share (see Figure 1). Choice of demand deposit accounts with a desired fee structure, along with the advent of new investment vehicles such as index funds, all fuel the banking customer's desire for new and better financial products.

In addition, consumers are moving away from the use of checks to other financial products, albeit slowly (see Figure 4). Consumers are also demanding variety of delivery channels available for their use (Table 3). In spite of the assertion that branch delivery is dead, most consumers still frequent the branch. In fact, there has been a rise in the number of branches, including supermarket-based locations (called "in-store branches") and kiosk-like branches found in many shopping malls. And, as can be seen in Figure 5, this trend to open new physical sites seems likely to continue. Furthermore, it is the "mixed channel consumer"—one who frequents multiple delivery points—that is the norm in the industry (Figure 6).

Consumers are demanding and receiving a larger variety of traditional and new banking products and delivery systems. The question, however, is how banks capture the value generated by this increase in variety. At present, one needs to look only at the controversy surrounding consumers' resistance to paying fees for various ATM transactions to understand that this increase in variety may be detrimental to a bank's profitability. Over decades, banks have invested heavily in ATM machines because of their cost advantage on a per-transaction basis (Table 4). The traditional teller transaction is almost an order of magnitude more expen-

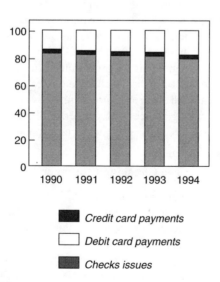

FIGURE 4 Use of various payment instruments (millions of transactions).
Source: Kennickell and Kwast (1997), Table 1.

TABLE 3 Percent of U.S. Households Using Various Delivery Channels

Delivery Channel	% of households
In person/ branch visit	86.7
Mail	57.4
Phone	26.0
Electronic transfer	17.6
ATM	34.4
Debit card	19.6
Direct deposit	59.6
Pre-authorized debit/ payment	23.6
PC banking	3.7

Source: Kennickell and Kwast (1997), Table 2.

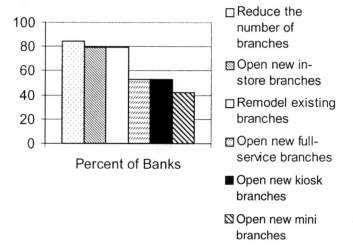

FIGURE 5 Branch activities planned over the period 1995-1998.
Source: Ernst and Young (1996), annual survey of major U.S. banks.

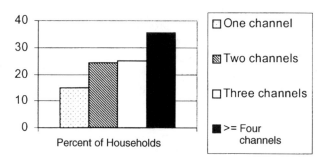

FIGURE 6 Percentage of U.S. households using various numbers of delivery channels.
Source: Kennickell and Kwast (1997), Table 2.

TABLE 4 Comparison of Cost per Transaction for Various Delivery Channels

Distribution channel	Cost per transaction
Teller	$1.40
Telephone (human operator)	$1.00
Telephone (automated voice response unit)	$0.15
ATM	$0.40

Source: Oliver Wyman and Company.

sive than ATM and automated phone systems. This has led banks to attempt to change consumer behavior through the addition of fees (the "stick") and a variety of rebates (the "carrot"). Despite these efforts, the total cost of serving certain customer segments has not changed significantly because customers, in the meantime, have altered their own transaction behavior. Some customers even use ATMs more frequently: a typical college student may use the ATMs once a day, for one $20 bill! Changing customers' behavior—to use ATMs instead of tellers, for example, or to use ATM's less frequently—would, in theory, yield the greatest benefit to banks in terms of cutting costs. However, in practice, this change in behavior will be difficult to achieve, as evidenced by the recent customer uproar over increases in ATM fees.

Thus, banks must continue to innovate in order to meet the changing needs and desires of the consumer. In particular, banks seek to leverage the developments in information technology to create new products and services. At the same time, banks must develop new fee structures to shift consumers away from high-cost delivery systems. This blend of innovation and behavior change lies at the heart of the modern banking organization. We now turn to the innovation mechanisms banks use to meet these challenges.

HOW DO BANKS INNOVATE?

How does a retail consumer bank innovate? To begin to answer this question, consider two important developments in banking—the emergence of the PC/electronic delivery of financial services (a product innovation) and creating new distribution channel designs (an organizational innovation). Both have had significant impact. In the case of PC banking, this innovation promises to revolutionize the cost structure of retail banking. However, such changes will not occur unless banking organizations can adapt their structure to exploit the new technology. Both technical and organizational innovation are crucial to retail banks.

A Product Innovation: PC Banking

Pushed by growing consumer demand and the fear of losing market share, banks are investing heavily in PC banking technology (Frei and Kalakota, 1997).

In collaboration with hardware, software, telecommunications, and other companies, banks are introducing new ways for consumers to access their account balances, transfer funds, pay bills, and buy goods and services—all without using cash, mailing a check, or leaving home. The four major approaches to home banking are, in historical order:

- *Proprietary Bank Dial-up Services.* A home banking service, in combination with a PC and modem, lets the bank become an electronic gateway to customers' accounts. This enables customers to transfer funds between accounts or pay bills directly to creditors' accounts.

- *Off-the-Shelf Home Finance Software.* This category is essential in helping banks cement relationships with existing customers and gain new customers. Examples include Intuit's Quicken, Microsoft's Money, and Bank of America's MECA software. This software market is also attracting interest from banks because it has steady revenue streams through upgrades, updates, and the sale of related products and services.

- *On-Line Services-Based.* This category allows banks to set up retail branches on subscriber-based on-line services such as Prodigy, CompuServe, and America Online.

- *World Wide Web-Based.* This category allows banks to bypass subscriber-based online services and reach the customer's browser directly through the World Wide Web. There are two great advantages: the flexibility to adapt to new online transaction processing models facilitated by electronic commerce, such as on-line bill paying, and the ability to eliminate the constricting intermediary or on-line service.

In contrast to packaged software that offers a limited set of services, the on-line and Web-based approaches offer further opportunities. As consumers increasingly purchase items in cyberspace with credit cards, debit cards, and newer financial instruments such as electronic cash or electronic checks, they will need software products to manage these electronic transactions and reconcile them with other off-line transactions. In the future, an increasing number of paper-based, manual financial tasks may be performed electronically on machines such as PCs, hand-held digital computing devices, interactive televisions, and interactive telephones. Banking software must have the capability to facilitate these tasks.

Home Banking Using Bank's Proprietary Software

On-line banking was introduced in the early 1980s when at least four major banks (Citibank, Chase Manhattan, Chemical, and Manufacturers Hanover) offered home banking services. Chemical introduced its Pronto home-banking services for individuals and Pronto Business Banker for small businesses in 1983. Its individual customers paid $12 per month for the dial-up service, which allowed them to maintain electronic checkbook registers and personal budgets, see account balances and activity (including cleared checks), transfer funds among checking and savings accounts, and—best of all—make electronic payments to some 17,000 merchants. In addition to home banking, users could obtain stock quotations for an additional per-minute charge. Two years later, Chemical teamed up with AT&T in a joint venture called Covidea meant to push the product through the second half of the decade. Despite the muscle of the two home-banking partners, Pronto failed to attract enough customers to break even and was abandoned in 1989.

Other banks had similar problems. Citicorp had a difficult time selling its personal computer-based home-banking system, dubbed Direct Access. Chase Manhattan had a PC banking service called Spectrum. Spectrum offered two tiers of service—one costing $10 a month for private customers and another costing $50 a month for business users, plus dial-up charges in each case. According to their brochure, business users paid more because they received additional services such as the ability to make money transfers and higher levels of security.

Banc One had two products, Channel 2000 and Applause. Channel 2000 was a trial personal computer-based home-banking system available to about 200 customers that was well received. Applause, a personal computer-based home-banking system modeled after Channel 2000, attracted fewer than 1000 subscribers. The trial was abandoned before the end of the decade, as the service could not attract the critical mass of about 5000 users that would let the bank break even. In each of the above instances, the banks discovered that it would be very difficult to attract enough customers to make a home-banking system pay for itself (in other words, to achieve economies of scale). Figure 7 describes a traditional proprietary system of banking.

On-line banking has been plagued by poor implementations from the early 1980s. Home-banking services lost too much from concept to reality. Many

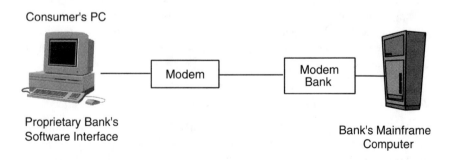

FIGURE 7 Proprietary software method for PC banking.
Source: Frei and Kalakota (1997).

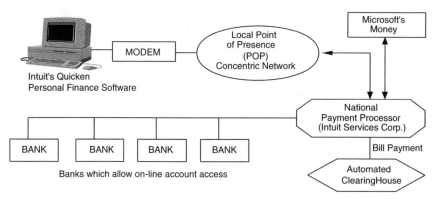

FIGURE 8 Banking with dial-up software.
Source: Frei and Kalakota (1997).

systems evolved gradually, which often meant that consumers who initially used the service and left dissatisfied could not be coaxed back into using it again.

Recently Citibank has revamped its Direct Access product, allowing consumers to dial in to Citibank's system and check their account balances, transfer money between accounts, pay bills electronically, review their Citibank credit card account, and buy and sell stock through Citicorp Investment Services. Although the underlying systems run in batch mode, Citibank has put together a middle-ware piece of software that makes consumers think that they are operating in a real-time environment. While this can work in a setting where Citibank is not interacting with third-party systems, there are potential difficulties with this batch/real-time mix if Citibank offers outside products and services such as insurance products. In addition, because consumers are interacting directly with Citibank's system, they have no way of performing household budgeting functions on their financial data. Clearly, Citibank will need to either provide this function itself or provide easy interface to the popular personal finance packages. However, it is important to point out that the new Direct Access represents the first major improvement in proprietary software home banking in 15 years, which is demonstrated by their explosive growth from 40,000 subscribers to 190,000 in 1996.

Banking with the PC Using Dial-Up Software

The main companies that are working to develop home-banking software are Intuit, the maker of Quicken; Microsoft, the maker of Microsoft Money; Bank of America and NationsBank, who acquired Meca's Managing Your Money software from H&R Block; and ADP, which acquired Peachtree Software. Banking with third-party software means that there is an intermediary between the bank and the consumer. In fact, as can be seen in Figure 8, it is easy to imagine how the banks can become back-end commodity providers in this system, with the third party controlling the customer interface.

Banking with On-Line Services

Although personal finance software allows people to manage their money, it only represents half of the equation. No matter which software package is used to manage accounts, information is managed twice: once by the consumer and once by the bank. If the consumer uses personal finance software, then both the consumer and the bank are responsible for maintaining systems that do not communicate. For example, a consumer enters data once into his or her system and transfers this information to paper in the form of a check, only to have the bank then transfer it from paper back into electronic form. In the instance where an electronic check is issued, the systems that receive the information rarely communicate automatically with bookkeeping systems.

Unfortunately, off-the-shelf personal finance software cannot bridge the communications gap or reduce the duplication of effort described above. However, a few home-banking systems that can help are beginning to take hold. In combination with a PC and modem, these home-banking services let the bank become an electronic gateway, reducing the monthly paper chase of bills and checks. The general structure of the on-line services banking architecture is shown in Figure 9.

How to Innovate with PC Banking

Although there is no clear choice as to the appropriate home-banking model, it is quite clear that very explicit trade-offs must be made. In addition to considering control of the interface, security, speed of access, and convenience, banks must consider the level of customer support required for each model. Basically,

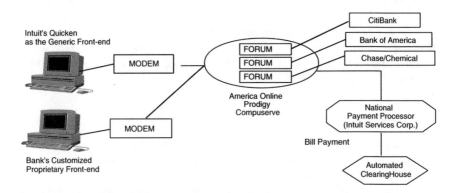

FIGURE 9 On-line services banking architecture.
Source: Frei and Kalakota (1997).

the larger the numbers of intermediaries, the higher the level of support the customer will need. Those banks that understand the technology, human resource, and process issues will have a better chance of coming out ahead in this innovation.

Thus, the fundamental challenges to innovation in PC banking are not technological per se but arise from the complex set of organizational choices to implement such a service for the consumer. Suppliers can provide not only the software needed to support a PC banking operation, but the back-office fulfillment processes as well. The basic innovation for the bank lies in its integration of these software and fulfillment processes to create the electronic banking service. To illustrate the fact that it is often organizational change that fuels innovation in banking, we now turn to an example of a bank that is in the process of re-creation.

An Organizational Innovation: Re-Creating a Bank

National Bank,[6] one of the larger U.S. commercial banks, with branches in many states, has a retail banking arm that is in many respects typical of the industry. Our research team has spent the past year studying the process of innovation at National, tracking the implementation of a major redesign of the retail delivery system.

Confronted by an increasingly competitive environment, National was challenged with improving the cost-efficiency of its far-flung retail delivery system, comprising hundreds of branches. At the same time, National sought to transform these numerous retail branches into sales-focused financial stores: ones that concentrated more directly on the sale of financial products and services. Our account of the continuing process of redesign at National illustrates a number of the observations made earlier in this chapter.

National's retail banking organization was quite decentralized. No single organizational unit in the bank had responsibility for retail operations. Rather, each of the major geographic areas served by the bank had its own management team. The challenge of redesigning the bank was heightened by the diversity across geographic areas. Some of the state-based operating divisions, and many of the branches, had been acquired from other banks and quickly folded into National, retaining many of their former employees and some of their technology and business processes. To carry out the redesign, therefore, National had to build from scratch a group responsible for its implementation. National assembled a re-engineering team of over 50 employees, drawn from a diverse set of geographic areas and functional backgrounds, and charged this team with spearheading the overhaul of the branch delivery system.

Initially, the redesign at National focused around very basic business process re-engineering in the branches. Over a period of decades, a huge number of

---

[6]National Bank is a pseudonym, used to protect confidential information.

administrative functions had accumulated in the branch systems, so that branch managers and service representatives spent a considerable amount of time on these activities rather than in contact with customers. Further, most of the time spent with customers was centered on simple, transaction-oriented and basic servicing of accounts rather than on activities that were thought to be likely to lead to sales opportunities. Leaders at National, recognizing these problems, engaged a leading consulting firm as a partner in the re-engineering of the branch system, and the consulting firm spent several months working with the implementation team to identify opportunities to streamline branch activities. The outcome of this partnership became known as the "pilot" redesign, and it was agreed that the redesign should be tested in a few small market areas before being rolled out across the bank more broadly.

From the start, both the consultants and the team conceived the redesign to require broad, systematic change. Effective innovation therefore required the participation of virtually all the functional areas within the bank, from information systems to marketing to human resources, with each of these areas represented on the implementation team. Anchoring the redesign was the streamlining of branch processes and the relocation of many of the administrative tasks and routine servicing of accounts to central locations outside the branch. To take one simple example, incoming telephone calls from customers were re-routed so that phones in the branch did not ring; rather, customers calling National and dialing the same number they had always used to contact the branch, would now find their calls routed to a central call center.

The innovation also required redesign of the physical layout of the branches. A goal of the redesign was to encourage more customers to use automatic teller machines and telephones for routine transactions. Customers entering the redesigned branch, therefore, were to be greeted by an ATM, an available telephone, and a bank employee ready to instruct them in the use of these technologies. The customer would be directed toward a teller or a service representative only if the customer insisted or when such personal attention was clearly necessary, for example, to deposit cash, to access a safe deposit box, or to meet with a sales representative about purchasing a product or service.

These technological innovations, along with the redirection of customers to alternative delivery channels, were intended to make operations more efficient. As an example of the expected efficiencies, early projections by the consulting firm, although later shown to be overly optimistic, envisioned a 65 percent decrease in the number of tellers required in the branch system. Over time, it was hoped that many customers would cease to rely on the branch and its employees for routine transactions and services. The re-engineering was also expected to transform service employees into sales personnel, allowing them to concentrate their efforts on activities that had potentially higher added value, such as customized transactions and the provision of financial advice coupled with sales efforts.

A clear requirement for effective innovation at National, then, was the participation not simply of the employees but also of the customers in the new service processes. In its design, National elected not to pursue some of the more notorious routes favored by other banks (such as charging fees to see tellers), but to lead customers somewhat more gently, by making customer relations a key feature of the redesigned retail bank. The redesign created a customer relations manager in each branch. It was this employee's responsibility to ensure that each retail customer who entered the branch was guided to a service employee or, alternatively, to a technological interface, such as an ATM, to receive the appropriate level of service.

The redesign also required a large degree of innovation in two further areas: the information system and the telephone call center. Changes in the information system were designed to help relocate and standardize a large number of routine types of account inquiries and transactions and to give National employees a fuller picture of each customer's financial position and potential. This more complete picture of the customer's portfolio was thought to enhance sales efforts, enabling service representatives to suggest a fit between customers and services, and to refer the customers to areas in the bank with expertise in a particular product. The retail bank branch would be turned into a sales-focused financial store.

Challenges in the IT area were heightened by the legacies of existing technologies and the requirement that customer service continue to be provided accurately and without interruption; customers are not patient with errors or delayed access to their own money. Over time, a large number of systems, laid one on top of the next, had accumulated in the bank. Further, the redesign had both the advantages and disadvantages of being introduced on the heels of a number of earlier, more piecemeal technological and sales initiatives aimed at the same goals. Both the marketing and IT functions had been continuously seeking to improve National's capabilities in these areas. Support for these initiatives, and their success, had been uneven across the various geographic areas. The Marketing and IT departments had also worked with a number of other outside vendors. It was not immediately obvious whether the more systematic redesign should complement or substitute for these earlier, more incremental changes in systems or whether these vendors would, or should, have a role in the redesign. Over time, however, these consultants and vendors came under increasing pressure to coordinate their efforts with those of the implementation team, and those who were unsuccessful in doing so were replaced.

The importance of the telephone call center raised a new set of challenges. National had lagged a number of its competitors in the sophistication of its telephone banking system; yet, through the redesign, it hoped to make telephone banking and eventually, PC or home-banking, cornerstones of its delivery system. Branch redesign, therefore, also required the construction of new call centers, staffing them as the customers began to be directed toward them, and developing an organizational structure not simply to run the call centers but to manage the relationship between the call centers and the branches. Even more consultants and vendors were required here. The delineation between the new redesign in the branch system and the specialized expertise of the vendors working with telecommunications technology was clearer, so that managing these continuing relationships raised fewer immediate problems than in the case of the branch-based vendors. However, and more recently, as implementation has continued, new challenges have emerged. The increasing importance of the telephone centers has increased the pressures on the call centers for accurate and effective service, even as the call centers struggle with much more basic issues around staffing and the physical implementation of the telecommunications systems.

Changes in the physical layout of the branches, in information systems, and in the design of key business processes therefore attracted the attention of the implementation team from the beginning of the innovation process. As planning for the implementation of the pilot redesign proceeded, however, it became increasingly obvious to many on the implementation team that the true anchor for the set of innovations was none of these factors. Most critically, the innovations relied upon significant changes in key jobs in the branch systems, on the human resource practices that supported these jobs, and on employees' reactions to these changes.

In order to reinforce the idea of standardization across the branch system, and to focus efforts toward sales and efficient delivery of services more clearly, the implementation team recommended that the redesign eliminate the position of local branch manager. In each branch, a customer-relations manager would coordinate customer service efforts, but this person would not have direct authority over the tellers and platform employees in branches. Rather, branch employees would report to supervisors by area: customer-relations employees, branch-sales specialists, and tellers each would be assigned to remote leaders. On the platform, a variety of specialized customer service and sales positions were to be consolidated into a position that was eventually titled "Financial Specialist." Local areas were also to be staffed with a few roving financial consultants who did not have specific branch assignments. Only the tellers were to remain relatively unscathed by the proposed changes.

With this design, the pilot was implemented in two small local markets. Most of the hundreds of administrative and servicing processes were removed from the branch. Telephones no longer rang in the branches. The financial specialists were freed to concentrate on sales activities and found themselves with time available to pursue sales opportunities prospectively rather than simply reacting to walk-in traffic. Most customers responded to the innovation positively, quickly migrating to the new technologies with few problems. The active roles played by the customer-relations managers, many of whom were former branch managers, helped this migration along.

The pilot implementation also revealed a number of problems in the design. First, employees and customers in a few of the most rural branch locations met the redesigned branch with great skepticism. After a period of wrestling with modifications to the design, and considering the benefits associated with the implementation of a single, standardized form of service delivery, the implementation team agreed to abandon the idea of a single best design. National Bank acknowledged that the characteristics of rural markets differed fundamentally from urban and suburban locations. Rural customers, and the way they expected banks and their employees to provide service, were not likely to be served effectively by the redesigned branch. A new task force was commissioned to explore this problem and to come up with a design that gained some of the efficiencies associated with standardization and re-engineering for rural branches while acknowledging the key differences.

A second critical problem was the slow implementation of new technology. Many of the new features of the technology needed to support the new design simply were not ready or did not work as promised. The implementation team, finding it necessary to push forward and being uncertain about when these features would be ready, moved ahead with the new design anyway once the team was assured that there would not be critical gaps or stoppages in the provision of services. Basic services were satisfactory. The remaining problems related chiefly to ease of use, performance measurement software, and databases and other systems that were intended to provide more support for sales.

Third, while most customers adjusted to the new arrangements quickly, and the new processes that were accompanied by supportive technology worked effectively, turning the retail bank branch into a sales-focused financial store proved more difficult. Financial specialists found it difficult to move from the idea of reacting to the sales opportunities that routine servicing occasionally provided to the more pro-active role that the redesign called for. Some even claimed that the redesign was responsible for decreased sales as a result of the streamlining. The implementation team wondered in turn how much of this difficulty could be attributed to the design and how much to skills deficits among the financial specialists.

A fourth problem was the difficulty in implementing human resource practices necessary to support the new organization. The deficit in skills raised further issues. For example, training was critical to the success of the implementation, yet the organization had little time to spend in developing the skills critical to the pilot's success. Further, it had been clear that the selection process for new employees would have to be adjusted to seek employees who were more likely to be effective sales agents, but the initial difficulties with the design made this even more imperative. And while incentive compensation systems were also changed to reflect the new goals of the redesign, these were experimental and required considerable fine-tuning. Perhaps most important, however, was that the new jobs had effectively destroyed career ladders in the pilot branches. No longer could tellers easily move to platform positions; these positions were now expected to require an entirely different skill set and, for new applicants, usually a college degree. The financial specialists, who in most cases had been platform employees, could no longer expect to be promoted to branch management positions: these positions had been abolished and many of the branch managers became customer-relations managers. In each functional area, the hierarchy was flattened. While this yielded efficiency gains, it left employees quite uncertain about their future in the organization.

The implementation team spent much of its time with the nuts and bolts of the new design. Technological and process-related problems with implementation, and the challenges associated with performance measurement, consumed the team's attention. However, the human resource problems raised serious concerns for the longer-range success of the redesign. Employee confusion and skepticism over the new design was emerging as an impediment to the success of the innovation—and this was occurring in an environment designed to soft-pedal such concerns. Because the team was concerned about the effectiveness of the technological, process, and architectural changes, they decided that the redesign would not be accompanied by any layoffs in the pilot branches. They also knew that, to achieve the eventual efficiencies they expected, some downsizing of the retail bank would be necessary. They did not expect that natural attrition, even in the relatively high-turnover retail bank, would yield the cuts in jobs that they hoped for. The team realized that, in future implementation, the insecurity generated by the job changes would be intensified by the layoffs that would accompany these changes.

Despite these problems, the redesign, with some modifications, moved forward. A second pilot redesign was implemented in urban and suburban markets in a geographic area distinct from the earlier pilot. More attention was paid to training and selection for the new positions; again, outside consultants were relied upon, this time to help identify employees with appropriate skills and to develop those skills. Some of the technological gaps and challenges had been addressed, yet some remained, yielding a new set of complications in the specifics of implementation. And the second pilot revealed a new set of problems. In this local area, the situation in the branches before the change differed considerably from those in the first set of pilots. In particular, these branches had already been sharply focused on sales opportunities, a reflection of the bank's strategy in this geographic area. While disruption of the status quo in the first set of pilots had been considered to be positive, the benefits of this disruption in the second group—which was already moving toward a system of sales-focused branches—were less clear to local managers. Consequently, they were more skeptical about the benefits of redesign and of a standardized model. Local managers consistently argued for local adaptation of the model, claiming that they knew best what sorts of processes, technologies, and job structures were likely to be most effective in their area.

The implementation team, while sympathetic to these claims, generally resisted the pressure to adapt but recognized a further difficulty. To argue that the redesigned model must be strictly adhered to was to admit that no further learning was to occur as a result of the innovation. They struggled to find ways to differentiate between local learning that truly represented a positive improvement to the design concepts and local arguments grounded more in resistance to changing established routines. They also sought principles for making these distinctions as the design was to be rolled out over a much wider area.

Currently the team is preparing to implement the new design across the remainder of National's retail bank system, with substantial modifications drawn from the lessons learned through the pilots and other issues that have emerged as the process of innovation has continued. Among these challenges are the problems associated with introducing yet another round of innovations in local areas that have already witnessed massive change in recent years as a result of the frantic pace of mergers and acquisitions in the industry. Some of the branches that will be the objects of the redesign will have had three parent banks in the past three years; each change has been accompanied by changes in jobs, processes, systems, and supporting human resource practices. Heaping yet more change onto these locations will be especially difficult.

A second challenge facing the implementation team stems from the current decentralized approach to management of the retail bank. While the details of the pilot redesign have not been formally disseminated across the various geographic areas, word that the bank of the future is soon to arrive has traveled widely. Some of the members of the implementation team have returned to management positions in their local areas. Smart local managers have already begun to identify the trends that the implementation team was charged with addressing and have begun to address these challenges locally with their own changes and strategies. Thus the implementation team will be trying to innovate not in a static or standard set of channels but in a wide array of varied and dynamic conditions: in short, against moving targets. Already some local managers have explicitly expressed a desire to get ahead of the game by proceeding with implementation of the features of the pilot redesigns they find most attractive. Left unanswered is how and whether the implementation team will be able to implement other features or how they will reconcile differences in the preemptive local redesigns with their own plan.

Appropriately configuring human resource practices to support innovative systems and process changes raises further, significant challenges. On the one hand, it is clear that simply changing job design and pay systems, and coupling these with other technological and system changes, will be insufficient. Attention must also be given to employee selection and promotion systems, training programs, appraisal systems, the use of flexible scheduling, and the bank's overall approach to employee involvement. However, contemplating such sweeping change severely taxes the organization. While piecemeal change in the human resource system is unlikely to yield the results desired, more comprehensive change raises significantly more challenges in implementation. At National, the hope is that investment in the redesign will improve several areas of performance simultaneously—sales effectiveness, productivity, and the quality of customers' relationships with the bank. In practice, this has proven difficult. The early, piloted version of the redesign was effective at serving customers efficiently: the bank streamlined processes and introduced new technological options. However, the effect of the redesign on sales performance and on the overall depth and quality of the customer relationship is not as clearly positive. In fact, some of the streamlining designed to supplement or improve employee-customer interaction may be replacing this interaction. This may mean missed sales opportunities and fewer chances for bank representatives to assess and attempt to meet customers' needs.

Because much of the change is held to be a necessary response to continuing competitive pressures, it is unlikely that the redesign will actually be evaluated in strict cost-benefit terms. Such an evaluation of these innovations, their costs and benefits, will require a longitudinal, sustained, consistent effort by the bank, even as members of the implementation team begin to rotate to other positions within the bank. It will also be difficult to decouple the effects of the redesign from other major changes in marketing and product offerings and from the results of continuing merger and acquisition activity.

Should the design prove successful, this itself will raise sequential challenges for National, which must further innovate to deliver on the promises raised by successful change. To the extent that customers are convinced to migrate to alternative, more efficient delivery channels, the bank must continue to develop its ability to manage those channels effectively. Such channels, particularly telephone and PC banking, are not only more technology-intensive but also raise new sets of organizational and human resource problems. As the use of such channels grows, and as their range of functions increases, questions over appropriate staffing, training, performance measurement, and reporting structures multiply. Innovation, both organizational and technological, may actually have to intensify as a result of the success of prior changes.

Where's R&D? The Process of Innovation in a Bank

The two examples given above highlight the complex organizational design issues involved in the innovation processes in retail banking. Simply put, most retail banks do not have something called an R&D group. If they do, these groups play an important, but small, role in the overall innovation practices of the organizations. Marketing, business units, information technology, and a complex web of information technology suppliers and consultants drive the innovation processes in banking.

Consider the case of National Bank, where there was no division devoted to thinking about or implementing innovation, no "research and development" or similar functional structure. Rather, pressure for innovation built incrementally as a result of numerous smaller initiatives by marketing, by those responsible for managing technological systems, and by line managers. Each area felt competitive pressure and began to develop responses. At National Bank, these responses were eventually, to some extent, collected and channeled through the implementation team, although they also maintained some momentum of their own.

At National Bank, translating this pressure to innovate into actual technological and organizational changes was greatly facilitated by the continuing presence of consultants and of suppliers of technology. Indeed, one way to understand at least part of the role of consultants is that they function as suppliers of the organizational technology required to leverage the potential gains from innovations in computing and telecommunications systems. While the organization continues to develop its capacity to learn and innovate, it explicitly recognizes that it has considerable distance to travel in order to exercise this capacity more independently.

One further lesson we take from National in the midst of this redesign is that changes in IT, and in technological capabilities can spark the desire for system-wide innovation and even shape its particular form. With the enthusiastic promotion of consultants and outside vendors, technology is perceived by retail banks to be a catalyst for change across the organization. Yet even where this technology is over-sold, poorly understood, or fails to deliver on its promises, the process of innovation may take on its own momentum.

In the case of PC banking, such organizational changes are heightened by the presence of external suppliers of technology, consumer access, and fulfillment services. As banks continue to grapple with the variety of choices for electronic delivery, new organizational forms and entities are sure to emerge. As an example, the Bank of Montreal recently created a direct bank called *mbanx,*[7] whose purpose is to be a non-branch-based deliverer of financial services that will directly compete with the existing Bank of Montreal delivery and sales organization. Such developments of new organizational systems for non-physical delivery are sure to accelerate in the next decade.

WHAT CONSTITUTES EFFICIENT INNOVATION?

To produce innovation in the banking industry, complex organizational structures are needed. Given this context, which banks are efficient at such innovation? To address this issue, Prasad and Harker (1997) consider the overall impact of information technology on productivity in the retail banking industry in the United States. Using a Cobb-Douglas production function, Prasad and Harker estimate the following equation with a combination of publicly available and proprietary data:

$$Q = e^{\beta_0} C^{\beta_1} K^{\beta_2} S^{\beta_3} L^{\beta_4} \qquad (1)$$

where Q = output of the firm,
    C = IT capital investment,

---

[7]For details on mbanx, see the following Web address: http://www.mbanx.com/.

K= non-IT capital investment,
S = IT labor expenses,
L = non-IT labor expenses, and
β_1, β_2, β_3, and β_4 are the associated output elasticities.

Using this function, the following three hypotheses were tested:

• IT investment contribution to output is positive (that is, the gross marginal product is positive);
• IT investment contribution to output is positive after deducting depreciation and labor expenses (that is, the net marginal product is positive); and
• IT investment makes no contribution to the firm's profits or stock market value.

Just what constitutes a bank's *output* is a subject of some discussion. Summing up the problem Benston et al. (1982) posit that, "output should be measured in terms of what banks do that cause operating expenses to be incurred." Various studies of productivity have taken various approaches to this question, and they may be classified into three broad categories—the assets approach, the user-cost approach, and the value-added approach (Berger and Humphrey, 1992). Prasad and Harker look at a wide variety of output measures, both financial and customer satisfaction. The most meaningful results from this analysis arise when total loan + deposits is used as the output of the institution; these results are summarized in Table 5.

From this table, it can be seen that the elasticities or coefficients associated with IT capital and labor are positive. However, the low significance associated with the IT capital coefficient implies that there is a high probability (0.93) that the elasticity of IT capital is zero. Thus, there is not sufficient evidence to support the hypothesis that IT capital produces positive returns in productivity. It is interesting to note that the elasticity of non-IT capital is at best zero (being not significantly different from zero), implying that IT capital investment is rela-

TABLE 5 Results of the Estimation of Equation 1 When Output = (Total Loans + Total Deposits)

Parameter	Coefficient	Standard Error	t-statistic	t-statistic significance (%)	Ratio to output	Marginal product
IT capital	0.00116	0.013	0.089	7	0.000452	2.56
IT labor	0.25989	0.031	8.34	100	0.0006	449.75
Non-IT capital	−0.02071	0.026	−0.79	57	0.00428	−4.84
Non-IT labor	0.53244	0.059	8.95	100	0.01475	36.10

Note: R^2 = 41% (OLS); 99% (2-step WLS)
Source: Prasad and Harker (1997).

tively better than investment in non-IT capital. However, since the marginal product of IT labor is $449.75, it can be concluded that IT labor is associated with a high increase in the output of the bank.

Since the first hypothesis cannot be supported for IT capital, the discussion of the stronger hypotheses, the second in the list, is restricted to the IT labor results. First, it can be seen that the marginal product for IT labor is very high. Since IT labor is a flow variable, then every dollar of IT labor costs a dollar. In view of this, the excess returns from IT labor can be computed to be $(449.75 - 1), or $448.75. Thus, this hypothesis cannot be rejected for IT labor. For the last hypothesis, one has

$$\beta_3 - (\text{IT labor expenses} / \text{non-IT labor expenses}) \times \beta_4 = 0.2390 > 0.$$

Thus, there is support for the claim that investment in IT labor makes a positive economic contribution.

As far as capital expenses are concerned, it can be seen that the marginal product of non-IT capital is negative. Further, given the standard errors of the estimation, it is asserted that IT capital is more likely to yield either slightly positive or no benefits, whereas non-IT capital will most probably have a negative effect, decreasing productivity. More formally,

$$\beta_1 - (\text{IT labor expenses} / \text{non-IT labor expenses}) \times \beta_2 = 0.0034 > 0.$$

Given the significance associated with the IT capital estimate, however, the last hypothesis failed to be rejected.

Thus, these results show no strong evidence of IT capital making a positive contribution to output. This result is significantly different from previous studies in the manufacturing sector (Lichtenberg, 1995; Brynjolfsson and Hitt, 1996), and seems to be more in conformity with those obtained by Parsons et al. (1993), the only formal study on IT in banking to date. While Parsons et al. report a slightly positive contribution to IT investment, this analysis demonstrates zero or slightly negative contributions.

IT labor presents a very different picture than does IT capital. IT labor contributes significantly to output; its marginal product is at least 10 times as much as that of non-IT labor. Rather than make the simplistic conclusion from this that a single IT person is equivalent to 10 non-IT persons, it is better perhaps to speculate that this may simply reflect the fact that there is a significant difference between the types of personnel involved in IT and non-IT functions. It is more interesting to compare the marginal product of IT capital versus IT labor. It is striking that while IT labor contributes significantly to productivity increases, IT capital does not. Thus, these results state that while the banks in our study may have over-invested in IT capital, there is significant benefit in hiring and retaining IT labor.

This result and interpretation is consistent with the idea that aligning capital, rather than throwing technology at problems, is what affects efficiency. IT personnel are likely to be much more effective at ensuring that the implementation of technology does what it is meant to do. The general point is that the management of IT has profound effects on efficiency. Banks that are able to manage their IT effectively are likely to be efficient. These results are consistent with our fieldwork experiences. They are also consistent with the fact that today's high demand for IT personnel is unprecedented in U.S. labor history. Figures from the U.S. Bureau of Labor Statistics show that while the overall job growth in the U.S. economy was 1.6 percent between 1987 and 1994, software employment grew in these years at 9.6 percent every year and jumped to 11.5 percent in 1995. The prediction is that, over the next decade, software employment will grow 6.4 percent every year (Rebello, 1996).

The problems are actually likely to be more subtle than our measures suggest. For example, IT personnel, while evidently valuable, may not be equally valuable. The point was driven home to us in a series of interviews at a major New York bank. A Senior Vice President there lamented the fact that, "The skills mix of the IT staff doesn't match the current strategy of the bank," and said that he "didn't know what to do about it." At the same bank, the Vice President in charge of IT claimed, "Our current IT training isn't working. We never spend anywhere near our training budget." IT labor is in very short supply, and issues as basic as re-skilling the workforce cannot be addressed given the lack of sufficient IT labor in banking.

Other researchers have observed this dependence and under-investment in human capital in technologically intensive environments. To quote Gunn's (1987) work in manufacturing, "Time and again, the major impediment to [technological] implementation . . . is people: their lack of knowledge, their resistance to change, or simply their lack of ability to quickly absorb the vast multitude of new technologies, philosophies, ideas, and practices that have come about in manufacturing over the last five to ten years." Another observation about the transitions firms need to make to gain from technology comes from Reich (1984), again in the manufacturing context: "The transition also requires a massive change in the skills of American labor, requiring investments in human capital beyond the capital of any individual firm."

The evidence also suggests that the effects of managing IT are being felt more broadly in a retail bank. Consider the *inclusive* model for managing branches. In this model, IT and process redesign (so-called re-engineering) combine to remove as many basic servicing tasks as possible from employees. These tasks—simple inquiries, transactions, and movement of funds—can be automated or turned over to customers. Re-engineering frees employees to concentrate more effort on activities that have potentially higher added value—customized transactions and the provision of financial advice coupled with sales efforts. Second, IT gives each employee a full picture of each customer's financial position and potential. This enhances sales efforts, enabling tellers and customer service representatives to suggest a fit between customers and services and to refer the customers to employee teammates with particular expertise in a product, as necessary.

Challenges under the *segmented* model are less acute, yet still present. In this model, technology is used to simplify the majority of the jobs—to make them easier to learn and, therefore, to make turnover less costly. Only those bank employees in high value-added, personal banking jobs have access to the broad range of information that might be useful in generating sales leads and opportunities.

In order for either model to function effectively, those responsible for designing IT must understand not only the purposes of the technology but the capabilities and propensities of the workforce and the likely effects of different choices in technology on employee and customer behavior. Further, IT staff must be able to assess the likely effects of different configurations of technologies and employment systems if they are to be able to contribute to strategic decisions around the deployment of IT.

Thus, our results are very consistent with Osterman's (1986) conclusion that ". . . as IT capital prices fall, production becomes increasingly information-worker intensive." Our results seem to confirm this; increasing IT investments in banks requires a substantial investment in IT labor. Further, IT labor is the most profitable of all four types of investment—IT and non-IT capital and labor—available to the bank. Accordingly, the biggest challenge facing banks with respect to efficient and effective innovation lies in the management of the "New Age Industrial Engineers" that must combine technological knowledge with process design in order to create the delivery systems of the future.

BANKING INNOVATIONS: LESSONS FOR THE STUDY OF SERVICES

Our study of banking innovation leads us to reconsider the basic model of innovation in the standard textbooks and readings in the field.[8] While the basic steps of the innovation process, such as those outlined by Marquis (1969), remain the same, the change arises in the combination of actors who perform these steps. The standard view is that R&D, operations, and marketing combine in a complex web of interactions to generate innovation (Figure 10).

However, as we have seen from our previous discussion, vendors that supply outsourced services and technology play a vital role in this innovation process.[9] Even more important in the development of innovations is the role of the "systems integrator"—the person or organization that pulls together not only the op-

---

[8]For example, the collection of readings in Tushman and Moore (1988).

[9]For a discussion on the strategic role of firms that supply outsourcing services, see, for example, Jonash (1996), Chesbrough and Teece (1996), and Rubenstein (1994). For the particular case of financial services, see Drew (1995).

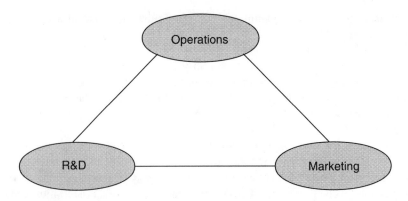

FIGURE 10 Basic relationship in innovation processes.
Source: Adapted from Galbraith (1982).

erations, IT, and marketing functions for a single innovation but also manages the portfolio of innovations in the organization. At National Bank, this systems integration role is played by an in-house re-engineering team in conjunction with their external consultant (see Figure 11).

Ultimately, it is this systems integration function that will make or break innovation efforts. Jonash (1996) argues that the systems integration function belongs in the hands of the chief technology officer who will coordinate the efforts of internal and external innovation efforts for the benefit of the organization. The discussion in the previous section about the critical role of the IT organization in the overall efficiency of the banks tends to support this view.

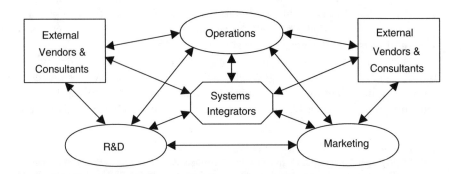

FIGURE 11 A new model for innovation involving expanded relations.

The role of the systems integrator is crucial for the future of retail banking. Frei et al. (1997), in summarizing their various analyses of retail banking efficiency based on the dataset described in the Appendix of this chapter, paint a picture of what makes an effective bank. They conclude that there is simply no single set of management practices, capital investments, and strategies that lead to success. Rather, it appears that the "devil" is truly in the details. The alignment of technology, human resources management, and capital investments with an appropriate production "technology" appears to be the key to efficiency in this industry. To achieve this alignment, banks need to invest in a cadre of "organizational architects" that are capable of integrating these varied pieces together to form a coherent structure. In fact, several leading financial services firms have realized the need for such talents and are investing heavily in senior managers from outside the industry—most notably, from manufacturing enterprises—to drive this alignment of technology, human resources management, and strategy. The challenge, therefore, is not to undertake any one set of practices but rather to develop senior management talent that is capable of this alignment of practices.

This is especially important because the future direction of the industry is subject to a tremendous degree of uncertainty. This uncertainty was revealed in the strategy-related data we collected as part of this study. We found that most banks simply could not articulate a consistent and coherent strategy for the future.[10] In numerous visits with the banks that were a part of the study, we would feed back the data that bank officials had given to us in order to check their validity. When we came to our survey's strategy-related questions, someone in the bank, usually at a senior management level, would state something along these lines: "This is wrong. This CAN'T be our strategy!" We would tell them who provided these data (always another senior manager), and we would become embroiled in a debate over defining the strategy of the bank!

This confusion reflects the tension between making investments to perfect today's strategy versus investing in a portfolio of alternative strategies for the future. This tension is both quite typical and quite real in the banking industry. Given the inability to control the use of the varied distribution channels, including ATMs and branches, banks are either investing in all channels simultaneously or undertaking fairly radical changes in their service offerings to deal with this proliferation of services. Thus, bank managers face a crucial decision as to missing the "correct" strategy for the future versus living with misaligned systems that they know to be inefficient.

Given this uncertainty, the removal of inefficient firms may take quite a long time. Furthermore, if we are correct in our assessment that a major cause of inefficiency in the industry is the misalignment of management practices, a major cause of persistent inefficiency in the banking industry may be the necessity for integrated financial services organizations to "hedge their bets" on the future.

---

[10]See Hunter (1996), in the context of human resources.

Clearly, alignment would be simpler and more rapid in an industry made up of many "niche" players, each focusing on a likely future scenario. Such movement to disintegrate financial services is already under way in most banking organizations; business units such as credit cards and trust divisions are now being run as completely separate operations.

The bottom line of this analysis is that service industries such as banking must develop a new generation of management talent to play the role of systems architect—one who can blend technical knowledge with complex organizational design issues to drive innovation through their firms.

REFERENCES

Akhavein, J.D., A.N. Berger and D.B. Humphrey. (1997). "The effects of megamergers on efficiency and prices: evidence from a bank profit function," *Review of Industrial Organization* 12:95-139.

Benston, G.J., G.A. Hanweck, and D.B. Humphrey. (1982). "Scale economies in banking: a restructuring and reassessment," *Journal of Money, Credit and Banking* 14:435-450.

Berger, A.N., D. Hancock, and D.B. Humphrey. (1993). "Bank efficiency derived from the profit function," *Journal of Banking and Finance* 17:317-348.

Berger, A.N., and D.B. Humphrey. (1992). "Measurement and efficiency issues in commercial banking," in *Output Measurement in the Services Sector: National Bureau of Economic Research Studies in Income and Wealth*, Z. Griliches ed., Chicago: University of Chicago Press.

Berger, A.N., W.C. Hunter, and S.G. Timme. (1993). "The efficiency of financial institutions: a review and preview of research past, present and future," *Journal of Banking and Finance* 17:221-250.

Berger, A.N., A.K. Kashyap, and J.M. Scalise. (1995). "The transformation of the U.S. banking industry: what a long, strange trip it's been," *Brookings Papers on Economic Activity* 2:55-218.

Brynjolfsson, E. and L. Hitt. (1993). "Is information systems spending productive? New evidence and new results," Working Paper, Coordination Laboratory, MIT Cambridge, MA.

Brynjolfsson, E., and L. Hitt. (1996). "Paradox lost? Firm-level evidence on the returns to information systems spending," *Management Science* 42:541-558.

Cates, D.C. (1991). "Can bank mergers build shareholder value?" *Journal of Bank Accounting and Finance* 6-7.

Chesbrough, H.W., and D.J. Teece (1996). "When is virtual virtuous? Organizing for innovation," *Harvard Business Review* 74(January-February):65-71.

Council on Financial Competition. (1996). *Letter from the Future: Beyond the Branch-Based Franchise,* Washington, DC: The Advisory Board Company.

Drew, S.A.W. (1995). "Accelerating innovation in financial services," *Long Range Planning* 28:11-21.

Ernst and Young. (1996). *Creating the Value Network 1996,* New York.

Frei, F.X. (1996). *The Role of Process Designs in Efficiency Analysis: An Empirical Investigation of the Retail Banking,* unpublished Ph.D. dissertation, The Wharton School, University of Pennsylvania. Philadelphia.

Frei, F.X., P.T. Harker, and L.W. Hunter. (1997). "Inside the black box: what makes a bank efficient?" Working Paper 97-20, Wharton Financial Institutions Center, The Wharton School, University of Pennsylvania, Philadelphia; also available at http://wrdsenet.wharton.upenn.edu/fic/wfic/papers.html

Frei, F.X., and R. Kalakota. (1997). "Frontiers of Online Financial Services," in *Banking and Finance on the Internet* M. J. Cronin ed., New York: Van Nostrand Reinhold Press.

Fried, H.O., C.A.K. Lovell, and P.V. van Eeckaut. (1993). "Evaluating the performance of U.S. credit unions," *Journal of Banking and Finance* 17:251-266.

Galbraith, J.R. (1982). "Designing the innovating organization," *Organizational Dynamics* (Winter):3-24.

Griliches, Z. (1992). *Output Measurement in the Services Sector: National Bureau of Economic Research Studies in Income and Wealth* Chicago: University of Chicago Press.

Gunn, T.G. (1987). *Manufacturing for Competitive Advantage,* Cambridge, MA: Ballinger.

Herring, R.J., and A.M. Santomero. (1991). "The role of the financial sector in economic performance," *Study Prepared for the Kingdom of Sweden's Productivity Commission,* Stockholm.

Hitt, L., and E. Brynjolfsson. (1996). "Productivity, business profitability, and consumer surplus: three different measures of information technology value," *MIS Quarterly* (June):121-142.

Huber, G.P., and D.J. Power. (1985). "Retrospective reports of strategic-level managers: guidelines for increasing their accuracy," *Strategic Management Journal* 6:171-180.

Hunter, L.W. (1996). "When fit doesn't happen: The limits of business strategy as an explanation for variety in human resource management practices," presented at the Academy of Management Annual Meeting, Cincinnati, Ohio, August 1996.

Jonash, R.S. (1996). "Strategic leveraging making outsourcing work for you," *Research-Technology Management* 39:19-25.

Kennickell, A.B., and M.L. Kwast. (1997). "Who uses electronic banking? Results from the 1995 survey of consumer finances," Working Paper, Division of Research and Statistics, Board of Governors of the Federal Reserve System, Washington, DC.

Leibenstein, H. (1966). "Allocative efficiency verses 'X-inefficiency," *American Economic Review* 56:392-415.

Leibenstein, H. (1980). "X-efficiency, intrafirm behavior, and growth," in *Lagging Productivity Growth,* S. Maital and N. Meltz eds., Cambridge, MA: Ballinger Publishing.

Lichtenberg, F.R. (1995). "The output contributions of computer equipment and personnel: a firm-level analysis," *Economics of Innovation and New Technology* 3.

Marquis, D.G. (1969). "The anatomy of successful innovations," *Innovation* (November).

"Mutual Fund Review," *Wall Street Journal,* April 1996.

National Research Council. (1994). *Information Technology in the Service Society,* Washington, DC: National Academy Press.

Oliver Wyman and Company. (1997). Personnel communication.

Osterman, P. (1986). "The impact of computers on the employment of clerks and managers," *Industrial and Labor Relations Review* 39:175-86.

Parsons, D., C.C. Gotlieb, and M. Denny (1993). "Productivity and computers in Canadian banking," in *Productivity Issues in Services at the Micro Level.* Z. Griliches and J. Mairesse eds., Boston: Kluwer Academic.

Peristiani, S. (1997). "Do mergers improve X-efficiency and scale efficiency of U.S. banks? Evidence from the 1980s," *Journal of Money, Credit, and Banking* 29:326-337.

Perrin, T. (1996) "Letter from the Future: Beyond the Branch-Based Franchise," Council on Financial Competition. Washington, DC: The Advisory Board Company.

Prasad, B., and P.T. Harker. (1997). "Examining the contribution of information technology toward productivity and profitability in U.S. retail banking," Working Paper 97-09, Financial Institutions Center, The Wharton School, University of Pennsylvania. Philadelphia; available at http://wrdsenet.wharton.upenn.edu/fic/wfic/papers.html

Pine, B.J. (1993). *Mass Customization: The New Frontier in Business Competition.* Boston, Harvard Business School Press.

Rebello, K. (1996). "We humbly beg you to take this job. Please," *Business Week* (June).

Reich, R.B. (1984). *The Next American Frontier.* New York: Penguin Books.

Rhoades, S.A. (1993). "Efficiency effects of horizontal (in-market) bank mergers," *Journal of Banking and Finance* 17:411-422.

Roach, S.A. (1993). *Making Technology Work*. Economic Research Unit, Morgan Stanley & Co., New York.

Rubenstein, A.H. (1994). "Trends in technology management revisited," *IEEE Transactions on Engineering Management* 41:335-341.

Shaffer, S. (1993). "Can mergers improve bank efficiency?" *Journal of Banking and Finance* 17:423-436.

Singh, H., and M. Zollo (1997). "Learning to acquire: knowledge accumulation mechanisms and the evolution of post-acquisition integration strategies," Working Paper 97-10B, Financial Institutions Center, The Wharton School, University of Pennsylvania. Philadelphia; available at http://wrdsenet.wharton.upenn.edu/fic/wfic/papers.html.

Tushman, M.L., and W.L. Moore, eds. (1988). *Readings in the Management of Innovation, 2nd Edition*. New York: Harper Business.

U.S. Bureau of Labor Statistics. (1992). *Employment and Earnings Report*, March.

APPENDIX: STRUCTURE OF THE
WHARTON/SLOAN RETAIL BANKING STUDY

This paper is partially a result of the work undertaken by the retail banking study at the Wharton Financial Institutions Center. The retail banking study is an interdisciplinary research effort aimed at furthering the understanding of competitiveness in the industry, where competitiveness means not simply firm performance but the relationship between industry trends and the experiences of the retail banking labor force.

In the exploratory first phase of a study of the U.S. retail banking industry during summer 1993 through fall 1994, a research team conducted open-ended and structured interviews with industry informants and shared its impressions with these informants at a number of conferences.

The team interviewed top executives, line managers in retail banking, human resource managers, executives responsible for the implementation of information technology, retail bank employees, and industry consultants. The first phase featured site visits to thirteen U.S. retail bank headquarters and interviews with numerous other managers and employees in remote and off-site locations. The interviews began with very general questions, and the questions increased in specificity as the research progressed. In this phase of the study, the team collected data through the use of two waves of structured questionnaires in seven retail banks. The team's analysis of the data in these questionnaires was then presented to management teams in six of the seven banks and was used as the basis for the second phase, a large-sample survey.

This detailed survey addressed technology, work practices, organizational strategy, and performance in 135 U.S. retail banks, chosen to yield the broadest coverage of trends in human resources, technology, and competitiveness in the industry. The survey focused on the largest banks in the country and was not intended as a random sample of all U.S. banks. In the end, the approach gained the participation of banks holding over three-quarters of the total assets in the industry in 1994. The process began by compiling a list of the 400 largest bank holding companies (BHCs) in the United States at the beginning of 1994. Merger activity, and the fact that a number of BHCs had no retail banking organization (defined as an entity that provides financial services to individual consumers), reduced the possible sample to 335 BHCs. Participation in the study was confidential but not anonymous, enabling the team to match survey data with data from publicly available sources.

Participation in the study required substantial time and effort on the part of organizations. Therefore, commitment to participation was sought by approaching the 70 largest U.S. BHCs directly; in the second half of 1994, we requested the participation of one retail banking entity from each BHC. Fifty-seven BHCs agreed to participate. Of these, seven BHCs engaged the participation of two or more retail banks in the BHC, giving us a total of 64 participating retail banks.

Multiple questionnaires were delivered to each organization in this sample. Questionnaires ranged from 10 to 30 pages and were designed to target the "most informed respondent" (Huber and Power, 1985) in the bank in a number of areas, including business strategy, technology, human resource management and operations, and the design of business processes. The team made a telephone help line available to respondents who were unsure of the meaning of particular questions. Questionnaires were delivered to four top managers: the head of the retail bank, the top finance officer, the top marketing officer, and the top manager responsible for technology and information systems. These banks received questionnaires for one manager of a bank telephone center and for one branch manager and one customer service representative (CSR) in the bank's head office branch, defined as the branch closest to the bank's headquarters. In addition, an on-site researcher gathered data about all business process flows in the head-office branch. Identical questionnaires were mailed to five more branch managers; the instructions to the bank were to choose the sample branches so that, if possible, data were received from two rural, two urban, and two suburban branches. Questionnaires were also mailed to CSRs in those branches. In these questionnaires, the CSRs themselves mapped processes associated with home equity loans, checking accounts, certificates of deposit, mutual fund accounts, and small business loans.

In order to facilitate the creation of process maps via the mailed survey, a worksheet was developed for the CSRs to fill out. These worksheets, a sample of which is shown by Frei (1996), list the majority of potential steps required in the process so that the CSR need only indicate the order of the step, the person responsible for its execution, the type of technology involved, and the amount of time the step takes. Adequate space was provided for the addition of steps unique to an institution.

In late 1994, survey questionnaires were mailed to top executives of the 265 next largest BHCs and followed with a telephone call requesting the participation of one of their retail banking organizations. Sixty-four of these BHCs agreed to participate in the study, and four of these engaged the participation of two or more retail banks in the BHC, so that a total of 71 retail banks participated in the mailed survey. For this group of banks, the head of the retail bank was surveyed, and many of the questions directed to the other top managers were consolidated into this survey. Prior interviews had suggested that, for banks of this size, the head of retail was able to answer this broader set of questions accurately. For this sample, questionnaires were mailed to one telephone center manager, one branch manager, and one CSR in the head office branch. The telephone help line was also available to respondents in this sample.

All together, the entire survey of retail banking covers 121 BHCs and 135 banks, which together comprise over three-quarters of the total industry, as measured by asset size. The scope and scale of this survey make it the most comprehensive survey to date on the retail banking industry.

Computing

TIMOTHY F. BRESNAHAN
Stanford University

The computer industry is remarkable for the pace of its technical change during the last half century and for the pace of its organizational change during the 1990s. From its inception in the 1940s, the industry has been characterized by rapid and sustained technical change. Major breakthroughs leading to new uses have punctuated continuous product innovation serving existing uses better each year. For decades, established sellers experienced success based on the persistence of key interface standards linking their proprietary technology to investments by users and by producers of complements. Much of that success arose from these firms' ability to coordinate and direct the wide variety of different technologies—components, systems, software, and networking—that make up computing. At the same time, the industry, opening up new markets, offered opportunities for entrepreneurial firms to pioneer new kinds of computers for new classes of users.

The industry is undergoing even more change in the 1990s, change of a different character. New technologies are being developed and introduced, many by new companies, in an industry organized in a radically new way. The entrepreneurial companies and the established firms no longer coexist but are in direct competition. Extraordinary returns to capital, and to highly skilled human capital—rents—are moving from vertically integrated firms expert in coordinating multiple technologies to clusters of loosely linked specialized firms. This is a revolution in systems of organizing innovation. At the moment, the "Silicon Valley" system of organizing innovation is on the ascent, and the "IBM" system appears to be fading. The change is so radical that one can speak of an old computer industry and a new one (Grove, 1996).

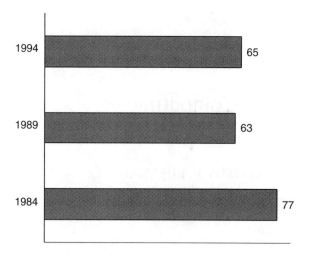

FIGURE 1 U.S. companies' share of worldwide revenues (%).

Despite all this change, one element of continuity is remarkable. Despite the decline of once-dominant IBM, U.S. firms continue to dominate the rent-generating portions of the industry, such as packaged software, microprocessors, and networking. Although the U.S. share of overall industry revenues is slowly falling (Figure 1), rents are staying put. Consider Microsoft, Intel, and Cisco, a troika that is small in revenue share but very large in rents and influence.[1]

This chapter examines the changing structure of the innovation process in computing alongside the enduring dominance of the United States. It examines the sources of the recent changes and the forces allowing a single country to earn most of the producer rents. The change in the character of the industry raises a set of serious questions about the persistence of international technological and competitive advantages in one country and a set of related questions about the origins of technological and commercial success for companies, countries, and regions within countries. Why was the United States, and not some other country, able to profit from the opportunities to become the world technological and competitive leader? What are the key performance characteristics of the IBM model of industrial organization versus the Silicon Valley model? The new computer industry rewards different kinds of technological skills, company organization, and inno-

---

[1]It would be useful but extremely difficult to turn these anecdotes about the persistence of rents in the United States into a systematic measure. A wide variety of sources, including the financial performance of U.S. and overseas firms, the export market penetration of products made in different countries, and the study of commodity vs. innovative products, strongly suggests that the rents have remained largely in the United States. Yet statistics on production and exports do not permit a systematic answer, partly because the most innovative products are the worst measured and partly because the portion of the industry that earns rents is shifting over time.

vation processes, but how is the United States able to persist in its leadership role despite the changing basis for success? What forces tend to make a single company, or region, a leader in the industry?

For answers one must look at the interrelationships among four very distinct areas:

- technology,
- firm and market organization,
- national support institutions, and
- demand and commercialization.

The necessity for congruence among the first three areas is by now familiar to most readers as a general observation about national success in an industry.[2] To understand the revolution in systems for organizing innovation in the computer industry, we need to understand the new technologies,[3] the new kinds of firms, the new market mechanisms for organizing technology from a wide variety of companies into useful computer systems, and the financial, legal, and educational infrastructure supporting the development of new firms and markets. Clearly the joint and mutually reinforcing development of technology, market structure, firms, and institutions is a source of national competitive advantage. In the computer industry, invention by users is very important. As a result, the forces of demand and commercialization must be included in any analysis of firm or national competitive advantage. Indeed, I categorize computer hardware, software, and networking firms not only by their technological capabilities but also by their marketing and commercialization capabilities. Overall competitive success is typically built upon joint and mutually reinforcing development in all four areas.

Within the United States, there have been several separate and distinct instances of this joint and mutually reinforcing development—separate clusters. On several occasions, a new technological breakthrough and a new kind of demand have combined to touch off new mutually reinforcing developments. The origins of the computer industry had that flavor. So, too, did the origins of the minicomputer industry, the microcomputer industry, and so on. With their different kinds of firms, different technologies, distinct relationships to support institutions, and distinct bodies of demand, each of these is a separate cluster of innovation. They even tend to be located in different regions—the original IBM-centric computer industry in and around New York; minicomputing in and around Boston; and much of microcomputing in California. In large part, the character

---

[2]See Nelson (1992, 1993) on this with particular regard to the United States.

[3]Indeed, there are some observers, using strongly technologically deterministic modes of explanation, who think that technical change is all we need to examine. Rather than review their arguments in detail in this chapter, I will let them die gently by ignoring them and making it obvious that other assets and factors have been very influential.

of each process, and thus of each cluster, was determined by the demand for the particular computers it made and by commercialization capabilities.[4]

These observations are directly relevant to the contemporary computer industry. Once again, computing is undergoing a revolution in its technical basis, firm and industry organization, and major applications. Networked computing involves transitions such as commercialization of Internet technologies and creation of electronic commerce—transitions that will certainly change the structure of firms and markets and may even mean the end of the vertically disintegrated Silicon Valley system. By their very nature, networked applications are integrative, drawing on technologies and commercialization capabilities from a variety of previously distinct clusters. Firms like Microsoft are proposing a new, more vertically integrated structure for the computing industry, with themselves in leading roles. Meanwhile, uncertainty about the new applications of networked computing opens up entry opportunities. It is a turning point.

The goal of this chapter is threefold. It lays out the structure of innovation in the computer industry, emphasizing the very wide variety in sources of innovation. It then shows how each of the main clusters has organized innovative activity, with an emphasis on the forces—some strong and some weak—that have caused innovative rents to flow to the United States. Finally, it discusses the radical reorganization of the industry and of innovation in the present and its implications for the future international allocation of producer rents.

THE EVOLVING STRUCTURE OF THE INNOVATION PROCESS

An initial look at the major features of the computer industry identifies some aspects that are relevant for the analysis presented in this chapter. The first feature is steady, rapid, and sustained technical progress. Fueled by fast-paced advances in the underlying electronic components as well as in computers themselves, computer hardware price and performance have both improved rapidly.[5] A wide variety of hardware categories have emerged—large and powerful computers such as mainframes, intermediate classes such as minicomputers and workstations, and classes with less expensive products such as personal computers. Technical progress has made the largest computers much more powerful and the smallest more affordable and has increased choice and variety in between. A few pioneering firms once supplied computers; now there are hundreds of successful suppliers of components, software, systems, services, and networks. Performance

---

[4]I am indebted to my long-time collaborators Shane Greenstein (see Bresnahan and Greenstein, 1995a) and Franco Malerba (see Bresnahan and Malerba, 1997) for much of this argument.

[5]For an extensive review of measurement studies of computer price-performance ratios, including discussion of alternative definitions of "performance," see Gordon (1989). On any definition of price/performance, improvements of 20-25 percent a year have been sustained over four decades. For a key class of electronic components, semiconductors, see Langlois and Steinmueller (1997) and Malerba (1985). For the complementary technology of software, see Mowery (1996).

increases and price decreases with dramatic improvements in all the different complementary technologies and considerable innovation and learning-by-using by customers. All of these factors woven together by firm, market, and other coordinating institutions have built a multi-billion dollar worldwide industry.

Complementarity: Multiple Technologies, Multiple Innovators

Computer systems draw upon a wide range of distinct, ever-advancing technologies. Computer hardware, software, and networking are capital goods used in a broad array of production processes. As with many general purpose technologies, investments in this capital lead only indirectly to valuable outputs. Without complementary innovations in other inputs or the complementary invention of new computer-based services, computers are useless.[6] At a minimum, to understand innovation in the computer industry, one must examine both invention by sellers of information technology and co-invention by buyers. Co-invention—users' complementary investments in human capital, new products, applications, business systems, and so on—has pulled computing into a wide variety of uses. The uses share invention but vary in co-invention.

Both invention and co-invention are complex processes combining innovations of many different forms. It is not a trivial problem to coordinate the direction of technical progress in invention with that in co-invention. A variety of market and commercialization institutions, ranging from the management and marketing functions in IBM to the markets and standards of open systems computing, have been used for this coordination. The innovation process in computing, dramatically oversimplified, consists of at least the elements described in Table 1.

The table begins with the familiar hardware technologies that most people tend to think of as "computers." Fueled by fundamental advances in materials and production processes, electronic components such as microprocessors and memory chips have seen steady and rapid technical progress. A tremendous amount of innovative effort lies behind empirical regularities such as Moore's law, by which the number of transistors on a cutting-edge integrated circuit doubles every 18 months.[7]

Another large and ongoing innovative effort is needed to bring these electronic components into useful electronic devices. A microprocessor may be the "brains" of a computer—or of a printer, a disk controller, or many other peripheral devices for that matter—but the design of computer systems and related hardware devices is a difficult and demanding piece of invention. It is not at all true

---

[6]See Bresnahan and Trajtenberg (1995) for analysis of general purpose technologies.

[7]See Langlois and Steinmueller (1997) for an analysis of this inventive process, which includes improvements in materials such as carefully doped silicon, equipment such as etchers and steppers, and the production process for integrated circuits themselves. On the complexities of the latter, see, for example, Hodges and Leachman (1996).

TABLE 1 Schematic of Technical Change in Computing

Invention	
Technologies (examples)	Coordination institutions for invention
Electronic components (microprocessor) Computer systems (mainframe, PC) Peripherals (disk drive, printer)	Vertical Integration
Systems software (operating system, network software, database management system) Applications software (accounts payable, computer aided design)	Interface Standards
	Coordination institutions for commercialization Vendor field sales and service Systems integrator Custom software house Consultant, VAR
Co-invention technologies (examples)	Coordination institutions for co-invention
Applications software (credit-card fraud detection system, spreadsheet macro)	MIS Department
New services (sorted checking account statement, instant account balance, frequent-flight bonus program)	Systems Analyst
New jobs and organizations (business process re-engineering, bank teller as sales representative)	CIO

that Moore's law for integrated circuits translates in any immediate and direct way into rapid declines in price-performance ratios for computers or other hardware. Computers themselves have exhibited smooth declines in price-performance ratios only because their designers have conquered a series of bottlenecks in different technical areas.[8] Progress in these areas may come in fits and starts, a choppiness that is smoothed out only when all the various subtechnologies are combined. Furthermore, a series of major technical discontinuities such as the founding of whole new classes of computers has punctuated the smooth advance.

For key peripherals, such as disk drives, major discontinuities and breakthroughs have characterized the process of technical advance.[9] Perhaps that fact should not be too surprising. Technical progress for peripherals is not purely electronic: a disk drive needs extraordinary precision in its reading and writing

---

[8]See Iansiti (1995) for an analysis of advances in computer systems technology along these lines and for discussion of management structures for dealing with rapid technical change in a variety of subcomponent technologies.

[9]See Christensen (1993) on how these discontinuities have led to major competitive turnover in the disk drive industry. Leading firms have fallen aside and been replaced by new firms. Also see Henderson's (1993) work on semiconductor equipment.

functions, for example; a printer needs a way to deposit ink. Computer hardware uses complex logic components and complex electromechanical components, both drawing on a wide variety of distinct subtechnologies.

All of computer hardware, taken together, draws on a wide array of distinct technologies. Overall, the rate of technical progress in computer hardware of a variety of types has been rapid. In the most innovative segments, technical progress reflects large investments in invention. There appears to be little difficulty with appropriability in this area, as the largest scale economy hardware technologies, such as microprocessors, tend to have quite concentrated industry structures as the mechanism for appropriability. The wide variety of product and process technologies in computer hardware means that some parts of computing have matured and become commodity businesses. In turn, production and to some extent invention, which were once largely confined to the United States, have now become global activities.

Software is a separate set of technologies that make computers and networks of computers useful in a variety of tasks. Systems software is best understood as a general purpose technology enabling a wide variety of distinct applications. Operating systems, network operating systems, communications controllers, and database management systems are as much a part of computers as the relevant hardware, but they are invented separately, and the total effort in their invention is very substantial.[10]

Applications software is a newer category as a market phenomenon (OECD, 1989; Mowery, 1996). For many years, applications were part of co-invention—almost all applications were custom-built for use in an individual company. Suppliers might be a management information systems (MIS) department or "end-user" departments. Now there are several important applications software markets in which software inventors sell their wares to user companies. The largest, in unit sales, are individual productivity applications (spreadsheets and word processors). Other important categories include general business software such as accounting, inventory management, and enterprise resource planning, and "vertical" applications software, which provides computing tailored to the needs of a specific industry. The transition from the co-invention of applications software to applications software markets is incomplete, so Table 1 lists applications software both at the top, under invention, and at the bottom, under co-invention.

To complicate the picture further, intermediaries can play the same role. A variety of commercialization institutions bridge the gap between invention and co-invention. Custom software, written for one customer at a time, existed as a market sector from the earliest days of computing. This service is sometimes

---

[10]The rate of technical progress in systems software is difficult to measure separately from the rate of technical progress in computer systems, because the two are such close complements. The measures of technological progress in computer systems shown in Table 2 should probably be interpreted as covering systems hardware plus operating systems but not other fundamental systems software.

sold as consulting, sometimes as explicit custom software, sometimes performed by systems integrators, and sometimes bundled with inventors' products as field sales and service. I list it here almost as an afterthought, as every one does. But that is an analytical mistake. By recent estimates, the "computer services" industry is larger in revenues than the software industry.

Why is this sector so large? It is difficult to plan, implement, and use computerized business systems; it is hard to manage computer and telecommunications departments in support of core businesses; high-quality computer personnel are scarce and expensive; and making strategic use of information technology is complex and sometimes fails. Using companies' Management Information Systems (MIS), departments of chief information officers (CIO) turn to the diverse computer services industry for support and help. Buyers choose between service firms whose initial competence was in the information technology arena, such as EDS or IBM's ISSC unit, in competition with others from the business consulting world, such as Andersen. Some users outsource their management information systems; others outsource their entire operational departments that use computers intensively, such as payroll processing (perhaps to ADP), or bank credit card (perhaps to First Data).

How Much? How Fast? How Well?

Table 2 presents some information on the sizes of the sectors just discussed and on their rate of technical progress. The main purpose of the table is to dramatize the wide variety in sources of innovation in computing. The table once again shows invention at the top, commercialization/intermediation in the center, and co-invention at the bottom.

The boundaries and definitions behind these tables are subject to some dispute, but a large message is clear from the size figures. A very substantial fraction of the activity in the industry is farther down the table, in commercialization or co-invention. The figures for the invention and commercialization segments reported in Table 2 are worldwide sales of those sectors.[11] They include both the costs of inventing new information technology and the costs of the goods, such as computers, in which that invention is embodied. Co-invention is harder to measure. The most objective part of it is programming personnel expenditures in computing departments, and this is the figure in Table 2.[12] The commercializa-

---

[11]This follows International Data Corporation (IDC) definitions and uses their 1996 report.

[12]The budget figures in Table 2 represent aggregate expenditures of using companies on their computer departments, less the products and services they buy and lease. These are primarily expenditures on programming personnel and cover the costs of writing applications programs in corporations, maintaining them, and so on. Some of the other expenditures counted here are training, planning systems, and the like. These costs count only the part of co-invention that is centralized and professionalized in MIS departments. If the finance department writes a spreadsheet macro on its own budget, or if the marketing department hires a webmaster on its own budget, it is almost certainly not counted here.

TABLE 2 Market Size and Rate of Technical Progress in Major Computing Industries

Technology	1996 Worldwide market size ($billion)(IDC)	Technical progress
Invention		
1. Electronic components (microprocessor)	(included in 2.)	Rapid—Dulberger (1993)
2. Computer systems (mainframe, PC)	261+	Rapid—Gordon (1989)
3. Systems software and tools (operating system, network software, DBMS	48+	(unstudied)
4. Applications software (accounts payable, CAD)	} 48+ combined	(unstudied)
5. Applications software (spreadsheets, word processors, etc.)		Slower—Gandal (1994)
Commercialization		
6. Computer and software services vendor (billed) services; systems integrator; custom software house; consultant	176+	(unstudied)
Co-invention		
7. Applications software (credit card fraud detection system, spreadsheet macro)	310+	Slow and difficult—Friedman (1989)
8. New services (sorted checking account statement, instant account balance, frequent flight bonus program)	} Together larger than 7 —Ito (1996) Very substantial—Brynjolfsson and Hitt (1996); Bresnahan and Greenstein (1995b)	Slow and difficult—Barras (1990); Bresnahan and Greenstein (1995b)
9. New jobs and organizations (business process re-engineering, bank teller as sales representative)		

*See footnotes 11 and 12 for International Data Corporation (IDC) sources and my calculations based on them.

tion services shown in the table are measured by their sales. Most of these charges are for activities, such as custom software and service, integration, or maintenance, that user companies might have done themselves.

An important part of co-invention is not measured at all in Table 2—the co-invention of new products and new processes based on computing. For example, much of the new product and process innovation in the services sectors is computer-based. The table measures only the part of this innovation that is explicitly computer programming. The equally important but unmeasured innovation is invention of new tasks for the computer. A new customer service—getting your balances by calling your bank on the telephone at night—will typically be invented and designed by marketing people. A new organizational structure—permitting the telephone bank operators to resolve certain account problems in those same account-query phone calls—is typically invented by operations managers, not the computer department. No firm accounts these activities as R&D, but they are an important part of innovation.[13]

Although these activities are not easily measured, a substantial anecdotal literature shows that inventing tasks for computers to perform and coordinating the efforts of a corporation's technical people with those of its marketing or operations people are difficult activities that consume a great deal of inventive energy.[14] There is also indirect quantitative evidence for the importance of these costs. Bresnahan and Greenstein (1995b) examined computer user's demand for a major new technology. Ito (1996) looked at major upgrades to existing computer systems. Brynjolfsson and Hitt (1996) considered the increases in sales per unit cost of companies that make new investments in computers. All these approaches reveal substantial costs of inventing the business side of computer applications.[15]

The last column in Table 2 offers a view of the rate of technical progress in the different portions of computer industry invention. As one reads down the column, the measured rates of technical progress fall. Hardware technical progress has been stunning, software technical progress has been rapid, and technical progress in co-invention has been slow. One implication of this variety is that the aggregate rate of technical progress in computing is slower than the rate of technical progress in hardware, dragged down by the large and slow co-invention sector.

The causes of the variety are important. In computing, technical progress in the applications sectors is slower than it is in the general purpose technologies. Technical progress in the general purpose technologies is difficult but has an excellent science and engineering base. Because of their generality and the growth in computer use, especially business computer use, general purpose technologies have huge markets that, partly as a function of available intellectual property protection tools, have provided strong if risky profits. Applications have a less well-developed science base, tending to draw on the business school's knowledge of organizations rather than the engineering school's knowledge of

---

[13]This is an example of the general problem of measuring innovation and productivity in services. For the computer-based production process, see Barras (1990).

[14]Friedman (1989) gives an interesting history of this.

[15]The substantial costs of co-invention have led some observers to doubt that companies' investments in computers and other information technology have been useful. This concern has led to a large and, in my judgment, hugely misguided literature on the "productivity paradox." Computers are useful. What is doubtful is the accuracy with which the output of the service sector is measured. Computers appear to be low productivity investments only when those measurements are assumed to be accurate.

circuits, bits, and bytes. The very nature of computer applications makes scale economies difficult to achieve. And bridging between technology and business purpose has proved conceptually difficult. The implications of the variety are also important. In this particular general purpose technology, the problem of innovation incentives does not arise in appropriability for the general purpose or generic technology. Instead, it arises in the applications sectors.

The Organization of Innovation

Managing the disparate technologies so that they work together has not been a trivial matter. Two main management structures have been deployed historically (see the right side of Table 1.) One is the *vertically integrated technology firm*, such as IBM. In it, a management structure is in place to coordinate the joint development of the many distinct technologies that make up computing. The other management structure is less centralized and explicitly coordinated. In the "Silicon Valley" form, distinct technologies are advanced by a wide number of different firms. Interface standards, cross-company communication, and markets have been used when supply is by a *group of vertically disintegrated specialty technology firms*.[16]

The emergence of applications software markets with independent "packaged" software vendors acting as suppliers was not merely a technological event.[17] It involved changes in industry structure and business models to be effective. Because software is a business with increasing returns to scale, the existence of a large number of computers on which the same program could run encouraged the emergence of software companies—first custom and then market. The invention of the computer platform by IBM in 1964 was a landmark event. Computers within the same platform have interchangeable components. Interchangeable components across computers of different sizes also permit growing buyers to use the same platform over time, avoiding losses on long-lived software. The invention of the platform and the creation of new platforms in minicomputing, personal computing, and so on improved the economics of software by permitting exploitation of scale economies. Some of these platforms are more "open" than others, so that control of the interface standards determining what software runs on the platform is spread out among many sellers, including "independent software vendors." The packaged software business that results is primarily American, with considerable invention and a large export market.[18]

---

[16]Although this form is named for the region that brought it to a high art, not all firms that participate in Silicon Valley innovation are located there. Many computer platforms with strong links to Silicon Valley nonetheless have key components supplied elsewhere in the United States (Washington State, Texas) or worldwide (Taiwan).

[17]See Steinmueller (1995) for a penetrating analysis.

[18]This stands in contrast to custom software, in which a great many sellers are local to particular countries. The importance of customer connections is the likely explanation. See Mowery (1996).

The trend so far has been to a more vertically disintegrated organization for innovation. This has had the considerable advantage of permitting specialization, including some international specialization. It has also increased the number of companies that can invent new and useful computer technologies. Yet the trend is by no means absolute. In the next section, we turn to the forces that have permitted the vertically integrated and vertically disintegrated systems of organization to exist in parallel for many years.

Co-invention has also been reorganized,[19] and much of co-invention has been shifted from end-user companies to commercialization or software companies. The latter trend parallels the vertical disintegration of invention and permits gains from specialization and scale economies.

This completes our tour of the different agents in the computer industry. The important message is that innovation in this industry is spread out over a wider range of economic agents than one might have imagined. This variety has been organized both by the IBM model— extended to influence over customers—and the Silicon Valley model.

DIFFERENT TYPES OF DEMAND

The variety in types of demand served by computing has had an important influence on the development of supplying firms and on markets. Several very different kinds of *demand* and *use* are important here. First is *business data processing in organizations*. Typically, the applications of computers in this domain involve changes in white-collar work across an organization. The demanders are senior managers seeking cost control or new ways to serve customers; they are supported by professionalized computer specialists. A second kind of demand is that for *business individual productivity applications* on PCs. More and more white-collar workers have seen their work at least partially computerized by these applications, but the span of the application tends to be a single worker. The third demand is for scientific, engineering, and other technical computation. Served by supercomputers, by minicomputers and later by workstations, and by PCs, factories, laboratories, and design centers do a tremendous volume of arithmetic. In total, the *technical computing* market size is roughly as large as each of the two kinds of commercial computing described above. Computer networks have changed technical computing through developments such as the Internet.

Computer networks have also changed business computing. *Interorganizational computing* links together firms or workers in distinct organizations. Applications such as networked commerce, electronic data interchange, and on-line

---

[19]Some of the organizational changes are shown in the Coordination Mechanisms column of Table 1. It is hard to say that this series of changes has much improved performance in co-invention. See Friedman (1989).

verification of a credit-card holder in a store, for example, are now the most rapidly growing part of computing.

This variety in demand has permitted the emergence of *different suppliers and markets*. More important, demand variety has permitted the emergence of new, entrepreneurial firms in parallel to established ones. In the next sections, we take up the distinct clusters of invention and co-invention that have arisen to serve the three oldest parts of demand and then turn to the question of American dominance of those very distinct clusters. The newest demand, interorganizational computing, involves both a blurring of the boundaries between the clusters and another set of reasons for American dominance, so current developments are treated in later sections.

ORGANIZATIONAL COMPUTING

The oldest part of the computer industry, mainframe computers, consolidated around a dominant firm (IBM) and a dominant platform in the 1960s, most likely because IBM had created the best combination of supplier firm, market organization, technical standards, and technical progress to support business data processing in organizations.[20] IBM emerged from an early competitive struggle to dominate supply, in the process determining the technologies needed for computing, the marketing capabilities needed to make computers commercially useful, and the management structures that could link technology and its use. Within organizational computing IBM managed both the cumulative and the disruptive/radical parts of technical change. Customers' learning by using and IBM engineers' learning by doing were focused on the same IBM computer architectures. IBM was not only the owner of established technology but also the innovator of the new.

From the mid-1960s onward, IBM dominated organizational computers with a single mainframe platform that began as the IBM System/360. This historical experience is important because it reveals some of the forces leading the computer industry to have only a very few platforms, even today. And the question of how open those platforms would be—that is, the extent to which a single firm would control and profit from direction of the platform—is timeless.

The key to IBM's invention of a platform was operating system compatibility across computers with different hardware. Because the same IBM software worked on all models, application software and databases on one system could be moved easily to another. IBM invented technical standards for how the products worked together and embedded them in its products. Further, the company had a

---

[20]The boundaries of the mainframe segment are not clear. Commercial minicomputers, which are not treated in this paper, eventually became much like mainframes. For international comparison purposes, the commercial minicomputer segment can be thought of as an extension of the mainframe segment.

field sales and service force to help users choose and configure computers and then make new, compatible, purchases when their use expanded.

Total investments by IBM and its customers in platform-compatible components were large, sunk, and platform-specific.[21] Those three features were then, and are now, powerful forces that limit the number of platforms that will serve any particular demand. "Platform-steering" vendors, such as IBM, create customer value as well as dominant positions for themselves.[22] IBM dominated organizational computing from its inception until the beginning of the 1990s, investing heavily in new technology, both hardware and software. When technical progress meant that existing IBM technologies were outdated, the company routinely abandoned its earlier investments in order to move forward. IBM's users decided, again and again, that it was better to stay with the established IBM platform than to switch to another. As a result, control of the direction of the platform and its standards remained completely centralized. Users did not have much in the way of competitive choice, but their investments were preserved. The inventive efforts of other participants in the platform—customers, user groups, and third-party providers of compatible components—made the platform better. A platform is a virtuous cycle of positive feedback, as more invention by customers, by third party providers, or by IBM encouraged further invention by the others. This was to all participants' advantage but notably to IBM's.

Economists often focus on the persistence of dominant firms in this industry. To understand the international industry structure, however, it is far more important to look at their origins. What was the origin of IBM's position, and what does it have to say about the sources of U.S. rents?

Despite substantial early enthusiasm about the potential for computers, there remained throughout the industrial world fundamental uncertainty on the technological development of the industry, the range of applications, and the potential size of the future market (Rosenberg, 1994). In particular, it was unclear whether the largest demand segments would be military, scientific and engineering, commercial, or something else altogether. These uncertainties in turn meant that the most important directions for technical progress as well as the nature of buyer-seller relationships and of commercialization efforts were unsettled.

As a result three distinct types of firms entered the early computer industry: office equipment producers, electronics firms, and new firms. Computers were a new electronics good that attracted several producers already active in other electronics fields. Similarly, some of the first applications of computers were in business, attracting firms with established connections to business data processing. This tension between technology-based and market-oriented firm organizations and competencies is ongoing (Davidow, 1986). The three groups of en-

---

[21]See Bresnahan and Greenstein (1995a) for more complete discussion and sources.

[22]Then and now, there has been a debate over the advantages of the customer value vs. the disadvantages of the dominant position.

trants had distinct capabilities and distinct strategies, but the capabilities and strategies of each group were similar in Europe and the United States (it was too early in the technological development of Japan for competition from that country to play much of a role). The electronics-based firms faced the challenge of either building or acquiring a business-equipment marketing capability—including a substantial field sales force—or finding a way to succeed without it. Firms with business equipment capabilities needed to add technological ones.

The combination of technical drive and customer focus required new management structures. IBM's success sprang from its major R&D investments and, more generally, its adherence to the Chandlerian three-pronged investment strategy (managerial capabilities, technology, and commercialization).[23] IBM rapidly became the world market leader because of its continuous R&D effort in developing new products, combined with advanced manufacturing capabilities, excellent marketing competence, and management structures keeping technology and market aligned. In Europe IBM's superiority in products and customer assistance was coupled with a local presence on the main markets. IBM Japan was for a long time first in revenues among "Japanese" computer companies. IBM used the "IBM World Trade" model, making itself everywhere as local a company as possible.

The Sources of U.S. Advantage in Organizational Computing

As the IBM experience illustrates, firm-level sources of national advantage have proved very important in organizational computing.[24] Of course, a wide variety of institutions and policies supported the emergence of IBM as the world market leader.

Universities

Universities in both the United States and Europe were active at the scientific and prototype levels before computers were commercialized.[25] In the United States, universities were less important for their scientific and engineering contributions that would be useful in computing than for their participation in computer projects for military and commercial sponsors. This created not only a body of knowledge but also a flow of trained people.[26] University research and university

---

[23]See Usselman (1993) and Chandler (1997).

[24]This is consistent with the simplest and most basic theory of international rent-steering in imperfectly competitive industries. If equilibrium in the industry leads only a few selling companies to earn rents, then the nations in which those companies are located will gain producer rents.

[25]This era is discussed in far greater detail in Bresnahan and Malerba (1997).

[26]See Flamm (1987) for a careful history of the American and European technology development efforts.

people did not provide the key competitive advantage to any country, but their absence would have created a bottleneck.

Government's Role

Substantial government backing for the early U.S. computer industry offered advantages to firms in the United States. Federal, especially military, research funding backed many of the purely technical capabilities needed to build computers (Flamm, 1987). Further, many early U.S. computer systems were themselves directly supported by federal funds. It was clear that the military was going to purchase many computers from domestic suppliers. All this encouraged development in the United States.

There is little support, however, for the view that the U.S. government "bought success" for IBM, and no support whatsoever for a "strategic trade policy" view of U.S. government actions (Bresnahan and Malerba, 1997). U.S. government actions were far removed from intentional strategic trade policy aimed at creating a "national champion." The ultimate national champion, IBM, was not an important part of the defense effort, nor was defense funding all that large a portion of IBM's commercial computing initiatives. The Defense Department spread a good deal of money around and let the supplying industry structure emerge in the marketplace.[27]

Antitrust Policy

U.S. antitrust policy worked actively to prevent IBM from emerging as the dominant firm in the fledgling computer industry. In particular, the U.S. Department of Justice systematically opposed IBM's strategy of strengthening commercialization and technical capabilities within the same firm. The department was not particularly anti-IBM nor anti-large-firm, but it did object to aspects of IBM's three-pronged strategy as anti-competitive. A 1956 consent decree between IBM and the government limited IBM's effectiveness as a commercialization company. A second antitrust lawsuit, brought in 1969 and contested for more than a decade, viewed IBM's service, sales, and support efforts as anticompetitive lock-in devices. The legislative branch also tilted procurement policy against IBM.[28] The point here, again, is that the industry structure leading to U.S. dominance was not the invention of the government.

---

[27]Usselman (1993) offers a very interesting argument that U.S. procurement policy favored IBM only because it took this form. IBM would not likely have been chosen as the national champion in the critical early phases, nor would a "supply side" procurement policy have led to the development of the IBM commercialization capabilities.

[28]There is an active debate on whether government agencies were able to evade the law and procure from IBM as they saw fit. See Greenstein (1993). There is no debate about the anti-IBM policy itself.

Albeit mostly by accident, antitrust policy was helpful in creating an independent software sector. The partial unbundling of IBM platforms meant that firms other than IBM could sell into IBM-using sites more easily. At the same time, the United States has had comparatively strong intellectual property protection for computer hardware and software. Countries with weaker protection have been less successful in software supply.

Intentionally in this case, the same antitrust policies led to a multicompany basis for the U.S. hardware sector, at least in the peripherals sector. The emergence of mass-storage device sellers (PCMs) other than IBM undercut IBM's ability to price discriminate using storage as a metering device. And it increased the variety and flexibility of hardware supply in the United States.

Competition from Other Countries

During the long period of IBM's hegemony, other companies sought a share of the worldwide market for organizational computing. In Europe and Japan, governments used trade and procurement policy to protect their weak domestic firms from IBM. These governments also influenced large quasi-governmental buyers in European and Japanese markets, principally banks and telephone companies, to buy domestic computers rather than IBM imports. In Europe this protectionism was no more than a barrier to exit, slowing the ultimate decline of European firms.

Japanese firms and the Japanese government came closer to creating an effective barrier to IBM's dominance.[29] Instead of a national champion policy, the best of half a dozen competing members of government-sponsored consortia would receive government support. The policy was technologically flexible. After some initial failures at making a purely Japanese computer, IBM compatibility became the focus. The effects of the consortia were to build a very substantial hardware technological capability within some Japanese firms, partially catching up to IBM. Developments on the software side were far weaker.

TECHNICAL AND PERSONAL COMPUTING: DISTINCT MODELS SUCCEED IN NEW SEGMENTS

A second set of U.S. computing successes served different markets and drew on different national capabilities and institutions. Although these two new areas, minicomputers and microcomputers, shared some fundamental technical advances with the existing mainframe segment, considerable innovation and entry characterized each new segment's founding. The two segments, minicomputers and microcomputers, developed new markets and organizational structures. The

---

[29]This very abbreviated treatment draws heavily on Anchordoguy (1989), Fransman (1995), and Bresnahan and Malerba (1997). See the last reference for a discussion of Japanese near-success.

"technologies," as engineers use that term, of each of these segments were distinct. These new kinds of computers served new kinds of demand. Accordingly, co-invention and commercialization in these markets were distinct. A very different form of firm and market organization characterized supply in the new segments, and supply was from new firms. With distinct invention, co-invention, and markets, these new segments formed new clusters of positive feedback and changed the face of competition within the industry.

Technical Computing

The first of these new segments was *minicomputers*, machines that from their start in the late 1950s were intended for scientific and engineering use.[30] Most of the firms in technical computing were entrepreneurial start-ups, many with their origins in universities. DEC, the largest of these start-ups with about one-third of minicomputer sales over many years, had its origins at M.I.T.'s Lincoln Laboratory. Other firms originated in the instrument business. Hewlett-Packard came out of both fields.

A series of reasons led this segment to be served by a distinct cluster. The appropriate seller commercialization model for manufacturing and scientific minicomputers was built on the fact that the relevant buyers were technically fluent. Software support came not from minicomputer producers but from "third parties," notably value-added resellers (VARs) and consultants, or from the end users themselves. The technical computing segments were focused on raw technology, low commercialization cost, and dramatic progress in computer price/performance ratios. The sources of national advantage for the United States came from universities and from the cluster of financial and other institutions supporting entrepreneurship. After a time, the minicomputer business itself became one of the supporting institutions, as later entrants were often spin-offs from existing minicomputer firms.

Personal Computing

The second new segment was the personal computer, which developed starting in the mid-1970s in the United States on the basis of hobbyist demand. Early entrants resembled those in minicomputers—established electronics (but not computer) firms and de novo entrants. Entrepreneurs entered not only computers, but also other hardware components and software. Positive feedback among these different entrepreneurs led not only to their success, but also to a pronounced regional advantage for the western United States, and the available support institutions for entrepreneurship in that region, such as venture capital, both rein-

---

[30]The boundary between organizational and technical computing is unclear. Early minicomputers and later supercomputers and workstations were not in competition with IBM, however.

forced and were reinforced by the new cluster of entrepreneurs.[31] The original location of personal computer suppliers in the U.S. was encouraged by forces similar to those in technical computing but quite dissimilar from those in organizational computing.

Few of those original firms in the PC business survive, and fewer still continue to earn rents. Yet the supplier rents in the PC business remain in the United States. What were the key competitive forces that permitted the replacement of early success stories yet left the rents within the same country?

Competition in Personal Computing

By the mid-1980s the structure of supply and demand in personal computing had unleashed three powerful new forces that affected competition.[32] These forces were unanticipated, so they presented very considerable opportunities for profit to the firms that first understood them. First, vertical disintegration of supply meant *divided technical leadership*. Second, large unit sales increased the importance of *scale economies*. Third, suppliers were capable of, and demanders of PCs were eager for, *speed-based competition*. These forces changed the way long-term features of computer market equilibrium, such as network externalities and positive feedback, played out in the PC market. These changes would matter not only in PC but also in networked computing. The technological leadership determining the direction of technical advance of the PC came to be divided among four distinct sectors of PC computing:

1) Makers of computers, of which IBM was the largest and most influential;
2) Intel, leading maker of the microprocessors in the PCs;
3) Microsoft, maker of the dominant operating system for PCs; and
4) Applications software makers such as Lotus, WordPerfect, and Ashton-Tate.

This divided technical leadership is striking for two reasons. First, it was remarkably effective at advancing the PC platform. Divided technical leadership meant rapid advance from specialists and the ability to take advantage of external economies among sellers and users (Langlois, 1990). Second, divided technical leadership is remarkably competitive. Each of the firms named above achieved, at least for a time, a position of dominance or even of near-monopoly in its pri-

---

[31]The early 1980s saw the aging of the original personal computer platforms and a standard-setting opportunity. IBM introduced the IBM PC and, for a while, controlled the new standard. IBM entered through external linkages with competent firms and with an open architecture. Although the PC was a product from long-established IBM, the majority of IBM's complementors were entrepreneurial, many university-based. Thus the invention of the IBM PC did not convert the personal computer business into one like the mainframe business.

[32]Several of these were the result of IBM's decisions at the time of its entry into the PC market. Yet nobody saw the competitive consequences at that time.

mary product. Yet each felt, or was substantially harmed by, competition from producers of complements. This new vertical competition—competition in which producers of complements attempt to steal one anothers' rents—is an important feature of PC markets and is likely to characterize all future computer markets. In most industries, competition comes from horizontal directions, that is, from firms selling substitutes.[33]

Vertical competition can be powerful, as demonstrated by the transition from the "IBM PC" (control of the PC business by IBM) to the "Wintel" (control by Microsoft and Intel). Attempts by Lotus and WordPerfect to earn operating system rents in their applications programs, spreadsheets and word processors, were vertically competitive initiatives. Operating system vendor Microsoft also engaged in vertical competition and succeeded in taking most of the rents of the IBM, Lotus, and WordPerfect products.[34] Microprocessor manufacturer Intel has destroyed the economic basis for many board-level products by including their functionality in new versions of the computer's "brain." There are many other examples in which firms in the PC industry whose products are complements in the short run are in competition for the same rents in the long run. The important elements of this vertical competition are standards stealing, time-based competition, and racing for rents.

Why is there vertical competition? First, vertical disintegration provides a source of competitors that is not available when a vertically integrated firm supplies the complements. Second, boundaries between vertical product segments are inherently malleable and thus subject to manipulation. Most important, the control of key interfaces directs the flow of producer rents. So producers compete for control of future boundaries between their products and thus future rent flows. The newfound importance of *scale economies* in PCs underscores this vertical competition. Unit sales in PC markets are very large by the standards of computer markets generally. Another force, less often mentioned but equally important, is the importance of a few key applications in PC use.

The final novelty in PC competition is perhaps its defining character—speed-based competition. The de facto standard-setting process favors early firms in the market for several reasons. First, once customers have made their co-investments, the standards in use tend to persist. Thus there is first-mover advantage, a substantial motivator for races. Races are particularly likely to occur when there are new opportunities, such as the Internet. Incentives to improve products, and to do it quickly, are very high.

---

[33]The most important example of horizontal competitive innovation in PCs came from IBM's entry, which destroyed the rents of preexisting sellers of CP/M computers and software.

[34]See Breuhan (1997) for an analysis of the transition from Microsoft DOS to Microsoft Windows as an example of vertical competition that freed customers from their lock-in to applications programs.

Second, PC customers are quite accepting of products rushed into the market. Personal computers are overwhelmingly used for applications that are not mission critical, a fact that removes much of the risk of system crashes. Customers will use partially tested and buggy new systems to gain access to new features or new performance. Researches show that successful software companies forgo quality to speed the time it takes to get the product to market.[35] Beta testing, once a long and carefully contracted process that linked a few lead customers to a vendor, is now a marketing tool, a way to get software in the hands of customers. Speed is king, in large part because customers tolerate change that would be "too fast" in other environments.

Vertical disintegration and divided technical leadership permit very rapid technical progress. Experts push each technology. Divided technical leadership that becomes vertical competition not only permits speed, it also forces a wide variety of competitive races. Enough desire for speed, in turn, demands specialization. Firms cannot master all the distinct technologies they need to bring new platforms, new standards, and so forth to market quickly. Thus speed and disintegration feed back to one another. As a result, for all the competitiveness of its structure, the PC business has had little difficulty in providing economic incentives for technical progress. If anything, the transition to a more competitive structure, notably a vertically competitive one, has led to a transition to an even faster pace of technical progress.

NEW SOURCES OF U.S. ADVANTAGE IN THE NEW SEGMENTS

The technical and personal computing segments of the American industry quickly dominated worldwide competition, but their sources of competitive advantage were different from those that applied in the mainframe computer age. Venture capital played a major role in supporting the entrepreneurial firms' entry and growth in both mini- and microcomputing, while universities played a new role as sources of scientific knowledge and entrepreneurship. Only Cambridge in the United Kingdom has played a role in Europe similar to, albeit weaker than, that played by M.I.T. for minicomputers and by Stanford and the University of Texas for microcomputers and workstations. Industry-specific government policies did not play a major role, while more general policies favoring education and skill development helped market development.

In PCs, the presence of strong complementarities and local knowledge externalities gave major international advantages to the United States or, more precisely, to Silicon Valley, where several firms were at the frontier in each market layer. Intense formal and informal communication and highly mobile personnel, together with the high entry and growth rates already present, exposed these firms early on to new experiments, knowledge, and technologies. These external econo-

---

[35]See Barr and Tessler (1997).

mies gave American firms major innovative advantages over competitors located elsewhere in the world.

In Europe few new mini- or microcomputer firms entered the industry, for several reasons. In both segments, American producers had first-mover advantage and rapidly took over the European market. Only limited spin-off from European universities took place, while badly developed venture capital markets limited financial support for new ventures. The protectionist measures, such as public procurement, that European governments used for mainframes could not be extended to the new markets.

In Japan the PC industry was focused on the local market; in a time of worldwide standards this local focus resulted in a fragmented market specific to Japanese needs. As a result, Japanese PC hardware exports were small, and PC software exports were near zero. The Japanese industry was thus unable to participate in the worldwide scale economies and substantial external economies associated with microcomputers. As a result, both European and Japanese suppliers were largely irrelevant in the world minicomputer and PC markets. The exceptions to these general observations serve mostly to underscore the analytical lessons. UK start-ups have had some success in niche hardware markets (handheld computers, for example) but have not been effective competitors in worldwide markets.

NETWORKED COMPUTING AND CONVERGENCE

The structure of the overall computer industry has changed dramatically in the 1990s. The segments that had once been separate are converging, bringing firms that were once separate U.S. successes into competition with each other. Fueled by advances in computer networking, convergence has permitted networks of personal computers and workstations to compete with minicomputers and mainframes.[36] Convergence has also enabled vertical competition between sellers in the previously separate segments, as their products are linked together in the same networks. Firms and technologies from personal computing now supply "clients," which are networked to products and technologies from organizational computing, now called "servers." Finally, convergence has led to the development of new technologies and new applications, so it has created important entry opportunities. All three of these changes are competitive, and all have led to reallocations of producer rents.

An important source of demand for networked computing in the early 1990s was as a new technology in organizational computing. Much of this involved replacing mainframes or commercial minicomputers with servers from technical computing. Organizational computing users, however, wanted their separate computing systems linked and also wanted to retain the positive features of personal computing, such as ease of use, and organizational computing, such as power and

---

[36]See Bresnahan and Greenstein (1995a) for an analysis of this horizontal competition.

the embodiment of business rules and procedures in computer systems. Accordingly, networked computing used hardware and software technologies from personal and organizational computing as complements.

Networked computer systems are highly complex and rich in opportunities in all their various components and dimensions. No single firm could innovate in all parts and subsystems. As a result, network computing has attracted a flood of new specialized entrants: these include technology-based spin-offs from established computer firms, science-based firms established by university scientists, and new firms with market or commercialization competencies. Consider Sun Microsystems, a strong workstation firm that has been a leader in converting workstations, a tool for engineers, into servers. In the process Sun is making a serious attempt to be a standard-setter in commercial computing. The point is that the entry wave in networked commercial computing in the 1990s is stunning in both the variety of sources of entrants and in the variety of offerings the entrants bring to the table.

Networked computing has also led to the very rapid growth of another class of computer applications, *interorganizational computing*. This includes electronic commerce, management of supply chains, electronic data interchange, and a host of other technologies and markets. Interorganizational computing is not new. Yet the opportunities for developing new interorganizational computing applications that stem from use of Internet and other networking technologies are substantial. At this writing, there is tremendous uncertainty about the nature of applications in interorganizational computing. Networked computing is changing rapidly, and it is changing in an unpredictable, constantly evolving direction.

Competition in Networked Computing

The divided technical leadership of networked computing, the uncertainty about invention in it and co-invention in interorganizational computing, and the likely large size of these new segments are a recipe for vertical competition. The rents associated with the control of future standards and technologies will be large, so there is a very large return to moving to control them now.

A stable market structure for networked computing has not yet emerged. Connectivity and compatibility have led to vertically disintegrated supply. Technical change is following a variety of directions with a rise in the number of potential technologies associated with the relevant platforms. Interdependencies and network externalities have increased. New entrants have pioneered many of these technologies. Firms are heterogeneous in terms of size and specialization, activity in various platform components, strategies, and modes of commercialization.

In 1998, neither the dominant design for a network of computers nor for a computer company in this environment is clear. Much of this lack of clarity stems, as it did in the past, from difficulties in forecasting the highly valuable uses of networked computing. Vertical competition seems certain to be a permanent feature of this platform, but the relative strengths of client-based and server-based strategies are highly uncertain. The continued rents of suppliers depend on the emerging structure of the computer industry. There will be technologies whose sellers earn rents and forms of organization of computer companies that deliver rents as well. Because of the competitive disruptions of the 1990s, however, the old rent-generating structures and technologies are threatened, and their replacements are not yet obvious. The market will now select from a wide variety of technological, company and industry structure, and commercialization initiatives. Because the uncertainty about the direction of technical progress is so large, it would be unwise for any particular firm to give up simply because it is behind in the race. Accordingly, a large variety of interesting racing initiatives are taking place throughout networked computing.

One of the more interesting races is for control of network interfaces. Those firms with strong positions on the server end of the business, such as Sun, Oracle, and IBM, have attempted various strategies to extend their control over clients, such as NC and Java, or to render them less influential. Microsoft, the firm with a strong position on the client side has defended itself against all comers, such as Browser, Java, and NC, while attempting to extend its control into the server side. As a result of all this maneuvering, there is widespread speculation that one or a few of the firms controlling key interfaces for connecting modular products will come to dominate networked computing, but no single firm has so far been able to govern change and coordinate platform standards. Clearly vertical integration will increase in the next few years, but substantial vertical competition will also continue. Thus the emergence of a new networked computing platform makes it possible for the U.S.-based firms to strike out for new rents, by innovating to compete for them. It by no means guarantees a position for any firm.

These standards races are struggles not only between distinct firms but also between distinct technologies. As a result, the technological basis of the future computer industry is difficult to forecast. Although this uncertainty makes it hard to predict which *firms* will earn the rents from networked computing, there is little doubt about which *country* will. Almost all the major initiatives to control the new industry rents are based in the United States.

Invention Incentives

Whether competition has changed the computer industry's R&D incentives is a question that cannot be answered quantitatively, for two distinct reasons. First, the boundary between R&D and other activities in the industry is very difficult to draw and thus to measure. The second reason is the dramatic increase in market opportunity facing the industry. The market size for stand-alone computers grew shockingly rapidly in the 1980s, as the personal computer became a far more successful product than anticipated. The market size for networked computing is doing the same in the 1990s, as the Internet is far more valuable as a commercial technology than anyone had anticipated. As a result the total amount of invention is rocketing upward, but it is very hard to determine how much should be attributed to demand-pull and how much to changes in supply, such as competition. Of course, in the long run supply, by opening up the new markets, has unleashed the demand forces. It still remains difficult to say what fraction of the increase in total invention should be attributed to competitive forces.

That portion must be considerable, however. Although the events of the 1990s have removed a great deal of monopoly power from the computer industry, they have not destroyed the return to innovation. Far from it. Now racing against competitors is the incentive to innovate, and it is a powerful and effective mechanism. The vast majority of computer industry competition is technological competition. Price competition might have destroyed the return to innovation; competition whose mechanism is constant racing to gain the next monopoly does not.

U.S. ADVANTAGE LIKELY TO CONTINUE

The uncertainties of networked computing notwithstanding, the United States seems likely to continue to dominate the worldwide computer industry. As computer hardware components and then entire systems became more and more divorced from the rent-generating software such as operating systems, it also became eligible for production at the worldwide cost-minimizing location. Accordingly, there has been a major reallocation of the industry's production of hardware—devices, components, and systems—out of the United States, notably to Asia. This has not posed a challenge to the continued dominance of the rent-generating segments within the United States.[37]

Many governments, notably in Europe, look at networked computing as being about the convergence of computing and telecommunications. Moreover, they are attracted to the idea of a top-down telecommunications-style regulation to direct the rents to their own national champions. This strategy will have some advantages, such as inducing markets to converge to unified and controlled standards more quickly than they would otherwise. In large part U.S. policy is the opposite, including as many alternatives for telephony and computing as possible and waiting for the market to select the winners. Everything in the recent development of computing suggests that the market will ultimately favor the U.S. approach over the European one.

Commercialization will, as always, play a large role in determining the ultimate technological and industry structure. The commercialization mechanisms by which this will occur are not at all clear at this time, as different firms use

---

[37]See Kraemer and Dedrick (1998) for a fuller treatment of this reallocation.

radically different commercialization strategies.[38] More nearly pure commercialization companies have opened a new international competitive front. Applications software companies for organizational computing, systems integrators, and customs software houses are all worth mentioning here. Many of these firms are American, but there is a healthy international supply in this area.[39] Commercialization, unlike technology, has a strongly local flavor. Many of the important new initiatives serve very specific bodies of customers, specific both to country and to industry. Finally, the uncertainty about applications in a networked environment has meant that there are entry opportunities in this area.

Coming all at once, change in the nature of competition, technologies, and relationships to customers as well as change in firm and industry structure has left traditional management doctrine out of date. Vertical disintegration implies the need to manage alliances, which not many firms know how to do.[40] The transition to effective speedy decision making has also been a difficult one for many companies.[41] There are many other contemporary examples of transition in management doctrine. This will play to the advantage of companies and regions that can experiment and change.

CONCLUSION

Useful business computer systems are complex. They draw upon a wide range of distinct technologies, each of which itself is advancing. Most of the technologies considered "technical"—microprocessors, networking equipment, and systems software, among them—advance rapidly. The less "technical" technologies—the organization of white-collar work and electronic commerce, for example—advance more slowly and are difficult to predict. These very different technologies are in a relationship of innovational complementarity; new kinds of technical capability, such as networked computing, are not much use without invention of new ways of organizing business, new ways of providing service to customers, or other "soft" technologies. The content of the technologies that makes up this system is variegated; microprocessors and advertising are not typically understood in the same way or by the same people. Moreover, seller inno-

---

[38]Some firms, such as IBM and Oracle, use the bilateral-customer-ties structure for commercialization. Using field sales forces, people-based support structures, and so on, these firms are better with larger customers. Others firms remain close to the PC market model, with only very distant connections to individual customers. This is the marketing and commercialization model of Microsoft, for example, which lacks the organizational capability to engage large customers in bilateral relationships. Like SAP, Peoplesoft, and others, they use commercialization specialists such as systems integration houses, consultants, and custom software firms.

[39]Cap Gemini Sogeti is a successful European systems integrator, while SAP and Baan are successful applications software houses. The service and sales forces of large computer companies, whether U.S. (DEC) or not, are important potential entrants here.

[40]See Eisenhardt and Schoonhoven (1996).

[41]See Brown and Eisenhardt (1997).

vation enables creation of new, but difficult to foresee, applications categories. Users' co-invention then takes time and cleverness. The overall innovation system is, as a result, extremely complex and unpredictable.

This means that market and firm organization are important in computing, especially for commercialization. Seller rents in computing have not gone to those companies or countries with purely technological capabilities. Instead, the sellers who have flourished are those who can align their own technological efforts with market needs and who can take advantage of the leverage implied by users' co-invention. The national institutions and infrastructure that support the development of firms and markets are as important as those supporting technology in explaining U.S. dominance.

A key element of commercialization in all regimes has been the use of the computer platform and associated compatibility standards. Platforms channel seller innovation; backward compatibility means that seller innovation does not outrun user needs. An important result of the use of platforms to organize innovation has been *punctuated equilibrium*. Once a platform standard is in place, technical progress within it tends to be rapid, mutually reinforcing, and focused on immediate market needs. Seller positions tend to be stable. Yet existing platforms do not always serve new needs, as the moves to PCs and networked computing demonstrated. The creation of new platforms is fundamentally disruptive and permits much seller entry. Rents are mobile.

The 1990s have seen three linked changes in computer industry structure and the workings of competition. The process of vertical disintegration, which had been historically confined to making each new market segment less integrated than the last, spread to all the segments. The locus of rent generation shifted downstream to software and applications developments. Computer hardware itself became more of a commodity. Finally, networked computing has brought a very wide list of old and new firms and technologies into a complex web of complementarities and competitive rivalries. Vertical competition appears to have become a permanent feature of the industry.

With all that change, it is natural to ask what has led to the long persistence of U.S. dominance in the industry. Some factors favoring American competitiveness *persisted over time*. First among these is the large size and rapid growth of the American market. Some of the growth is related to the U.S. macroeconomy; the rest is related to education in computer technologies and a highly skilled labor force in information technology. U.S. tax, antitrust, and legal policy has not been supportive of computing, but it has not been dangerously hostile either. U.S. universities, always a source of entrepreneurship, have been highly receptive to the launching of new scientific fields and academic curricula. Finally, there is the tendency for dominant firms and technologies to persist for a long time within the industry's established segments.

Other sources of American competitive advantages have been *changing over time*. In mainframes, for example, the major sources of American advantages were linked to *a single firm's advantages;* IBM presented a unique commitment to R&D policies and to the Chandlerian three-pronged investments in management, production, and marketing. No other firm in the world was able to match IBM's capabilities and investments. In mini- and microcomputers, U.S. advantages were related to *favorable entry and growth conditions* for new firms in new market segments and to the creation of open multifirm platforms that created local knowledge externalities. In computer networks, U.S. advantages are related to the presence of *local knowledge externalities* and strong complementarities between various components of the multifirm standard platform. The creation of each of these new segments involved very substantial entry opportunities for new firms.

Some of these advantages were *transmitted from segment to segment.* For example, the success of venture capital in supporting early computing entrepreneurs as well as other microelectronics and unrelated ventures led to the availability of abundant venture capital in microcomputers and computer networks. Moreover, some of the entrepreneurs important in founding new segments came from established U.S. computer firms. These are weak transmission links. A stronger link was the technologies that network computing drew from established U.S. firms in the already existing segments.

The geographic location of the competencies supporting American success has several times shifted within that large country. In mainframes, American advantages were related to the areas of IBM location of R&D and production, centered in New York but widely dispersed. For minicomputers, the sources of competitive advantage were mainly centered in the eastern part of the United States, with important exceptions such as Hewlett Packard. In microcomputing, and even more so in computer networks, there has been a regional shift from areas in the eastern part of the United States westward toward Silicon Valley. This shift implies the need to consider carefully the unit of analysis of competitive advantages—the division or department, the firm, the region, or the country (Saxenian, 1994). The United States is a large country; as one company or region declined, another grew.

Perhaps the most important advantage, however, has been the flexibility of the U.S. computing industry—its ability to abandon old competencies in favor of new ones. As an example, consider the decline of the centralized vertically integrated large firm in the 1990s. Changing market conditions meant that a new kind of firm was more likely to be successful. With no barriers to exit, the previously highly successful IBM model, not to mention the highly successful IBM, declined. This flexibility and variety has been the hallmark of the U.S. national innovation system. At each critical turn, when large rents were to be earned by an unknown form of computer firm and an unknown technology, the United States has brought forth a wide variety of distinct initiatives. Thus the United States has maintained its leading position not by protecting the old but by seizing the new.

BIBLIOGRAPHY

Anchordoguy, M. (1989). *Computers Inc.: Japan's Challenge to IBM*, Harvard University: Harvard University Press, Cambridge (MA) and London.

Barr, A., and S. Tessler. (1997). *A Pilot Survey of Software Product Management*, in the Proceedings of the Software Engineering Process Group '97 Conference, San Jose, CA.

Barras, R. (1990). "Interactive Innovation in Financial and Business Services," *Research Policy* 19(3):215-237.

Bresnahan, T., and S. Greenstein. (1995a). "Technological Competition and the Structure of the Computer Industry," Working Paper 315, CEPR, Stanford University, Stanford.

Bresnahan, T., and S. Greenstein. (1995b). "The Competitive Crash in Large-Scale Commercial Computing," in *Growth & Development: The Economics of the 21st Century*, R. Landau, N. Rosenberg and T. Taylor, eds. Stanford University Press.

Bresnahan, T., and F. Malerba. (1997). "Industrial Dynamics and the Evolution of Firms' and Nations' Competitive Capabilities in the World Computer Industry," forthcoming in *The Sources of Industrial Leadership*, D. Mowery and R. Nelson, eds. Cambridge University Press.

Bresnahan, T., and M. Trajtenberg. (1995). "General Purpose Technologies: 'Engines of Growth'?" *Journal of Econometrics* 65(1):83-108.

Breuhan, A. (1997). *Innovation and the Persistence of Technological Lock-In*, Ph.D. Dissertation, Stanford University.

Brown, S.L., and K.M. Eisenhardt. (1997). "The Art of Continuous Change: Linking Complexity Theory and Time-paced Evolution in Relentlessly Shifting Organizations," *Administrative Science Quarterly* 42:1-34.

Brynjolfsson, E., and L. Hitt. (1996). "Paradox Lost? Firm-Level Evidence on the Returns to Systems Spending," *Management Science* 42(4):541-558.

Chandler, A.P. (1997). "The Computer Industry: The First Half-Century," in *Competing in the Age of Digital Convergence*, D. Yoffie, ed. Harvard Business School Press.

Christensen, C.M. (1993). "The Rigid Disk Drive Industry: A History of Commercial and Technological Turbulence," *Business History Review* 67(4):531-588.

Davidow, W.H. (1986). *Marketing High Technology: An Insider's View*, New York: Free Press, London: Collier Macmillan.

Dulberger, E.R. (1993). "Sources of Price Decline in Computer Processors: Selected Electronic Components," in *Price Measurements and their Uses*, M. Foss, M. Manser, A. Young, eds. Chicago: University of Chicago Press.

Eisenhardt, K.M., and C.B. Schoonhoven. (1996). "Resource-based View of Strategic Alliance Formation: Strategic and Social Explanations in Entrepreneurial Firms," *Organization Science* 7:136-150.

Flamm, K. (1987). *Targeting the Computer: Government Support and International Competition*, Washington, DC: The Brookings Institution.

Fransman, M. (1995). *Japan's Computer and Communication Industry*, Oxford: Oxford University Press.

Friedman, A.L. (1989). *Computer Systems Development: History, Organization, and Implementation,* New York: Wiley.

Gandal, N. (1994). "Hedonic Price Indexes for Spreadsheets and an Empirical Test for Network Externalities," *RAND Journal of Economics* 25(1):160-171.

Gordon, R.J. (1989). "The Postwar Evolution of Computer Prices," in *Technology and Capital Formation*, D. W. Jorgenson and R. Landau, eds. MIT Press.

Greenstein, S. (1993). "Did Installed Base Give an Incumbent any (Measurable) Advantages in Federal Computer Procurement?" *RAND Journal of Economics* 24(1):19-39.

Grove, A.S. (1996). *Only the Paranoid Survive: How to Exploit the Crisis Points that Challenge Every Career and Company*, New York: Doubleday.

Henderson, R. (1993). "Underinvestment and Incompetence as Responses to Radical Innovation," *RAND Journal of Economics* 24(2):248-270.

Hodges, D. and R. Leachman. (1995). "Benchmarking Semiconductor Manufacturing," Berkeley Competitive Semiconductor Manufacturing Progress CSM-27.

Iansiti, M. (1995). "Technology Integration: Managing Technological Evolution in a Complex Environment," *Research Policy* 24(4):521-542.

IDC. (1996). International Data Corporation *Worldwide IT Industry Survey*. Framingham, MA: IDC Press.

Ito, H. (1996). "Essays on Investment Adjustment Costs," Ph.D. Dissertation/Master's Thesis, Department of Economics, Stanford University.

Kraemer, K., and J. Dedrick. (1998). *The Asian Computer Challenge*. Oxford University Press.

Langlois, R. (1990). "Creating External Capabilities: Innovation and Vertical Disintegration in the Microcomputer Industry," *Business and Economic History* 19:93-102.

Langlois, R., and E. Steinmueller. (1997). "The Evolution of Competitive Advantage in the Worldwide Semiconductor Industry, 1947-1996" forthcoming in D. Mowery and R. Nelson, eds. *The Sources of Industrial Leadership,* Cambridge University Press

Malerba, F. (1985). *The Semiconductor Business*, Madison: University of Wisconsin Press.

Mowery, D.C. (1996). *The International Computer Software Industry: Evolution and Structure*, New York: Oxford University Press.

Nelson, R. (1992). *U.S. Technological Leadership: Where Did It Come From and Where Did It Go?* Ann Arbor: University of Michigan Press.

Nelson, R. (1993). *National Innovation Systems: A Comparative Analysis*, New York: Oxford University Press.

OECD. (1989). *The Internationalisation of Software and Computer Services*, Organization for Economic Cooperation and Development.

Rosenberg, N. (1994). *Uncertainty and Technological Change*, Stanford: Stanford University Press.

Saxenian, A. (1994). *Regional Advantage: Culture and Competition in Silicon Valley and Route 128*, Cambridge, MA: Harvard University Press.

Steinmueller, E. (1995), *The U.S. Software Industry: An Analysis and Interpretative History*.

Usselman, S. (1993). "IBM and its Imitators: Organizational Capabilities and the Emergence of the International Computer Industry," mimeo, University of North Carolina, Charlotte.

Semiconductors[1]

JEFFREY T. MACHER
DAVID C. MOWERY
DAVID A. HODGES
University of California, Berkeley

INTRODUCTION

Often called the "crude oil of the information age," semiconductors are the basic building blocks of many electronics industries. Declines in the price/performance ratio of semiconductor components have propelled their adoption in an ever-expanding array of applications and have supported the rapid diffusion of products utilizing them. Semiconductors have accelerated the development and productivity of industries as diverse as telecommunications, automobiles, and military systems. Semiconductor technology has increased the variety of products offered in industries such as consumer electronics, personal communications, and home appliances.

Global production of semiconductor components grew from roughly $19 billion to $137 billion (in 1997 dollars) during 1980-1997, an annual growth rate of more than 12 percent (see Figure 1).[2] Nevertheless, the U.S. semiconductor industry, which had pioneered the commercial development of this technology,

[1]The research on which this paper is based was supported by the Alfred P. Sloan Foundation. We are indebted to our fellow participants in the Berkeley Competitive Semiconductor Manufacturing Research Program and especially to its director, Professor Rob Leachman, for invaluable data, advice, and support. We also appreciate the assistance of Jerry Karls and Howard Dicken of Integrated Circuit Engineering, Inc., Doug Andrey and Lynn Lehsten of the Semiconductor Industry Association, Dan Hutcheson of VLSI Research, Inc., and Jodi Shelton and Debra Scoggin of the Fabless Semiconductor Association in providing data for this paper.

This paper has benefited greatly from the comments of Melissa Appleyard, Rose Marie Ham, and Bill Spencer. The authors are solely responsible for any errors or omissions.

[2]Market share data presented in this paper represent the dollar amount of billings as reported by member semiconductor firms to World Semiconductor Trade Statistics (WSTS), Inc.

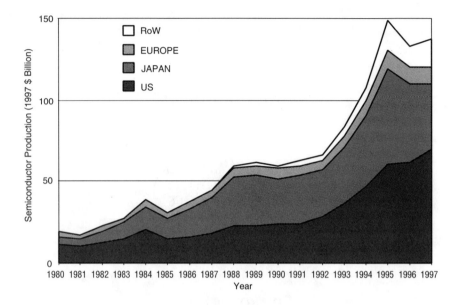

FIGURE 1 Worldwide semiconductor production, 1980-1997.
Source: SIA 1997 Annual Databook; ICE Status: A Report on the Integrated Circuit Industry, 1980-1998.

experienced wide swings in its competitive performance, especially relative to that of Japan, during this period. The 1980s opened with an expanding Japanese presence in memory components, products widely viewed as essential "technology drivers" for advances in semiconductor manufacturing processes. U.S. firms steadily lost market share to Japanese firms in memory components during the 1980s, and Intel, now among the most profitable semiconductor manufacturers in the world, nearly collapsed in the 1984-1985 industry recession. At the end of the decade, the M.I.T. Commission on Industrial Productivity (1989) suggested that:

> The traditional structure and institutions of the U.S. [semiconductor] industry appear to be inappropriate for meeting the challenge of the much stronger and better-organized Japanese competition.... The technological edge that once enabled innovative American companies to excel despite their lack of financial and market clout has disappeared, and the Japanese have gained the lead.

By 1989, however, this dismal picture had begun to brighten, and the market position and profitability of U.S. firms have since improved, especially relative to that of Japanese firms. Stronger U.S. performance is revealed in gains in global market share that rest in part on improvements in product quality and manufacturing process yields. Improved performance also reflects the withdrawal by most U.S. firms from the fiercely competitive DRAM segment of the semiconductor industry. During and after the late 1980s, U.S. firms shifted to logic and microcomponent products,[3] where foreign competition was less intense and they could pursue new product opportunities, many of which drew on their proximity to developers of computer software and other complementary products. Japanese firms, facing less progressive domestic computer hardware and software industries, were less successful in product innovation. Other U.S. firms, such as the so-called "fabless" semiconductor firms,[4] have entered the industry successfully as specialists in innovative device designs. Meanwhile, the Japanese firms that dominated DRAMs now face a domestic recession and entry by South Korean and Taiwanese firms with low costs and high manufacturing productivity. Entry by non-Japanese semiconductor manufacturers also has expanded export markets for U.S. producers of semiconductor equipment. Although the Asian economic crisis that began in 1997 is likely to depress global demand for semiconductors in the near term and erodes the financial performance of U.S. producers, the relative performance of U.S. semiconductor firms remains strong, as their continuing leadership in global market share indicates.

Much of the "renaissance" of U.S. competitive advantage in semiconductors thus reflects exploitation by U.S. firms of long-standing strengths in product innovation. Many of the new opportunities that appeared in the late 1980s for such product innovation reflected developments in other industries such as telecommunications and computers, in which U.S. firms demonstrated renewed innovative and competitive vigor. The repositioning of U.S. semiconductor firms was if anything aided by the U.S. industry's fragmented structure, criticized by the M.I.T. Commission and others (e.g., Florida and Kenney, 1990). U.S. semiconductor firms' exploitation of new opportunities for product innovation built on an unusual industry structure that distinguishes this industry from its Western European, South Korean, and Japanese counterparts. The U.S. semiconductor industry is dominated by merchant producers[5] rather than by subsidiaries of large, diversified electronics firms.

A number of federal government initiatives, ranging from trade policy to financial support for university research and R&D consortia, played a role in the industry's revival, but the specific links between such undertakings as SEMA-TECH[6] and improved manufacturing performance are difficult to measure. Col-

---

[3]Microcomponents include microprocessors, microcontrollers, DSP devices, and microperipheral devices.

[4]Fabless semiconductor firms design new microelectronic products but subcontract out the manufacture of these products to firms ("foundries") specialized in their fabrication.

[5]Merchant semiconductor firms sell most of their production on the open market, in contrast to captive semiconductor firms who produce semiconductor devices principally for internal "parent" systems divisions.

[6]The SEmiconductor MAnufacturing TECHnology (SEMATECH) consortium was created in 1987 to develop semiconductor manufacturing technology, using a combination of industry and federal government funding.

laboration between equipment and manufacturing firms contributed to improved manufacturing performance, but the size of this contribution as well as the factors that produced higher levels of collaboration in an industry long known for its fierce interfirm competition remain uncertain. Industry managers are virtually unanimous in emphasizing that the crisis of the 1980s forced them to devote much more attention to improving their development and management of manufacturing process technology. But we do not know how much of the overall improvements reflect this renewed focus by managers, nor do we understand why poor performance was tolerated for so long.

In a complex industry such as semiconductors, no single explanation for improved U.S. performance is likely to suffice. All of the factors discussed above have contributed to this industry's revival, and it is futile to attempt to assign weights to individual causes. At the same time, the foundation for this competitive revival is fragile. U.S. producers' success in repositioning their product lines and developing innovative products does not guarantee enduring dominance. The M.I.T. Commission's grim diagnosis of the "structural crisis" of the U.S. semiconductor industry does contain important insights; and at least some of the negative consequences of the U.S. industry's unusual structure have not been addressed. Many of the large corporations that supported much of the basic research that propelled the semiconductor industry's early growth have reduced the scope of their in-house basic research, and public funding for long-term R&D is more uncertain in the wake of the Cold War. Without a clearer understanding of the factors that gave rise to it, maintaining interfirm collaboration may prove difficult.

This paper surveys the competitive performance of the U.S. semiconductor industry since 1980. The following gives a description of the industry's decline and revival, focusing on measures of financial and manufacturing performance. In Section III, we discuss the changes in technology management that contributed to this revival. Section IV discusses the non-technological factors that affected the U.S. and global semiconductor industries. A short summary and concluding comments are presented in Section V.

INDUSTRY PERFORMANCE, 1980-1997

Our discussion of industry performance begins with a summary of the development of the global semiconductor industry, highlighting trends in the market shares of U.S. and non-U.S. semiconductor manufacturers and semiconductor equipment suppliers during the 1980-1997 period. Our market-share data are measured in terms of revenues and therefore confound trends in output quantity and the price per unit of that output. This effect is not entirely undesirable; one of the primary factors behind the resurgence of U.S. manufacturers' market share in semiconductors is precisely the higher average selling prices of their output during the 1990s. But we also wish to discuss trends in manufacturing performance, and therefore present data in subsequent sections on product quality, yield, and productivity that exclude these price effects.

Market Share

U.S. Dominance Prior to 1985

The first transistors, and subsequently the first integrated circuits (ICs), were developed and manufactured in the United States primarily for U.S. military and space programs. By the mid-1960s, the computer and communication industries surpassed the U.S. military as the predominant markets for semiconductors, and the market for semiconductor components has been dominated by commercial applications ever since (Tilton, 1971; Braun and MacDonald, 1978). From the invention of the IC in 1959 through 1985, the combined market share of U.S. producers exceeded that of firms from all other nations (see Figure 2).

A combination of unusual circumstances, including abundant venture capital, widespread licensing and cross-licensing of key patents, and the willingness of U.S. military and space agencies to purchase semiconductor devices from relatively new firms, produced an industry structure that by the 1960s contrasted with those of the Japanese and Western European semiconductor industries. The leading commercial producers of semiconductors in the U.S. included a number of "merchant" firms that specialized in semiconductor manufacture. Many of these firms were relatively young, having been founded during the 1950s and 1960s with venture capital financing. In contrast, the Japanese and Western European semiconductor industries were and continue to be dominated by subsidiaries of large, diversified firms in the electrical equipment industries.

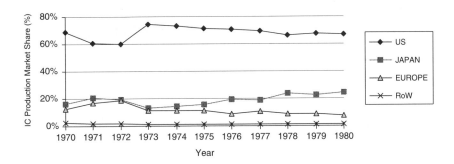

FIGURE 2 Worldwide IC production market share, 1970-1980.
Source: ICE Status: A Report on the Integrated Circuit Industry, 1976-1982.

Japanese Growth and Dominance, 1985-1990

The dominant position of U.S. firms was challenged by foreign semiconductor producers in the late 1970s. Japanese firms had been active in the semiconductor industry since the 1950s but lagged behind U.S. firms in product and process technology. But in the mid-1970s, MITI, NTT (at the time, Japan's state-owned telecommunications firm), and Japanese producers of semiconductor devices and manufacturing equipment launched several research programs to improve the semiconductor manufacturing capabilities of domestic firms. These initiatives included the well-known VLSI Program overseen by MITI and a parallel program for its semiconductor suppliers sponsored by NTT. Paradoxically, the VLSI Program sought to improve Japanese semiconductor capabilities in order to strengthen the international competitiveness of Japan's computer industry.

These technology development programs focused on memory devices, which were important in computer systems and whose relative design simplicity facilitated the testing of new process technologies. The market outlook for these devices appeared to be favorable, a projection that was amply borne out by subsequent events. Finally, "Moore's Law"[7] provided a clear "roadmap" of the path of future developments in DRAM technology, enabling Japanese firms to focus their efforts to "catch up" in semiconductor technology. In 1977, Japanese semiconductor producers gained a foothold in 16K DRAMs; by 1979, Japanese producers accounted for almost 42 percent of global DRAM sales (ICE Status, 1980).

Japanese producers became the dominant suppliers of memory devices in the industry by the mid-1980s, and U.S. firms' market share in memory products plummeted from 75 percent in 1980 to less than 20 percent in 1990 (see Figure 3). U.S.-Japanese competition in DRAM production took on the characteristics of a

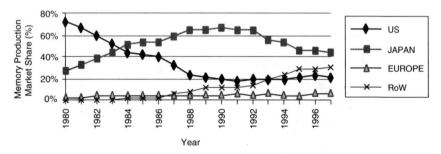

FIGURE 3 Worldwide memory production market share, 1980-1997.
Source: ICE Status: A Report on the Integrated Circuit Industry, 1980-1998.

---

[7]Moore's Law was articulated in 1965 by Dr. Gordon Moore, one of the founders of Intel, who pointed out that the number of transistors integrated on semiconductor devices tends to double every 18 months.

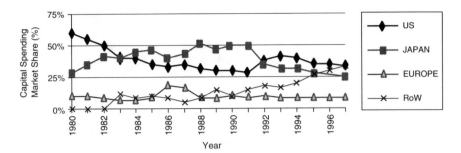

FIGURE 4 Worldwide semiconductor capital spending share, 1980-1997.
Source: ICE Status: A Report on the Integrated Circuit Industry, 1980-1998.

"capacity race"—firms in each nation invested aggressively in production capacity for next-generation products. Aided by their superior access to internal sources of finance, Japanese semiconductor manufacturers were able to dominate this investment competition. The U.S. share of capital spending in the world semiconductor industry declined from nearly 60 percent in 1980 to roughly 30 percent in 1990 (see Figure 4). During 1979-1990, Japanese producers were first to market and increased their overall market share with each new product generation (see Table 1). The enormous capital requirements of the investment capacity race, combined with fierce price competition in DRAMs and a U.S. industry recession, forced many U.S. merchant firms, with the notable exceptions of Texas Instruments and Micron Technology, out of the DRAM market by 1985. By 1990, Japanese firms accounted for 98 percent of sales of 4-megabit DRAMs, then the most advanced memory product.

Reflecting their declining fortunes in memory devices, U.S. merchant semiconductor producers lost considerable market share during this period (see Figure 5). From a leading share of almost 62 percent in 1980, U.S. chipmakers lost roughly 25 percent of the global market over the next nine years, declining to a

TABLE 1 Maximum Market Share by Device Type

Device type	Volume production	Maximum market share (%)	
		U.S.	Japan
1K	1971	95	5
4K	1974	83	17
16K	1977	59	41
64K	1979	29	71
256K	1982	8	92
1M	1985	4	96
4M	1990	2	98

Source: Dataquest, cited by Methé (1991) and Langlois and Steinmueller (1998).

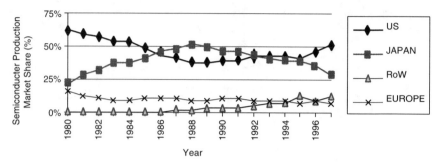

FIGURE 5 World semiconductor production market share, 1980-1997.
Source: SIA 1997 Annual Databook; ICE Status: A Report on the Integrated Circuit Industry, 1980-1998.

low point of 37 percent by 1989. Japanese semiconductor firms by 1989 accounted for more than half of global semiconductor revenues.

The Japanese semiconductor manufacturing equipment industry also enjoyed rapid growth during the 1980-1990 period (see Figure 6). Indeed, the trends in Japanese firms' share of overall capital spending and Japanese semiconductor equipment market share parallel one another closely, since many Japanese semiconductor firms purchased most of their manufacturing equipment from domestic suppliers. Japanese firms held less than 50 percent of the equipment market in Japan in 1980, but their share increased to 84 percent by 1991 and remains near 75 percent in 1997 (VLSI Research, 1998). Japanese semiconductor equipment manufacturers increased their global market share from less than 20 percent in 1980 to almost 50 percent in 1990, largely at the expense of U.S. equipment firms, whose market share declined from roughly 75 percent to less than 45 percent during the same period (VLSI Research, 1998). The rapid growth of Japa-

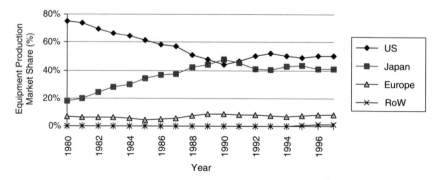

FIGURE 6 Worldwide semiconductor equipment production market share, 1980-1997.
Source: VLSI Research Semiconductor Equipment Consumption and Production by Region, 1998.

nese equipment firms appears to be attributable to the growth in investment spending by their major customers, rather than MITI initiatives such as the VLSI Program (Langlois and Steinmueller, 1998). Equally important, however, was the superior performance and reliability of Japanese equipment.

U.S. Revival, 1989-1997

Japanese firms' advances in DRAMs produced widespread concern within the U.S. semiconductor industry and among government policymakers. This dire competitive situation nevertheless began to change in the late 1980s. U.S. producers reversed their global market share decline in 1990 for the first time since 1975 (ICE Status, 1976-1991). But this reversal in market share took place in areas other than memory products, where U.S. firms' global market share has grown only slightly since 1990 (see Figure 3).

Much of the improvement in market share resulted from the efforts of U.S. firms to shift their product mix away from low-margin products such as DRAMs in favor of products that enabled them to exploit their strengths in product innovation. Having largely exited the DRAM market by 1985, U.S. semiconductor manufacturers in the 1990s focused on logic devices and "mixed-signal" and other digital signal processor (DSP) components for the burgeoning market in computer networking equipment. Strong demand for these "design-intensive" components propelled U.S. chipmakers to market share leadership in the global semiconductor industry by 1993 (see Figure 5). By 1997, U.S. producers controlled over 50 percent of the global semiconductor market, well above the 29 percent held by Japanese firms. Contradicting the predictions of analysts who argued that DRAM production was an indispensable "technology driver" for semiconductor manufacturing, U.S. firms' enduring market share losses in DRAMs did not prevent this revival in their competitive fortunes.

New Competition in DRAMS, 1992-1997

The post-1990 decline in Japanese firms' global market share reflected the revival of U.S. firms in new, more profitable product lines, as well as entry by South Korean and Taiwanese firms into the DRAM market. South Korean firms began DRAM production in 1984, and Taiwanese firms had entered large-scale merchant production of DRAMs by 1994.

Rather than shifting to logic products, Japanese firms remained in the DRAM business and sought to be technology leaders in introducing next-generation DRAM devices. But a global recession in the early 1990s and the subsequent prolonged domestic recession in Japan depressed demand for next-generation memory products. The weakness of the Japanese counterparts of the U.S. industries (e.g., computer networking, Internet applications, and packaged software) that sparked innovation in the U.S. industry also contributed to Japan's misfor-

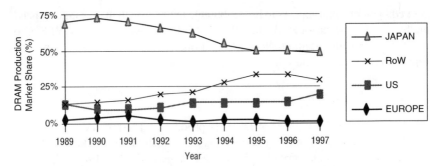

FIGURE 7 Worldwide DRAM production market share, 1989-1997.
Source: ICE Status: A Report on the Integrated Circuit Industry, 1980-1998.

tunes. The situation was made worse by the appreciation of the yen, which placed Japanese firms' memory chips at a competitive disadvantage vis-à-vis those of other DRAM producers in foreign markets. Japanese firms' market share in DRAM products has declined from roughly 70 percent in 1990 to less than 50 percent in 1997 (see Figure 7). Moreover, more intense price competition has reduced the profitability of DRAMs. Their loss of market share therefore understates the financial damage to Japanese semiconductor firms from their focus on DRAMs.

DRAMs now are essentially commodity products, and Japan, Taiwan, and South Korea are engaged in a global battle for market share based on low production costs and high yields. Japan no longer dominates the memory market as it did in the 1980s, having lost market share to Korean and Taiwanese semiconductor firms. The Korean semiconductor firm Samsung now holds the largest share of the global SRAM and DRAM markets, and Korean semiconductor firms occupy three of the top six spots in DRAM sales (see Table 2).

TABLE 2 Worldwide DRAM Merchant Market Sales (Million Dollars)

Company	Country	1995	1996	1997
Samsung	Korea	6,462	4,805	3,550
NEC	Japan	4,740	3,175	2,510
Micron	U.S.	2,485	1,575	2,003
Hitachi	Japan	4,439	2,805	1,950
Toshiba	Japan	3,725	2,235	1,750
Hyundai	Korea	3,500	2,300	1,650
LG Semicon	Korea	3,005	2,005	1,580
Mitsubishi	Japan	2,215	1,400	1,150
Texas Instruments	U.S.	3,200	1,600	1,100
Fujitsu	Japan	2,065	1,350	1,050
Others		4,999	1,880	1,505
TOTAL		40,835	25,130	19,798

Source: ICE Status: A Report on the Integrated Circuit Industry, 1996-1998.

The rapid growth of Japanese firms' market share during the 1980s relied in part on their reputation for high-quality products. Similarly, the revival of U.S. firms' market share in the late 1980s and 1990s rested in part on improvements in the quality of their products. Although the data on product quality are reasonably reliable, the causes of these trends are less easily discerned. We discuss both the trends and the available evidence on factors that lay behind them in the following section.

Product Quality

The Quality Challenge From Japan

In the early 1970s, Japanese firms recognized that improved quality in their semiconductor products could aid entry into the U.S. and global markets. These chipmakers targeted global firms such as IBM and Hewlett Packard, who needed high-quality components for their advanced electronic systems products. Drawing in many cases on practices they had long followed in their other manufacturing businesses, Japanese semiconductor manufacturers incorporated statistical process control (SPC), total quality management (TQM), and total preventive maintenance (TPM) into their semiconductor operations.[8]

By the mid-1970s, Japanese firms were applying SPC methods to semiconductor processes in fabrication and assembly in order to reduce process variance and defects—quality control practices that U.S. semiconductor firms did not pursue until well into the 1980s. Japanese semiconductor firms implemented TQM concepts through extensive training of line operators and selective automation of manufacturing to improve process control, material handling, and data processing and feedback. Japanese firms also improved the reliability of their semiconductor equipment through preventive maintenance and strengthened their relationships with systems-level customers, semiconductor equipment manufacturers, and materials vendors.

These internal management practices produced significant quality differences between Japanese and U.S. semiconductor products. Users of U.S. and Japanese devices discovered Japanese memory products had defect rates that were one-half to one-third those of comparable U.S. memory products (Barron, 1980). In 1980, leading Japanese memory producers averaged 160 defect parts per million (PPM) while U.S. semiconductor firms averaged 780 PPM for the same devices (Finan, 1993). Their skills in managing the development and introduction of new process technologies also enabled Japanese semiconductor manufacturers to "ramp" output of new products more rapidly than their U.S. counterparts. Faster achievement of high production volumes gave Japanese firms advantages in defining

---

[8]See Finan (1993) for a more extensive discussion.

product standards for leading-edge memory devices, facilitating more rapid market penetration (Finan, 1993).

U.S. industrial consumers of semiconductor devices publicized these U.S.-Japanese differences in product quality. Hewlett Packard presented data at a 1980 Electronics Industries Association of Japan (EIAJ) conference that showed Japanese memory products had average defect rates that were an order of magnitude lower than those in U.S. products. A 1989 SEMATECH survey revealed a preference among both U.S. semiconductor manufacturers and U.S. equipment and material suppliers for partnerships with Japanese firms (rather than U.S. firms) because of Japanese firms' commitment to quality and effective management of their supply chains (Erickson and Kanagal, 1992).

U.S. Semiconductor Firms' Response

By the mid-1980s most of the leading U.S. semiconductor firms recognized the strategic importance of quality and had initiated quality improvement programs. Some U.S. semiconductor firms devoted considerable effort to learning from Japanese firms; and those with operations in Japan, particularly TI and Motorola, were among the first to apply Japanese quality management techniques. Confronted with evidence of improvement in the performance of these domestic competitors, other U.S. firms began to emulate their practices. A survey in 1990 by the National Institutes of Standards and Technology (NIST) of 11 U.S. semiconductor firms' quality assurance investments revealed a doubling of the share of spending related to quality in the previous five years (Finan, 1993).[9] The estimated share of total company outlays directly or indirectly allocated toward achieving higher quality averaged 10-20 percent during 1980-1985 but increased to 20-35 percent by 1990. In addition, industry managers argue that the formation of SEMATECH supported more effective collaboration between U.S. manufacturers and equipment suppliers on quality and reliability problems.

These and other efforts were associated with a reduction in average product defect rates to less than 400 PPM by 1986, according to the Semiconductor Industry Association (SIA), the U.S. semiconductor manufacturers' trade group (Finan, 1993). Defect rates continued to decline through the rest of the decade, and by the early 1990s leading U.S. firms had matched Japanese memory producers' defect levels at less than 100 PPM (see Figure 8).

Other Measures of Manufacturing Performance

In addition to improving their product quality, U.S. semiconductor firms strengthened their performance in manufacturing process management. Data from

---

[9]Finan measures this increased spending on product quality as a doubling in the share of total firm expenses (operating, R&D, and capital outlays) devoted to quality improvement programs.

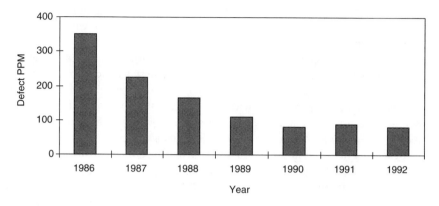

FIGURE 8 U.S. IC defective PPM, 1986-1992.
Source: SIA Quarterly Quality Survey (1992), cited in Finan (1993).

the U.C. Berkeley Competitive Semiconductor Manufacturing (CSM) Program[10] suggest that the U.S. firms have improved manufacturing "yield" and direct labor productivity in some product lines since the early 1990s,[11] although they still lag behind Japanese and other Asian firms in most of these performance measures. Nevertheless, narrowing this gap in manufacturing performance appears to have been sufficient, in combination with U.S. firms' product innovations and strategic repositioning, to improve their overall competitive performance.

A key measure of semiconductor manufacturing performance is die yield, the number of usable die per silicon wafer that emerge from the manufacturing process. Die yield is a measure of "process quality" that differs in at least one important respect from the product defect data discussed earlier. The number of defective PPM reported earlier referred to defects among products released to the market. A significant portion of the reductions in defective PPM in U.S. firms' commercial output reflects more intensive inspection of chips after manufacture and before distribution to the market. Our measure of die yield, however, is not directly affected by such inspection procedures. Instead, die yield is sensitive to

---

[10]This multi-year research effort is a joint project of the College of Engineering, the Haas School of Business, and the Berkeley Roundtable on the International Economy at U.C. Berkeley. The project has been supported by the Alfred P. Sloan Foundation and semiconductor producers from Asia, Europe, and the United States. Program directors are Dave Hodges and Robert Leachman of U.C. Berkeley's College of Engineering.

[11]In a CSM working paper (CSM-40) entitled *National Performance in Semiconductor Manufacturing*, Robert and Chien Leachman report fab performance in logic and memory devices for submicron CMOS processes using eight different performance metrics. National statistics are tabulated based upon fab location rather than the nationality of the owner firm, but the data reported here contain no "transplants." The next several paragraphs draw on Leachman and Leachman (1997), and the interested reader is referred to it for a more thorough discussion.

the execution of the numerous steps involved in the production of a new component, and its improvement reflects improved manufacturing methods, in many
respects a more difficult achievement.

The level of technical sophistication of semiconductor manufacturing processes typically is defined as the size of the smallest feature on a chip that is
manufactured with the technology. "State-of-the-art" manufacturing processes
now can produce chips with linewidths as small as 0.18 micron.[12] But this technological frontier is continually moving, and comparing manufacturing performance over even a brief length of time requires a choice of a single linewidth
category that has been in use within the U.S. and Japanese industries throughout
the period of comparison. Accordingly, our analysis of manufacturing performance uses data for devices with minimum linewidths of 0.7-0.9 micron for the
1989-1994 period.[13] We present data only for logic products because we lack a
sufficient number of observations for U.S.-located, domestically owned memory
production capacity to support a comparison of performance in U.S.- and Japa-
nese-owned memory production facilities for this period. Japanese defect density
data for logic products are available only for 1993, but during this period their
defect densities were far lower than U.S. or other firms. Nevertheless, U.S. firms
reduced their defect densities from as many as 2.5 fatal defects per square centimeter in 1991 to levels comparable with the 1993 performance of Japanese fabs
by 1994 (see Figure 9).[14] During this period, Taiwanese firms achieved similar
improvements in defect density.

Along with Finan (1993), Leachman and Leachman (1997) attribute improvements in U.S. manufacturing performance to increased use of quality management techniques. Widespread adoption of SPC methodologies by U.S. firms appears to have lowered defect densities and improved die yields. In addition, U.S.
firms improved the speed of collection, the reliability, and the accessibility of
data on manufacturing performance, all of which enabled faster identification and
diagnosis of problems in manufacturing yields. These steps included the use of
"end-of-line" yield analysis that relies on rapid transmission of data from probe
tests of wafers to engineers, the increased use of data collection systems that
provide statistical correlation of in-line data on process steps and lot characteristics with end-of-line yield tests, the increased automation of manufacturing pro-

---

[12]One micron is 1/1000th of a millimeter.

[13]Die yield is affected by particulate contamination of the silicon wafer's surface, among other
things, and reported die yield therefore is sensitive to the average size of die on a wafer. In order to
control for differences in average die size, the measure of die yield that is reported here is "defect
density," the number of fatal defects per square centimeter on a wafer.

[14]The figure reports defect density for "CMOS" logic manufacturing processes, which are the largest single category of MOS manufacturing processes. During the period of the sample, CMOS represented more than 90 percent of MOS technology used in all IC manufacturing (ICE Status, 1989-
1994).

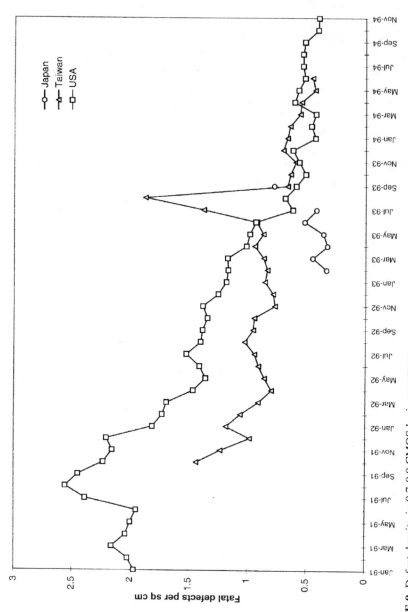

FIGURE 9 Defect density in 0.7-0.9 CMOS logic processes.
Source: CSM-40: ""National Performance in Semiconductor Manufacturing," Leachman and Leachman (1997).

cess information,[15] and the automated collection of equipment performance and control parameters.

Other measures of die yield indicate significant improvement in U.S. semiconductor firms' manufacturing performance by the end of the 1980s. According to the U.S. General Accounting Office (U.S. GAO), U.S. firms had fallen behind Japanese firms in "probe yield"[16] by 1981 (U.S. GAO, July 1992),[17] but U.S. firms' performance in this quality measure improved significantly during the 1986-1991 period. By 1991, U.S. manufacturers had narrowed but had not eliminated the gap between their performance and that of Japanese manufacturers (see Table 3).

Although U.S. semiconductor firms have narrowed the gap with Japanese firms in die yield for some devices, they continue to lag in other areas, such as direct labor productivity. Our data support comparisons of U.S. and Japanese productivity performance, measured in terms of the number of wafer layers per operator per day. This measure captures differences in "physical productivity;" the value of output per worker is not captured by this measure.[18] As such, differences in the price per die on wafers produced in different fabs may partially or entirely offset much of the financial consequences of differences in this measure of performance.

At CMOS logic fabs, U.S. firms' direct labor productivity improved during the 1991-1994 period but still lag behind Taiwanese and Japanese firms (see Figure 10). The importance of scale economies in semiconductor manufacturing means that the smaller average size of U.S. fabs, relative to those in Japan, South

TABLE 3 Average Probe Yield: U.S. and Japanese Semiconductor Manufacturers (1981-1991)

Country	Average probe yield (%)						
	1981	1986	1987	1988	1989	1990	1991
U.S.	55	60	60	67	74	80	84
Japan	45	75	79	81	85	89	93

Source: U.S. General Accounting Office (July, 1992).

---

[15]Such as "downloading" of recipes for specific device types to operators, helping to reduce errors.

[16]Probe yield is the percentage of good die on a silicon wafer after the last electrical test for functionality before semiconductor devices are cut from the wafer, packaged, and assembled. It is similar to defect density, although it does not control for variation in die size.

[17]The GAO study cited unpublished data from VLSI Research in this assessment.

[18]Significant differences within the sample in fab organization and relationships with other corporate functions, such as R&D, process development, and the like mean that the amount of "indirect" labor—i.e., engineering and management staff—is likely to vary among fabs in this sample. "Direct" labor productivity should reduce the influence of these differences in the comparison of fab-level productivity.

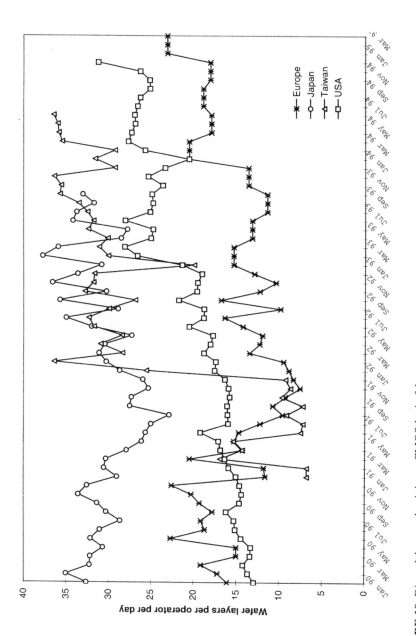

FIGURE 10 Direct labor productivity at CMOS logic fabs.

Source: CSM-40: ""National Performance in Semiconductor Manufacturing," Leachman and Leachman (1997).

Korea, and Taiwan, depresses U.S. firms' performance in this comparison (Leachman and Leachman, 1997). Nevertheless, as was noted above, the consequences of these differences in physical productivity are mediated by the prices and margins per die in each region's product mix. U.S. firms have specialized in relatively high-margin products and therefore are somewhat insulated from the financial consequences of their relatively low physical productivity. But as and if non-U.S. firms strengthen their capabilities in product innovation and shift their output mix to become direct competitors with U.S. firms in these newer, more design-intensive products, the relatively low direct labor productivity of U.S. fabs could produce more serious competitive difficulties.

During the late 1980s and 1990s, U.S. firms also improved their management of the development and introduction of new process technologies into high-volume manufacturing.[19] Increases in the design complexity of semiconductor devices during the 1980-1997 period meant that the process technologies necessary to produce them became more complex. Greater design complexity made it much more difficult to predict the performance of new process steps and equipment through simulation or laboratory-scale experimentation and complicated the "debugging" of new process technologies in the manufacturing environment. At the same time, intensified competition in many product lines meant that rapid expansion of the output of new products was essential to maximize sales in the increasingly brief period prior to entry by competitors. A prolonged period of "learning," during which yields are low and/or product quality is unreliable, reduces profits. The difficulties associated with new process development and introduction thus have grown simultaneously with the competitive and financial penalties of a poorly managed introduction. Evidence from the CSM study and other sources suggests that U.S. firms were slow to respond to these new realities until forced to do so by Japanese competition.

There is no single "best practice" for managing the development and introduction of new process technologies. Many U.S. firms have expanded their use of "development facilities," which are similar in many respects to pilot process plants in the chemicals industry. These facilities support the development and debugging of new process technologies and equipment in an environment that is insulated from the demands of high-volume manufacturing yet is designed to reproduce as many characteristics of that environment as possible. Intel's integrated process development facility in northwest Oregon doubled manufacturing yields from the 1980s to the mid-1990s and accelerated the "ramping" of production of new device designs (Cole, 1998).

Duplication of the manufacturing environment requires that the development facility duplicate production equipment and materials to the maximum extent, which can be costly and in some instances delays the adoption of the latest production technologies. Firms also rely on multifunction teams for the develop-

---

[19]See Hatch and Mowery (1998) and Appleyard et al. (1997) for further discussion of these issues.

ment and introduction of new manufacturing processes, and members of the process development staff participate directly in the introduction and debugging of a new manufacturing process. Improved management of new process development and introduction also requires an integrated approach to product and process development. Introducing a radically new manufacturing process (for which 75-90 percent of the hundreds of individual steps are new) and attempting to simultaneously begin large-scale production of a new product design with this process is formidable. Many U.S. firms instead introduce incremental advances in manufacturing processes and debug these modified processes on new versions of existing product designs—for example, a smaller version or "shrink" of an established logic or memory chip. This more incremental approach to new process development and introduction requires close coordination among product design, process development, and equipment procurement over multiple generations of existing and new products.

Other Factors Affecting U.S. Semiconductor Firm Performance

By the early 1990s, the global semiconductor industry had coalesced into three broad product categories—memory, logic, and microcomponent products. Memory has traditionally been the largest single segment of the semiconductor industry. The logic market includes both application-specific integrated circuits (ASICs) and other general and special-purpose logic devices. ASICs are logic devices produced for a specific customer and not, as the name implies, devices for a specific application. U.S. semiconductor firms' total ASIC market share declined from 1988 to 1991 but has rebounded. U.S. firms controlled almost 50 percent of the worldwide total ASIC market in 1995 and dominate the programmable-logic, analog array, and standard cell segments (ICE ASIC Outlook, 1987-1988, 1990-1998).

The microcomponent market of the semiconductor industry includes microprocessors, microcontrollers, DSPs, and microperipheral devices. Microprocessors, "computers on a chip," have continuously improved in functionality, complexity, and processing speed. Microcontrollers are somewhat simpler and less powerful than microprocessors and have their main applications in automotive, factory and industrial automation, and processing machinery.

Digital signal processors are a rapidly expanding segment of the microcomponent market because of their applications to the computer networking and communications industries. The global DSP market has expanded from roughly $340 million in 1991 to just under $3.4 billion in 1997 and is dominated by U.S. semiconductor firms who produced more than 90 percent of the DSP products sold in 1997 (ICE ASIC Outlook, 1990-1998). U.S. semiconductor firms' strength in this segment of the industry reflects their presence in the most dynamic end-user markets for these applications as well as their ability to exploit their proximity to U.S. systems producers that dominate these end-user markets.

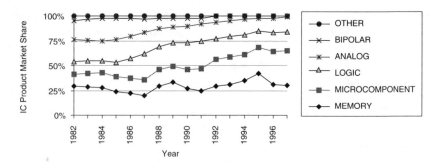

FIGURE 11 Worldwide cumulative IC product market share, 1982-1997.
Source: ICE Status: A Report on the Integrated Circuit Industry, 1982-1998.

In both Western Europe and Japan, the much slower growth of end-user markets has hampered European and Japanese semiconductor firms' entry into DSPs and related products.

Sales of logic, microcomponent, and memory products accounted for over 80 percent of worldwide IC market revenue in 1997 (see Figure 11). These product markets have benefited from strong end-use demand, most notably from the computer industry, which consumes almost two-thirds of the memory and microcomponent devices output. Memory sales in 1996 amounted to $36 billion, but declined in 1997 to $31 billion and are expected to continue to decline as a result of price competition (ICE Status, 1997-1998). Sales of microcomponent devices reached almost $40 billion in 1996, exceeding memory product revenues for the first time, and were greater than $50 billion in 1997 (ICE Status, 1997-1998).

Four distinct process technologies are used to manufacture semiconductor devices—discrete, bipolar, analog, and metal-oxide semiconductor (MOS). During the 1980s, semiconductor producers shifted from discrete and bipolar process technologies to MOS, while analog technologies retained roughly 15 percent of the overall IC market. From just under 55 percent of worldwide sales in 1984, MOS devices have grown to more than 80 percent of the overall IC market by 1997 (see Figure 11). For much of the 1980s and 1990s, DRAMs pioneered in the development of process technologies. Indeed, concern within the U.S. industry and U.S. government over loss of DRAM market share reflected the view that these products "drove" advances in manufacturing methods for many products. In recent years, however, DRAMs have lost their position as the "technology drivers" in MOS products as microprocessor manufacturing technologies have placed even greater demands on process limits and controls. In 1997, for example, Intel introduced large-scale production of portable Pentium microprocessors using 0.25-micron process technology, exceeding the then current state-of-the-art production technology of memory components (0.35 micron). At least for the near term, microcomponent devices, where U.S. firms retain a leadership position, appear to have assumed a role as "technology drivers" for manufacturing processes.

Since 1985, U.S. semiconductor companies have shifted away from the commodity memory business, concentrating instead on "design-rich" semiconductor market segments that have more specialized and demanding product design requirements than memory devices. U.S. producers such as Intel and AMD now dominate the microcomponent market and have increased the U.S. share of microcomponent device production to more than 70 percent in 1997 from 50 percent in 1989 (see Figure 12). The shift by U.S. firms to the microcomponent and logic product market segments has significantly changed the competitive positioning of firms in the semiconductor industry. The industry titans of the 1970s and 1980s—National Semiconductor, Motorola, Intel, and TI—are still around, but their product portfolios have changed. Whereas memory products were an integral part of their product portfolios in the 1970s and 1980s, logic and microcomponent devices now are the single most important source of revenues for all. Another major change in the U.S. industry is the growth of specialized design firms. During the 1980s and 1990s, a significant number of U.S. entrants into the semiconductor industry focused exclusively on the design and marketing of semiconductor devices, relying on third-party foundries for the manufacture of these devices (see below).

Shifts in the growth of demand and profitability of semiconductor market segments, along with the entry of new producers, have affected regional capital investment trends in the global semiconductor industry. U.S. firms have signifi-

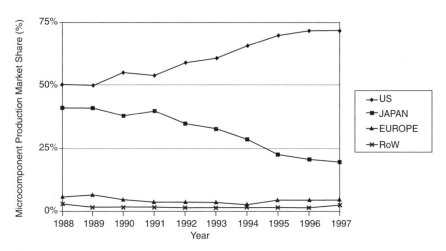

FIGURE 12 Worldwide microcomponent production market share, 1988-1997.
Source: ICE Status: A Report on the Integrated Circuit Industry, 1988-1998; ICE Microprocessor Outlook, 1997-1998.

cantly increased their capital spending since the early 1990s and since 1992 have accounted for the largest single regional share of overall investment in plant and equipment (see Figure 4). This represents a considerable change from the 1984-1991 period, during which Japanese semiconductor companies were responsible for nearly half of the capital expenditures made by IC manufacturers. Despite increases in their capital investment since 1993, Japanese producers accounted for only 25 percent of total industry capital and equipment investment in 1997. South Korean and "RoW" ("rest of world," largely Taiwanese, firms) have doubled their share of capital spending since 1991, and since 1996 combined investment in fabs and equipment by South Korean and RoW semiconductor firms has exceeded that of Japan (see Figure 4).

These investment trends have contributed to a revival of the U.S. semiconductor manufacturing equipment industry, which increased its global market share from roughly 45 percent in 1990 to more than 50 percent in 1997, while Japanese equipment firms have lost market share (see Figure 6). The market share gains by U.S. semiconductor firms after 1989 aided U.S. equipment suppliers because U.S. equipment suppliers maintain a position in the U.S. market that is only slightly less dominant than that of Japanese equipment firms in the Japanese domestic market. From 1980 to 1997, U.S. semiconductor equipment manufacturers supplied at least 75 percent of the equipment demanded by U.S. semiconductor firms in each year (VLSI Research, 1998). The development of the Taiwanese and Korean foundry industries has also played a part in the U.S. semiconductor equipment industry's recovery. These foundries typically manufacture U.S.-designed semiconductor products, which require multiple metal layers and advanced equipment for chemical vapor deposition (CVD) and thin-film sputtering, areas in which U.S. semiconductor equipment firms have traditionally excelled.

Summary: Factors behind U.S. Decline and Revival

The performance of the U.S. semiconductor industry during the 1980-1997 period reflected shifts in both product and process technology management. In contrast with the Japanese firms that during the mid-1980s appeared to pose a serious competitive threat, U.S. firms proved to be relatively agile in repositioning their product portfolios to emphasize new products that were relatively design intensive. At the same time, however, U.S. firms improved their manufacturing performance, which enabled them to exploit their long-standing strengths in product innovation more effectively. From a position of substantial inferiority in the development and management of semiconductor process technologies in the early 1980s, U.S. chipmakers narrowed the gap between U.S. and Japanese manufacturing capability and productivity in some product lines by the end of the decade.

Both repositioning and improved manufacturing performance almost certainly were necessary; neither was sufficient. Improvements in both of these dimensions of performance reflected improved technology management practices, where these practices are defined to include management of process technologies on the shop floor as well as improvements in the development and adoption of new process and product technologies. In addition to these changes in their internal management of innovation and production, U.S. firms expanded collaboration among one another, with equipment firms, and with non-U.S. firms. Finally, the entry of specialized design firms into the U.S. semiconductor industry signaled the development of new approaches to the organization of the innovation process that involved greater reliance on specialization and arms-length arrangements.

Although U.S. semiconductor firms' performance during the 1993-1997 period has been impressive, it has been aided in part by Japanese and South Korean semiconductor firms' failure to shift their product portfolios away from DRAMs to design-intensive components. The 1998 industry downturn caused by the continued economic problems in Asia and excess capacity in DRAMs, may bring new competitors to semiconductor markets that have traditionally been dominated by U.S. producers. Indeed, some Taiwanese DRAM producers have recently entered product markets led by U.S. semiconductor firms, such as flash memory (Takahashi, 1998). Drawing on their experience in operating "foundry" production facilities, other Taiwanese firms now are able to switch from memory to advanced logic components, depending on market conditions. The flexibility gives them an advantage over South Korean and Japanese semiconductor firms and may foreshadow the development of a formidable competitor in the years to come. U.S. firms will be challenged by foreign firms for the foreseeable future, placing a premium on their ability to innovate and shift to profitable new activities and products.

CHANGING TECHNOLOGY STRATEGIES

In this industry, like other U.S. high-technology industries, perhaps the greatest single change in the innovation process since 1980 has been the increased reliance by U.S. firms on collaborative strategies. Collaboration has been both "vertical," linking suppliers of equipment with semiconductor manufacturers, and "horizontal," linking semiconductor manufacturers with one another. Collaboration has also been both domestic and international; it has been supported by public and by private funds, and in a number of cases it has been associated with increased specialization by firms in different phases of the development and manufacturing process.

A central reason for collaboration is the higher costs and risks of new product development and the spiraling costs of new production capacity. Electronic system suppliers are demanding specialized chips that incorporate more features and provide more functionality, but these semiconductor components require chip facilities that cost more than $1 billion per plant (see Table 4). Many semiconductor firms have found it impossible to invest in new products or manufacturing capacity without some arrangements for risk-sharing.

TABLE 4 Fabrication Facility Production Costs

Year	Capital cost (million dollars)	Linewidth (micron)
Early 1970s	$ 20	3.00
Early 1980s	100	1.00
Early 1990s	300	0.70
Late 1990s	1,200	0.35
Late 2000s	12,000	0.10

Sources: Cost Effective IC Manufacturing (1998-1999, 1997); National Technology Roadmap for
     Semiconductors: Technology Needs (1997).

Producer—Designer Collaboration

One strategy to reduce financial risks that has been adopted by recent entrants into the U.S. semiconductor industry is specialization in design. By the late 1980s, the rapidly escalating costs of manufacturing facilities, along with burgeoning opportunities in less capital-intensive sectors for venture capitalists, had reduced the flow of venture-capital financing for new firms in semiconductor manufacturing. The declining opportunities for entry into semiconductor manufacturing, however, created other possibilities and financing for specialized design firms. These so-called "fabless" semiconductor firms design semiconductor components but rely on specialized "foundries" for the production of their designs.

Access by fabless firms to foundry capacity was aided by the rise to dominance within the semiconductor industry of MOS manufacturing processes, which effectively standardized manufacturing technologies for commercial semiconductor devices. The diffusion of MOS production technology facilitated the division of labor between device designers in fabless firms, who were able to operate within relatively stable rules and constraints, and foundries, who were able to incrementally improve their process technologies to accommodate a succession of new device designs. The fabless firm is largely a North American phenomenon; more than 300 of the worldwide population of 500 fabless firms were located in North America in 1998.[20] By contrast, most state-of-the-art foundries are located in Asia.[21]

---

[20]The estimate of North American fabless firms was provided by the Fabless Semiconductor Association (FSA) and Integrated Circuit Engineering (ICE) Inc. The estimate of the worldwide population of fabless firms was provided by the FSA through personal communication on August 1, 1998.

[21]Major "pure-play foundries" include TSMC and UMC (both Taiwan), Chartered Semiconductor (Singapore), and Tower Semiconductor (Israel). New pure-play foundries include Anam (a Korean startup) and WSMC (a Taiwanese startup). The prevalence of Southeast Asian pure-play foundries is subsiding as merchant semiconductor producers from all nations are converting older facilities or dedicating entirely new facilities to provide foundry services to this industry. IBM Microelectronics (U.S.), LG Semicon (Korea), Samsung (Korea), Winbond (Taiwan), and VLSI (U.S.) are notable examples.

Outsourcing of manufacturing is not new to the semiconductor industry; many producers of electronic systems have relied on third-party manufacturing of custom components for in-house designs. But the fabless business model involves outsourcing of production to specialized third parties and relies on skills that differ from those exploited by the traditional merchant semiconductor firms. Fabless firms concentrate on design R&D, utilizing design tools and architectures that must be compatible with the process requirements of the foundry in order for the designs to be manufacturable. The foundries that work with the fabless firms must be able to manage small production runs, support and modify their process technologies for a diversity of products, and provide short prototyping and good cycle times. Although the manufacturing capabilities of most advanced foundries lag behind those of merchant semiconductor firms, this gap is expected to close in 1998 (see Figure 13).

Fabless firms serve a variety of fast-growing industries, especially personal computers and telecommunications, and seek to dominate their markets by offering more innovative designs and shorter delivery times than merchant firms. Constant-dollar industry revenues have grown at an average annual rate of 32 percent since 1991, almost twice the average for the global semiconductor industry as a whole (see Figure 14). The fabless industry's trade association estimates 1997 fabless industry revenues at $7.8 billion (FSA, 1997), and Dataquest forecasts fabless industry revenues will grow to $11.7 billion in 2000 and 40 percent of the world's chip production by 2010 (Semiconductor Business News, 1998).

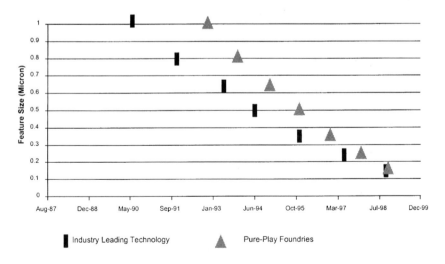

FIGURE 13 Trends in process technology migration, 1987-1999.
Source: FSA State of the Fabless Business Model (Sept. 1997), UBS.

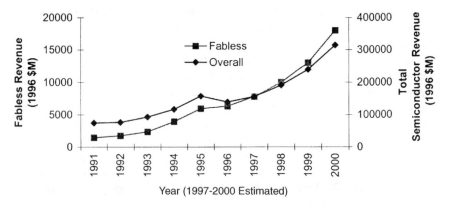

FIGURE 14 Fabless and total semiconductor revenue, 1991-2000E.
Source: FSA State of the Fabless Business Model (Sept. 1997); ICE Status: A Report on the Integrated Circuit Industry, 1990-1998.

International Collaboration

U.S., Japanese, and European multinational manufacturers have increased offshore R&D spending since 1980 but all still perform the vast majority of firm-financed R&D in their home regions. Examples of departures from this pattern by U.S. firms are long-established, relatively small R&D organizations operated by TI in Bedford, England, by IBM in Zurich and Tokyo, by Motorola in Hong Kong, and by Intel in Israel. These U.S.-based firms seek access to overseas talent, better understanding of competition and of foreign market needs, and improved access to local and regional markets. Overall, however, there is no evidence of major growth in offshore R&D by U.S. semiconductor firms.

During the past 20 years, several non-U.S. firms have established R&D facilities in the United States and other developed nations outside their home region.[22] These foreign R&D investments are motivated by the same factors that drive U.S. offshore R&D investment, although many foreign firms are especially interested in tracking new semiconductor product and process developments in the U.S. market. This desire to monitor technological change also has led a number of European and Japanese firms to support U.S. university research and education in the semiconductor field.

Although U.S. semiconductor firms have not significantly expanded their foreign R&D operations, alliances among U.S. and non-U.S. semiconductor firms have grown rapidly since 1980. Many alliances focus on specific product or process development projects and often involve some exchange by U.S. firms of product technology for foreign, usually Japanese, expertise in process technol-

---

[22]Philips, Siemens, NEC Hitachi, and Fujitsu are examples.

ogy. Such partnerships also facilitate access to international markets that are otherwise impeded by tariffs or political mechanisms. Manufacturing partnerships are driven by the same considerations as R&D partnerships, along with the escalating costs of new production facilities. Many of these international collaborative agreements focus on a single product area, such as nonvolatile memory or microprocessors, and many involve no U.S. partners. Coming from a "catchup" position in semiconductor design and manufacturing technology, firms based in Taiwan, Korea, and Singapore have actively sought relationships with more advanced firms in Japan, Europe, and the United States.

Domestic Collaboration

International collaboration in the semiconductor industry has been paralleled by expanding domestic collaboration, some of which is supported with public funds. Japanese firms' growing domination of the global market for semiconductor memory chips in the late 1980s reflected concern within both the U.S. industry and the U.S. government over the future viability of an industry that supplied critical components for defense applications.[23] This possibility led to an unusual initiative, spearheaded by the Defense Department, to strengthen U.S. semiconductor firms' commercial-device manufacturing capabilities.[24] SEMATECH was formed in 1987 by 14 U.S. semiconductor manufacturing firms that together accounted for more than 80 percent of U.S. semiconductor manufacturing capacity[25] and was financed jointly by member firms and the federal government.[26] SEMATECH's defense-related funding and sponsorship, along with broader political concerns, led to the decision to exclude non-U.S. firms from membership. In addition to paying dues totaling $100 million per year—matched by $100 million from federal sources—the member firms contributed roughly two-thirds of SEMATECH's 300-member research staff through temporary, usually two-year, rotation of "assignees" at the consortium. Concurrently with the foundation of SEMATECH, U.S. semiconductor materials and equipment (SME) suppliers formed SEMI/SEMATECH to facilitate linkages between U.S. SME suppliers and SEMATECH. SEMI/SEMATECH has more than 100 members who account for more than 85 percent of U.S. SME sales.

---

[23]This discussion of SEMATECH draws on Grindley et al. (1994).

[24]Concerned by the implications for national defense of U.S. dependence on foreign semiconductors, the Defense Science Board (an advisory committee within the Department of Defense) developed a competing proposal that recommended creation of a manufacturing facility jointly owned by government and industry to produce semiconductor components (McLoughlin, 1992).

[25]SEMATECH's founders included the following firms: Advanced Micro Devices, AT&T, Digital Equipment Corporation, Harris Corporation, Hewlett Packard Company, Intel Corporation, IBM, LSI Logic, Micron Technology, Motorola, National Semiconductor, NCR, Rockwell International, and Texas Instruments. Three of the founding members of SEMATECH (Harris Semiconductor, LSI Logic, and Micron) left the consortium in 1991.

[26]SEMATECH's federal funding ceased in 1996.

SEMATECH's original objectives—improving member firms' semiconductor manufacturing process technology—underpinned its decision to build a large-scale fabrication facility in Austin. But SEMATECH had difficulty developing a research agenda that could exploit this research facility and eventually altered its research agenda to one that sought to improve the technological capabilities of U.S. suppliers of semiconductor manufacturing equipment through "vertical" cooperation between U.S. suppliers and U.S. users of semiconductor process equipment (Katz and Ordover, 1990; U.S. Congressional Budget Office, 1990). This research focus could benefit all members without threatening their proprietary capabilities. In the words of William Spencer, former SEMATECH CEO, "We can't develop specific products or processes. That's the job of the member companies. SEMATECH can enable members to cooperate or compete as they see fit" (Burrows, 1992).

In many respects, SEMATECH now resembles an industry association, diffusing information and best-practice techniques, setting standards, and coordinating generic research. Like many Japanese cooperative research projects, SEMATECH is concerned as much with technology diffusion as with the advancement of the technological frontier. SEMATECH also has focused on medium-term, rather than long-term research, with the typical time horizon for R&D investments targeted at three to five years.

SEMATECH's formation and operations coincide with improvements in U.S. semiconductor manufacturing performance and increased market shares for U.S. semiconductor equipment suppliers. It is difficult if not impossible, however, to find direct cause-and-effect links between SEMATECH's activities and these developments. In the case of semiconductor manufacturing equipment, for example, a significant portion of the improved market share of U.S. suppliers reflects the decline in Japanese manufacturing firms' capital investments, which has depressed the growth of equipment demand in a market that was long dominated by Japanese equipment firms. U.S. equipment producers have not increased their share of the Japanese market significantly during the 1990s but have benefited from the rapid growth in the South Korean and Taiwanese markets, which were far easier to penetrate. Nevertheless, SEMATECH member firms have continued to support and participate in SEMATECH since the cessation of federal support, a strong signal that industry managers believe that the consortium has produced important benefits. Indeed, this continued support suggests that a smaller amount or shorter period of federal support might have sufficed to launch and sustain this consortium.

Even if its specific contributions to improved industry performance cannot be isolated definitively, the survival and evolution of SEMATECH suggest some important lessons for future consortium design. Industry leadership in the design and establishment of the research agenda, joint industry and public funding, staffing the consortium by employees of member firms, flexibility and adaptiveness in the research agenda, and the consortium's focus on "vertical" rather than "horizontal" collaboration all have contributed to its success. At the same time, however, the exclusion of non-U.S. firms from membership in the consortium did not prevent foreign firms from benefiting from its activities. Many member firms developed collaborative relationships with non-U.S. firms in related manufacturing areas, and the original restrictions on equipment firms' export of products embodying SEMATECH research results have been relaxed. Indeed, non-Japanese foreign firms now are active in a recent SEMATECH initiative that seeks to define equipment and performance standards for 300-mm wafer processing.

Another collaborative research initiative that predates the formation of SEMATECH is the Semiconductor Research Corporation (SRC), supported by industry firms, the Defense Department, and more recently, by SEMATECH. The SRC supports university research in order to bolster an important portion of the U.S. research infrastructure, attract faculty and students to work on problems of relevance to industry, and attract high-quality students to seek employment opportunities in the U.S. semiconductor industry. State-level programs, such as the California MICRO program, pursue similar objectives through a combination of public and industry support. Finally, the Microelectronics Advanced Research Corporation (MARCO) is a new industry-financed collaborative research initiative that will support long-range, university research on silicon IC technology. The academic R&D focus of MARCO is intended to complement the efforts of the SRC and SEMATECH.

Fabless semiconductor firms also are pursuing collaborative R&D in two consortia sponsored by FSA. The first is a 0.35-micron wafer level reliability project that seeks to standardize test structures and test methodologies and evaluate their usefulness. Dozens of fabless firms and foundries are participating in this endeavor. The second project, which involves five foundries, is developing a standard test chip that will improve manufacturing efficiency in 0.25-micron, five-level metal MOS processes and develop standards for process performance.

Publicly and privately funded R&D collaboration has expanded significantly within the U.S. semiconductor industry since 1980. Domestic collaborative ventures focus primarily on near-term or mid-term R&D rather than joint manufacturing or long-term basic research. Most ventures also are quite young, and their ultimate effects on U.S. industry performance are difficult to predict. In view of the limited experience of U.S. managers with such undertakings, both failures and successes are likely, and the essential point is to try to capture sufficient knowledge from each to improve performance. Collaboration is not a panacea, but it may offer some solutions to the competitive weaknesses associated with the fragmented industry structure cited by the M.I.T. Commission.

Who Will Fund and Perform Basic Research?

However useful, collaborative R&D in the U.S. semiconductor industry thus far has supported little long-term research. The large U.S. corporate laboratories of the 1950s and 1960s, most notably those of AT&T, GE, and IBM, performed much of the fundamental research that underlies today's mainstream semiconductor technology. Those laboratories now focus on near-term corporate goals and applied research, and no U.S. organization has emerged to fund the basic research needed for the future. Federally funded R&D in the U.S. semiconductor industry, mostly in defense-related applications, has declined from nearly 25 percent of total R&D spending in the industry (imperfectly defined in this case as SIC 367, "electronic components") in 1980 to slightly less than 7 percent in 1992 (National Science Foundation, 1996). Defense-related R&D funding is likely to continue to decline in the aftermath of the Cold War.

Although the leading U.S. merchant semiconductor firms, such as Intel, TI, Micron, and AMD, spend 10-15 percent of revenues on R&D, the bulk of these expenditures focus on new product development. Intel has announced its intention to expand its long-term research program, but few other semiconductor manufacturing firms conduct much R&D beyond development of next-generation products. None of the new leaders in digital communications maintains any internal semiconductor R&D; instead they focus their efforts on product definition, system design, and marketing of their end products. These smaller firms rarely perform much fundamental research, instead pursuing product development using sharply targeted technical teams stocked with Ph.D. engineers and scientists, categories of professional staff rarely employed by such small firms in earlier times.

By contrast, the major non-U.S. semiconductor manufacturers such as NEC, Hitachi, Toshiba, Philips, and Siemens still conduct considerable long-range R&D. These firms are integrated from materials and components to system-level products, and their varied internal customers for semiconductors allow them to extend the productive life of their semiconductor production facilities. Their other businesses produce generous cash flows that help to offset the heavy R&D and investment costs of the capital-intensive semiconductor business. Despite these apparent advantages over their smaller U.S. merchant and fabless competitors, these large, diversified foreign firms thus far have been relatively slow or ineffective in exploiting new opportunities for innovative products. Particularly in Japan, internal R&D has been applied to problems in manufacturing processes, and capital resources have enabled rapid capacity expansion for new generations of products that follow a well-established "trajectory" of technological development. Nevertheless, non-U.S. firms could re-emerge as formidable competitors in product lines in which the pace of product innovation has slowed or assumed a more incremental and capital-intensive character, as was the case in DRAMs during the 1980s.

What institutional mechanisms for supporting long-term research exist within the U.S. economy? U.S. research universities, even if they should receive expanded research funding, can fill only part of the gap left behind by the downsizing of U.S. corporate laboratories. Moreover, efforts by many U.S. research universities to expand their patenting of scientific advances in areas such as biotechnology that formerly were placed in the public domain could, if expanded into research in electronics and software, restrict the dissemination of critical research results to industrial and other practitioners. The large network of public laboratories in the United States also may be of limited use for this purpose, as only Sandia Labs has contributed significantly to the recent advancement of semiconductor technology.

The reconfiguration of the semiconductor industry described above merits detailed study. More public support for research may be needed to ensure U.S. leadership in semiconductor technologies over the long term, but the political rationale and institutional vehicles for such an initiative are uncertain at present. Publicly funded research might rely on partnerships among industry, universities, and government, extending and elaborating recent experiments in collaboration discussed above (Rosenbloom and Spencer, 1996). But any such arrangements would require change in the historical roles played by all three of the institutional partners in this industry.

THE ROLE OF NON-TECHNOLOGICAL FACTORS IN U.S. SEMICONDUCTOR INDUSTRY'S REVIVAL

Our discussion of the decline and revival of the U.S. semiconductor industry has emphasized technological factors, such as the improvements in U.S. firms' manufacturing performance and renewed emphasis on product innovation. But these factors cannot be considered in isolation from non-technological factors or the broader economic, institutional, and policy environment. Moreover, a central concern of the Board on Science, Technology, and Economic Policy project on which this book is based is the interaction between these non-technological or external influences and the competitive performance of U.S. industry. At least some of the factors associated with the industry's revival were facilitated by non-technological factors that were hardly exogenous. Indeed, the development by the U.S. semiconductor industry of an influential political voice in the SIA was partially responsible for some significant changes in U.S. government policy. Accordingly, this section considers the influence on the U.S. semiconductor industry's improved performance since 1980 of the environment for capital formation, the role of government antitrust policy and trade policy, and the changes in intellectual property protection. Our treatment is necessarily brief, and we conclude that non-technological factors have been helpful and necessary but are by no means sufficient in explaining the industry's revival since 1990.

Capital Formation

Any discussion of the cost and availability of capital in competitive performance within the semiconductor industry confronts a paradox. On the one hand, this issue is important in some segments of the industry, where the costs of a single commercial-scale production facility significantly exceed $1 billion. Indeed, we noted earlier that U.S. firms faced significant handicaps in the DRAM "capacity races" that developed in the early 1980s. On the other hand, capital expenditures by U.S. merchant semiconductor producers amounted to more than $14 billion in 1996 and $13 billion in 1997 (ICE Status, 1998). Investments of this size suggest that constraints on the supply of capital to U.S. firms are scarcely binding. In addition, the recent competitive performance of the large firms active in both the Japanese and South Korean semiconductor industries, especially those specializing in DRAMs, suggests that one can have too much of a good thing. Low-cost capital has been associated with overinvestment in manufacturing capacity for commodity products that yield low profits.

Historically, the U.S. semiconductor industry has faced an abundant supply of venture capital (VC). VC funds have supported the foundation of literally hundreds of semiconductor firms since this industry's inception four decades ago. The data in Figure 15 suggest that this high "birthrate," which has contributed significantly to the industry's technological dynamism, shows few signs of declining. Especially interesting is the sharp upsurge in new-firm formation during 1983-1985, a period of severe industry recession. Excluding this period, an average of eight semiconductor startups appear annually (Figure 15); the majority (70 percent) of these are fabless firms. The VC community has continued to support new fabless semiconductor endeavors, but has been less generous toward semiconductor ventures that include manufacturing.

Although there are few reliable estimates of the risk-adjusted cost of capital in the U.S., Japanese, South Korean, and other semiconductor industries, U.S. firms may well face a higher cost of capital. Nevertheless, any such differential has not deterred the foundation of new U.S. firms, nor has it deterred large-scale capital investments by U.S. firms that have developed successful competitive strategies that rely on their strengths in product design and innovation. For established U.S. semiconductor firms, competitive success appears to lead to abundant

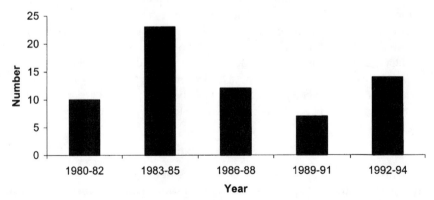

FIGURE 15 U.S. semiconductor start-ups, 1980-1994.
Source: FSA *Fabless Forum* (1995) V.2, n.1.

capital for investment in plant and equipment rather than vice versa. A higher cost of capital may contribute to the low level of investment in long-term, basic research by many U.S. semiconductor firms. Nevertheless, given the competitive realities of this industry, especially the short product cycles and high costs of R&D for maintaining near-term competitiveness, the risk-adjusted cost of capital would have to be very low indeed to produce higher levels of such investment.

Trade Policy

Among the several government initiatives emerging from the semiconductor industry's turmoil of the 1980s was a sector-specific trade agreement, the effects of which continue to be debated. The Semiconductor Trade Agreement of 1986 responded to accusations by U.S. firms that Japanese DRAM producers were "dumping" their products in the U.S. market.[27] By preventing the imposition of heavy antidumping duties on U.S. imports of DRAMs, the agreement sought to avoid a policy that would drive up the domestic prices of components that were essential to U.S. manufacturers of electronic systems, creating strong incentives for them to shift production to foreign locations. A system of "fair market value" prices for DRAMs was created under the terms of the agreement that was intended to prevent dumping in the U.S. and third-country markets. The agreement also included an "understanding" that foreign-sourced components would achieve a 20 percent share of the Japanese domestic market within five years. An extension of the agreement in 1991 retained the market share language but dropped the price-monitoring system.

Although the agreement was negotiated in response to the competitive crisis facing U.S. producers of DRAMs, its effects on these firms' activities in DRAM production were limited. Most of the major U.S. DRAM producers had exited from this product line by 1985, well before the agreement was finalized.[28] The agreement's price floors and the associated implementation by MITI of controls on production and capacity investment by Japanese DRAM producers, however, had several interesting effects, few of which directly benefited U.S. semiconductor manufacturers or were foreseen in 1986.[29] Higher prices for DRAMs pro-

---

[27]Flamm (1996) provides the most objective account of the agreement, and these paragraphs draw on his analysis.

[28]The agreement's "price floor" nevertheless may have aided the remaining U.S. domestic producer of DRAMs, Micron Corporation.

[29]The period following the agreement was also associated with severe shortages of 256K DRAMs, then a vital component of personal computer and other electronic systems. U.S. computer producers, among others, blamed the agreement and the informal, MITI-guided domestic production cartel that oversaw the agreement's implementation within Japan for the shortages. Concern over DRAM shortages and the alleged Japanese cartelization of the DRAM market (a condition to which U.S. policy, in the form of the bilateral trade agreement, arguably had contributed) led to the proposal by a group of U.S. computer manufacturers to jointly fund the creation of a DRAM manufacturing consortium, U.S. Memories. As supplies of DRAMs became more abundant, this proposal was abandoned in early 1990.

vided an opportunity for South Korean firms to expand their production of these devices, sowing the seeds for more intense competition in this product line in the future. Flamm (1996) argues that similar restrictions on production of electronically programmable memory chips (EPROMs) reduced Japanese exports of these devices and enabled U.S. producers of EPROMs to remain in this product line.

Although the Semiconductor Trade Agreement may have provided some benefits to U.S. EPROM producers, the effects of its pricing provisions seem to have had little effect on the overall U.S. industry. These provisions did not attract U.S. manufacturers back into DRAM production and imposed heavy short-term costs on major U.S. consumers of DRAMs. The market-share provisions of the 1986 and 1991 agreements, however, were eventually followed by a significant increase in U.S. semiconductor manufacturers' market share in Japan, and the agreement is viewed as a key factor in expanded Japanese imports of foreign components. In 1992, the foreign share of Japan's domestic consumption of semiconductor components increased beyond 20 percent, and recent data suggest that this share now is at roughly 25 percent (SIA Annual Databook, 1997). According to Flamm (1996), this increase cannot be attributed solely to growth in Japanese consumption of devices (such as microprocessors) in which U.S. firms have a strong competitive advantage but includes significant growth in other product areas. In other words, U.S. producers increased their Japanese market share in products where they historically had been relatively weak. The agreement's market-share provisions thus contributed to the revival of U.S. semiconductor firms after 1990, but the timing of this revival is such that the lack of such an import target would not have prevented the U.S. industry's recovery, which was well under way by 1990.

Antitrust Policy

U.S. antitrust policy played an important role in the earliest years of the semiconductor industry, as Bell Laboratories' liberal licensing of the original transistor and related patents was motivated in part by concern over the outcome of the federal government's antitrust suit against the firm that was settled in 1956. The 1956 settlement also led AT&T to manufacture semiconductor devices solely for internal consumption rather than entering the commercial market. These early actions by the technological pioneer in semiconductors powerfully influenced the subsequent development of the U.S. semiconductor industry.

The competitive crises of the semiconductor and other U.S. industries contributed to a far-reaching shift in U.S. antitrust statutes and enforcement policy in the 1980s. U.S. antitrust policy was widely criticized in the late 1970s for discouraging R&D collaboration. The U.S. Justice Department issued guidelines in 1980 that were intended to clarify the antitrust statutes and the Department's enforcement philosophy toward R&D collaboration, in order to remove impediments to such collaborative undertakings. Nevertheless, continuing industry and

Congressional dissatisfaction resulted in the 1984 passage of the National Cooperative Research Act (NCRA). The NCRA has been credited with facilitating the formation of SEMATECH, among other industry-wide collaborations, and R&D collaboration appears to have aided the revival of the U.S. semiconductor industry. The act was amended in 1993 to extend its coverage to joint production ventures.

An evaluation of the "real" effects of the NCRA and the broader shift in antitrust enforcement policy on the U.S. semiconductor industry's decline and revival is difficult without some clearer specification of the counterfactual situation. Would SEMATECH have been formed without the NCRA? Has R&D collaboration contributed to increased market power and/or poorer industry performance? Given the size of the firms that joined together to create SEMATECH and the sustained acquaintance of several of them with the federal antitrust authorities, the legislative endorsement of R&D collaboration under the terms of the NCRA almost certainly did aid in the creation of this consortium. The semiconductor industry's performance suggests that R&D collaboration need not result in cartelization and a weakening of competitive forces, although the large share of the U.S. semiconductor equipment market represented by SEMATECH member firms means that this consortium's vertical relationships deserve continued monitoring. Indeed, collaboration may provide one mechanism for combining the benefits of the U.S. industry's atomized structure and technological dynamism with those flowing from closer user-supplier relationships. Nevertheless, very few production joint ventures have been formed since the passage of the 1993 amendments to the NCRA, suggesting that this policy shift thus far has had little effect.

Intellectual Property Rights

Since 1980, the U.S. semiconductor industry has experienced considerable change in another important aspect of the public policy environment, intellectual property rights. Shifts in U.S. policy toward intellectual property rights began with the 1982 legislation that established the Court of Appeals for the Federal Circuit (CAFC), which strengthened the protection granted to patent holders.[30] The U.S. government also pursued stronger international protection for intellectual property rights in the Uruguay Round trade negotiations and in bilateral venues. These shifts in federal policy toward intellectual property rights involved both stronger international and domestic enforcement and a somewhat more favorable attitude in the judiciary and antitrust enforcement agencies toward patent

---

[30]According to Katz and Ordover (1990), at least 14 Congressional bills passed during the 1980s focused on strengthening domestic and international protection for intellectual property rights. The Court of Appeals for the Federal Circuit created in 1982 has upheld patent rights in roughly 80 percent of the cases argued before it, a considerable increase from the pre-1982 rate of 30 percent for the Federal bench.

holder rights. The shift was particularly significant for the U.S. semiconductor industry, because of the relatively limited economic role historically occupied by formal instruments of intellectual property protection such as patents. As we noted in our summary of the industry's development, the unusual circumstances of the industry's founding years, especially the extensive cross-licensing by Bell Labs and the Defense Department's requirements for "second-sourcing" of many devices, meant that knowledge flowed relatively freely among firms in the industry. Interfirm knowledge flows were further enhanced by high levels of labor mobility, by reverse engineering by firms of one another's chip designs, and by diffusion of advances in process technology through equipment suppliers.

In addition to these shifts in federal policy affecting all U.S. industries, the semiconductor industry was the beneficiary of a law designed to strengthen protection of industry-specific intellectual property. The Semiconductor Chip Protection Act (SCPA) of 1984 established protection for the design or "mask work" used in semiconductor manufacturing (Stern, 1986).[31] Passage of the SCPA established a new form of semiconductor intellectual property—a "chip design right," best described as a sui generis mode of protection that combines elements of patent and copyright principles with elements of trade secret law (Brown, 1990). The SCPA extends protection from copying to the three-dimensional images or patterns formed on or in the layers of the semiconductor component—that is, the "topography" of the chip—and provides for a reverse engineering clause whereby a competitor may reproduce a mask work for the purpose of analyzing it.[32]

Although it is an interesting experiment in sui generis protection of new forms of intellectual property, the SCPA's economic significance appears to be limited. Only one case has ever been litigated under its provisions.[33] The SCPA's unanticipated insignificance appears to be one result of the increasing complexity of manufacturing process technologies in the semiconductor industry. Copies of a device design and mask work are necessary but by no means sufficient to enable large-scale production of infringing products (Kasch, 1993). As a result, semiconductor firms during the 1980s and 1990s continue to rely on trade secrets and patents, the value of which has increased as a result of the policy shifts noted above.[34]

---

[31]Mask works represent the three-dimensional pattern of the layers (the topography) of a semiconductor component.

[32]Mask works may be reproduced for the purpose of teaching, analyzing, or evaluating the concepts or techniques embodied in the circuitry, logic flow, or organization of components. Legitimate reverse engineering may incorporate the results without infringement into another mask work to be produced and distributed.

[33]This case, *Brooktree Corporation v. Advanced Micro Devices*, resulted in the award of $26 million in damages for AMD's infringement under the SCPA and several patents.

[34]The registration of mask works under the SCPA provisions has advantages over patent filings, which require the disclosure of proprietary information and a time-consuming search through prior art to assert validity. Mask work filing provides immediate registration at minimal cost without a time-consuming search.

The SCPA nevertheless may play an important, albeit unintended, economic role in the fabless segment of the U.S. semiconductor industry. In order to maintain short design cycles, fabless firms must extensively reuse design data, "porting" designs from one product to another, or contracting with another firm for all or some part of the design. Reusable design components are generally referred to as "intellectual property (IP) blocks," and protection for these IP blocks under the SCPA facilitates the licensing process. The growth of licensing of IP blocks has supported further specialization by some design firms in specific components of overall device designs. These "virtual companies" operate by licensing their proprietary designs and architecture to other semiconductor design firms that produce an integrated design, contract with a foundry, and, in many cases, market the final product.

The broader shift of federal policy toward stronger enforcement of patent holder rights has been associated with a dramatic increase in patenting and licensing among integrated semiconductor manufacturers in the U.S. industry.[35] Licensing has become an important component of profits for some leading manufacturers. The royalty income of Texas Instruments has grown from roughly $200 million in 1987 to more than $600 million in 1995 (Grindley and Teece, 1997). Other firms, such as Intel, IBM and AT&T, now rely on licensing to generate revenues and protect product and process technologies.

The historic strengths of U.S. firms in product design and rapid innovation should be reinforced by stronger enforcement of patents and trade secrets. The distribution of these benefits within the industry, however, is less clear. Stronger intellectual property protection appears to have benefited established firms. Intel's strong position in its microprocessor product line relies in large part on the firm's intellectual property rights. Another historic strength of the U.S. industry, however, is the ease with which new firms can enter. The effects of stronger intellectual property rights on rates of new-firm formation and entry are less clear. On the one hand, new firms with strong patent positions often find it much easier to attract financing. On the other hand, the costs, in terms of litigation and patent prosecution expenses, of establishing such a patent position are very high. The empirical evidence on the social benefits from stronger intellectual property protection is thin and equivocal. Certainly, the increased litigiousness of established U.S. semiconductor firms has attracted criticism from other U.S. semiconductor producers. In the semiconductor industry, as in others, the U.S. is conducting an experiment in the effects of stronger intellectual property protection, and the implications of these new policies for long-term industry performance are surprisingly uncertain.

---

[35]The number of patents granted in the category "Semiconductor Devices and Manufacture" increased from 1655 in 1981 to 5427 in 1994 (U.S. Department of Commerce: Patent & Trademark Office, 1995).

CONCLUSION

Forecasts of the impending demise of the U.S. semiconductor industry in the late 1980s were considerably overstated. After declining through much of the 1980s, U.S. semiconductor firms undertook corrective actions on several fronts. They exited from product lines in which their historic skills at product innovation provided limited competitive advantage and their foreign competitors' superior access to capital made long-term competition difficult. U.S. firms also improved their product quality and appear to have enhanced their manufacturing performance, narrowing the gaps between them and foreign competitors, rather than moving ahead. The results of these steps have been dramatic. The U.S. semiconductor industry has regained its formerly dominant global market share, and the financial performance of U.S. semiconductor manufacturers now outstrips that of their South Korean and Japanese competitors. Moreover, the revival of the U.S. semiconductor manufacturing industry has reinvigorated the U.S. semiconductor equipment industry. Simultaneously, the South Korean and Japanese firms that specialize in the production of DRAMs are experiencing serious financial losses.

In many respects, the revival of the U.S. semiconductor industry relied on the elements of its structure that were the target of criticism in the 1989 report of the M.I.T. Commission on Industrial Productivity. The structure of the U.S. semiconductor manufacturing industry remains very different from that of the Western European or Japanese industries, although the structure of the emergent Taiwanese semiconductor industry is based on the U.S. model and still bears a passing resemblance to it. Populated by numerous, comparatively small, highly innovative firms, and exposed to competition by new entrants pursuing new product opportunities and new approaches to the semiconductor business, the U.S. industry remains adept at product innovation and rapid strategic repositioning. In addition, U.S. firms have relied on collaboration among semiconductor manufacturers, and between manufacturing firms and suppliers of equipment, to improve their manufacturing performance. The links between the collaborative initiatives of the 1980s and 1990s and the industry's improved performance remain elusive, however, and further research on these issues is essential if the current strengths of U.S. manufacturers and equipment producers are to be maintained.

Although the M.I.T. Commission's overall prognosis of the industry's future prospects was inaccurate, its analysis of the U.S. industry's weaknesses in manufacturing and long-term R&D investment highlighted other issues that could lead to future competitive difficulties. The very best U.S. semiconductor manufacturers appear to be capable of matching the yield and productivity of the best non-U.S. producers, but there is little evidence of consistently superior U.S. manufacturing performance. As a result, U.S. firms are likely to do best in periods of rapid innovation, especially because of their ability to exploit their presence in one of the world's most dynamic markets for applications of new products that use semiconductor components. But U.S. firms may have trouble competing on the basis of their manufacturing skills alone and therefore are likely to face challenges in future periods where they and foreign competitors are pursuing incremental innovations within a well-defined technological "trajectory." The U.S. industry is enormously effective in exploiting scientific advances for rapid commercialization but may underinvest in the basic research supporting these advances. This is a serious issue for debate, although the recent performance of the much larger Western European and Japanese firms in this industry that have made such investments suggests that simply creating large, diversified firms is an ineffective solution to this problem.

From its very inception, the U.S. semiconductor industry has had close relationships with federal government agencies in charge of R&D and procurement programs. Like other post-war U.S. high-technology industries, the U.S. semiconductor industry benefited from large-scale investments in defense-related R&D in both industry and academia as well as the procurement programs of federal military and space programs in the 1950s and 1960s. Federal policies in other areas, such as antitrust and trade policy, also have affected this industry throughout its history. But during the 1980s, apart from steel and automobiles, the semiconductor industry was almost without peer in the attention devoted to its welfare and competitive prospects by federal policymakers.

The record and legacy of federal intervention in this industry during the 1980s has been criticized by many observers. Nevertheless, one of the most remarkable features of federal policy in semiconductors was the rejection of some alternatives that almost certainly would have been far worse for the industry's competitive prospects. For example, consider the costs and consequences of a public-private venture like U.S. Memories, specializing in DRAMs, during the 1990s. Policymakers and industry managers might well have faced some very unpleasant choices between erecting trade barriers against competing imports or allowing this venture to slide into insolvency. The proposal of the National Advisory Commission on Semiconductors for a government-backed Consumer Electronics Capital Corporation, which would have been charged with financing the revival of a U.S. industry to consume the products of the domestic semiconductor industry, experienced an even more rapid and fortuitous demise. In hindsight, the avoidance by federal policymakers in the Executive and Congressional branches of government of programs that would involve the support with public funds of specific designs of commercial products was wise and consistent with well-established principles of technology policy.

The revival of the U.S. semiconductor industry is an impressive feat, for which government policymakers and industry managers, engineers, and researchers should share in the credit. But the unexpected nature of this revival, its rather complex causes, the contributions to it of cyclical factors, and the fragility of its foundation all suggest that competitive strength in this industry cannot be taken for granted. Indeed, some foreign producers, notably Taiwanese semiconductor firms, now are entering markets traditionally dominated by U.S. producers, a development that will intensify pressure on U.S. firms and increase the importance of manufacturing performance for competitive leadership. In other words, U.S. semiconductor firms must maintain their strategic agility and strength in product innovation while avoiding significant erosion in their manufacturing capabilities in order to maintain their strength. This task will require imagination and collaboration among government, industry, and academia.

REFERENCES

Appleyard, M.M., N. Hatch, and D.C. Mowery. "Managing New Process Introduction in the Semiconductor Industry," forthcoming in *Corporate Capabilities and Competitiveness,* G. Dosi, R. Nelson, and S. Winter, eds. London: Pinter.

Barron, C.A. (1980). "Microelectronics Survey: All That Is Electronic Does Not Glitter," *Economist* 1 March.

Braun, E., and S. MacDonald. (1978). *Revolution in Miniature: The History and Impact of Semiconductor Electronics,* Cambridge: Cambridge University Press.

Brown, H. (1990). "Fear and Loathing of the Paper Trail: Originality in Products of Reverse Engineering Under the Semiconductor Chip Protection Act as Analogized to the Fair Use of Nonfiction Literary Works," *Syracuse Law Review.*

Burrows, P. (1992). "Bill Spencer Struggles to reform SEMATECH," *Electronic Business* May 18.

Cole, R.C. *Managing Quality Fads: How American Business Learned to Play the Quality Game,* New York: Oxford University Press, forthcoming.

Erickson, K., and A. Kanagal. (1992). "Partnering for Total Quality," *Quality,* Sept.

Fabless Semiconductor Association. (1997). "State of the Fabless Business Model," mimeo, September.

Finan, W. (1993). "Matching Japan in Quality: How the Leading U.S. Semiconductor Firms Caught Up With the Best in Japan," M.I.T.-Japan Working Paper.

Flamm, K. (1996). *Mismanaged Trade? Strategic Policy and the Semiconductor Industry.* Washington, DC: Brookings Institution.

Florida, R., and M. Kenney. (1990). "Silicon Valley and Route 128 Won't Save Us," *California Management Review* 33(1).

Grindley, P., D.C. Mowery, and B. Silverman. (1994). "SEMATECH and Collaborative Research: Lessons in the Design of High-Technology Consortia," *Journal of Policy Analysis and Management.*

Grindley, P., and D.J. Teece. (1997). "Managing Intellectual Capital: Licensing and Cross-Licensing in Semiconductors and Electronics," *California Management Review.*

Hatch, N.W., and D.C. Mowery. (1998). "Process Innovation and Learning by Doing in Semiconductor Manufacturing," *Management Science,* forthcoming.

Integrated Circuit Engineering Corporation (ICE). (1987, 1988, 1990-1998). "ASIC Outlook: An Application Specific IC Report and Directory," Integrated Circuit Engineering.

Integrated Circuit Engineering Corporation (ICE). (1997). "Cost Effective IC Manufacturing 1998-1999," Integrated Circuit Engineering.

Integrated Circuit Engineering Corporation (ICE). (1997). "Memory 1997," Integrated Circuit Engineering.

Integrated Circuit Engineering Corporation (ICE). (1998). "Memory 1998," Integrated Circuit Engineering.

Integrated Circuit Engineering Corporation (ICE). (1997). "Microprocessor Outlook 1997," Integrated Circuit Engineering.

Integrated Circuit Engineering Corporation (ICE). (1998). "Microprocessor Outlook 1998," Integrated Circuit Engineering.

Integrated Circuit Engineering Corporation (ICE). (1976-1998). "Status: A Report on the Integrated Circuit Industry," Integrated Circuit Engineering.

Kasch, S. (1993). "The Semiconductor Chip Protection Act: Past, Present and Future," *High Technology Law Journal.*

Katz, M.L., and J.A. Ordover. (1990). "R&D Cooperation and Competition," *Brookings Papers on Economic Activity,* Washington, DC: The Brookings Institution.

Langlois, R., and W.E. Steinmueller. (1998). "The Evolution of Competitive Advantage in the Global Semiconductor Industry: 1947-1996" in *The Sources of Industrial Leadership,* D.C. Mowery and R.R. Nelson, eds. New York: Cambridge University Press.

Leachman, R., and C. Leachman. (1997). "National Performance in Semiconductor Manufacturing," University of California, Berkeley Competitive Semiconductor Research Program working paper (CSM-40).

McLoughlin, G.J. (1992). "SEMATECH: Issues in Evaluation and Assessment," Congressional Research Service. Science Policy Research Division, #92-749SPR. Washington, DC.

M.I.T. Commission on Industrial Productivity. (1989). *Working Papers of the Commission on Industrial Productivity,* Cambridge: M.I.T. Press, two volumes.

Methé, D.T. (1991). *Technological Competition in Global Industries: Marketing and Planning Strategies for American Industry,* Westport, Conn.: Quorum Books.

National Science Foundation. (1996). *National Patterns of R&D Resources,* Washington, DC: National Science Foundation.

Rosenbloom, R.S., and W.J. Spencer (1996). "The Transformation of Industrial Research," *Issues in Science and Technology* 12(3):68-74.

Semiconductor Business News. (1998). "Foundries may Build 40 percent of World's Chips by 2010."

Semiconductor Industry Association. (1997). "1997 Annual Databook".

Semiconductor Industry Association. (1992). "SIA Quarterly Quality Survey."

Semiconductor Industry Association. (1997). "The National Technology Roadmap for Semiconductors: Technology Needs," SEMATECH, Inc.

Stern, R. (1986). *Semiconductor Chip Protection,* New York: Law & Business.

Takahashi, D. (1998). "Chip Makers Enter Slump; Sales Fall 13 percent," *Wall Street Journal* July 6.

Tilton, J. (1971). *International Diffusion of Technology. The Case of Semiconductors,* Brookings Institution: Washington, DC.

U.S. Congressional Budget Office. (1990). "SEMATECH's Efforts to Strengthen the U.S. Semiconductor Industry," Washington, DC.

U.S. Department of Commerce: Patent & Trademark Office. (1995). "Technology Profile Report: Semiconductor Devices and Manufacture: 1/1969-12/1994," February.

U.S. General Accounting Office. (1992). "Federal Research: SEMATECH's Technological Progress and Proposed R&D Program," July.

VLSI Research. (1998). "Semiconductor Equipment Consumption and Production by Region," mimeo.

Hard Disk Drives[1]

DAVID McKENDRICK
University of California, San Diego

The hard disk drive (HDD) industry represents an interesting exception to received wisdom about American industrial competitiveness. Until recently, scholars have been pessimistic about the competitive prospects of much of U.S. industry, observing that "[i]t is too late for the United States to regain its position as the exemplar of best practice in the world" (Kogut, 1993). Florida and Kenney (1990) concluded that America may be good at generating new industries but is bad at sustaining them as they become more mature. American industry in general was said to have "attitudinal and organizational weaknesses" leading to "shortcomings in the quality and innovativeness of the nation's products" (Dertouzos et al., 1989). Yet, the experience of the disk drive industry suggests that these characterizations of American industry need not be its paradigmatic form. Like many industries that emerged in the twentieth century, the disk drive industry was dominated by American firms during its early years. Unlike other industries, however, the United States never relinquished its leadership. American companies hold more than 85 percent of the global market, an even greater share than they did in the late 1970s.

Why has the United States been so consistently successful in this industry? This paper argues that the industry's globalization was an important factor in

[1]This research was supported by the Alfred P. Sloan Foundation, grant numbers 95-6-13 and 97-1-10. The author is grateful to Allen Hicken, John Richards, Peter Gourevitch, Roger Bohn, Frank Mayadas, and David Mowery for careful and insightful comments on an earlier draft. For assistance with data collection and compilation, the author thanks Allen Hicken. He also thank James Porter, president of Disk/Trend, Inc., who not only reviewed this paper but also has generously shared his data, time, and knowledge about the disk drive industry. Mark Geenen, president of TrendFOCUS, Inc., kindly provided data on the media and heads segments of the industry.

sustaining the American competitive advantage. This is not to say that other, less global factors often invoked to explain the success of certain nations in particular industries do not apply to the disk drive industry. Home market demand, form of industrial organization, innovative capabilities, and the role of institutions, such as universities, government agencies, business associations, and other regional and national entities—all play or have played some role in the industry's evolution. The principal point here is that foreign investment not only complements innovation, style of industrial organization, and the favors conferred by historical chance, but it is also critical to *sustaining* industrial performance.

The disk drive industry offers a fascinating context for charting globalization as well as industrial evolution more broadly. American disk drive firms in particular have accumulated the organizational skills necessary for managing the geographic separation of R&D, production, and distribution to achieve economies of location. This kind of dispersion has been increasing in other industries as well, but, because the HDD industry is farther ahead than most in globalizing its activities, its experiences may provide a glimpse of what may come for other parts of the American economy and a touchstone for the maintenance of industrial leadership.

NATIONAL EMBEDDEDNESS AND PATH DEPENDENCE

The home market confers advantage upon national firms, and the success of customer industries confers success upon their suppliers. This national embeddedness is especially true for nascent industries. An industry emerges through the cumulative interactions of entrepreneurs and organizations. Interdependencies are established through the sharing of information and resources, and clusters of firms begin to form (Van de Ven and Garud, 1989). Clusters are largely national or regional phenomena, with firms serving national customers before growing through foreign trade and investment (Chandler, 1990). Ties among national firms persist as they expand into international markets. In his study of competitive advantage, Porter (1990: 138) finds that "a group of internationally successful domestic firms, selling worldwide, channel[ed] global demand to the domestic supplier industry."

The Origins of the Disk Drive Industry

One possible explanation for the success of the American HDD industry, therefore, is American success in the computer industry. This explanation seems reasonable on the face of it. At the time IBM shipped the first rigid disk drive in 1956, the United States was already the world's dominant computer producer and exporter. Although Europe contributed enormously to the technical development of the early computer industry, American firms led the world in computer installations, and many of these same firms developed their own HDDs. General Electric, Control Data, Burroughs, and Digital Equipment followed IBM's entry into HDDs in the 1960s. Some independent companies, such as Bryant Computer Products and Data Products, also emerged in the early 1960s to develop disk drives for sale to computer manufacturers that had not yet made their own, notably Sylvania, RCA, Honeywell, and Univac. In the late 1960s, after IBM secured its position as the clearly dominant mainframe maker, a new wave of independent companies emerged to make disk drives that were "plug compatible" with IBM systems: Memorex, Potter Instrument, Marshall Laboratories, and Information Storage Systems. Without incurring IBM's R&D expenses, the plug compatible companies were able to offer disk drives identical to or better than IBM's at a much lower price. Plug compatibility was not limited to IBM systems but extended to systems made by other computer manufacturers as well.

A parallel trend, but on a smaller scale, was evident in Japan and Europe. In Japan, the principal computer companies made their own disk drives: NEC, Fujitsu, Hitachi, and Toshiba all entered in the mid-to-late 1960s. Only in the 1970s did Japanese companies attempt to market disk drives to non-Japanese customers in the U.S. market; until then the size of the market for Japanese computers limited the market for their disk drives. A smaller domestic market also meant fewer independent Japanese disk drive companies entered in the 1970s as alternative sources of supply; the principal ones were Mitsubishi and Hokushin Electric Works.

In Europe Siemens and Philips made disk drives for their own computer systems, while Data Recording Instruments and BASF produced for the original equipment market (OEM). Data Recording Instruments was Europe's first firm to ship HDDs in 1968. Honeywell-Bull (later CII-Honeywell Bull and then Bull Peripherals) engaged in both captive and OEM production. In Eastern Europe COMECON organized the computer industry in such a way that DZU of Bulgaria was designated as the principal disk drive supplier for all computers in the region and became the most vertically integrated producer in the world. Only in rare cases did European disk drives find their way into American or Japanese computer systems.

Thus, throughout the 1960s and 1970s, the relative positions of the U.S., Japanese, and European disk drive industries could be explained by incorporation of their products into the systems manufactured by their respective national or regional computer industries.[2] During the 1970s captive production remained the largest channel for disk drives, though the relative importance of the original equipment market grew. Led by Control Data, Diablo Systems, CalComp, and Memorex, the OEM segment reached $631 million in sales revenues in 1979 but was still well below the $2.8 billion associated with captive production (Disk/

---

[2] A major exception to this general tendency was the success of Control Data in selling to European computer manufacturers. It claimed the bulk of the world's shipments of "noncaptive" drives in the 1960s and still almost half by the late 1970s. IBM's disk drives were solely for IBM computers.

Trend, 1980). In 1979, American firms had 81.1 percent of the global HDD market, Japan 14.3 percent, and Europe the remainder. Between them, IBM and Control Data controlled just short of 40 percent of the market.

The Personal Computer and the Desktop Disk Drive

Up to this point the story conforms strongly to an explanation of competitive advantage through path dependence and increasing returns; the large U.S. market for mainframes, and later minicomputers, gave the American disk drive firms an unassailable long-term advantage. But it does not account for the divergence in the fortunes of the American disk drive and computer industries after 1980 when both came under greater global competitive pressures.

For the computer industry, a watershed event was the debut of the IBM PC in 1981. The PC defined the dominant design in the industry for many years (Langlois, 1992; Anderson, 1995). In addition to setting the standard for what a desktop computer should look like, it featured an open architecture that attracted the entry not only of some of IBM's established mainframe and minicomputer rivals but de novo start-ups that set out to manufacture IBM clones. Compaq and Dell became two of the most important American entrants, but more interesting are the many new clone makers that emerged outside the U.S., especially Taiwan, Korea, and Japan. Daewoo, Epson, Hyundai, Acer, and scores of other smaller companies collectively dispersed the production of computers. As a result, the global market share of U.S. computer makers steadily eroded during the 1980s and early 1990s. The U.S. share of the worldwide computer market, including mainframes, fell from 88 percent in 1983 to around 56 percent by 1992. During the same period Japanese market share in the computer industry increased from 8 percent to 30 percent.[3]

The same open architecture that attracted the new clone manufacturers also stimulated entry into peripheral equipment. Where mainframe and minicomputer manufacturers made many of their own peripherals and components, the assemblers of personal computers almost entirely outsourced their production. Japanese, Korean, and Taiwanese producers of keyboards, floppy disk drives, monitors, DRAMs, and motherboards displaced U.S. firms in peripherals and components even more dramatically than American companies had been displaced in the PC market.

Given these trends, and the development of national clusters of computer-related capabilities in these countries, one might have expected other Asian companies to erode America's position in HDDs. Much the same competitive dynamics faced the HDD industry as disk drives were adapted to fit into a PC. Drive sizes decreased from 14-inch and 8-inch diameters in the 1970s to 5.25 inches in 1980 and 3.5 inches in 1983. An explosion of some 100 new entrants,

---

[3]Global computer market shares were calculated from the Datamation 100 for various years.

intense competition, and shakeouts occurred between 1980 and 1996. By 1996 fewer firms made disk drives than at any time during the previous 20 years (Figure 1). The HDD landscape became littered with the graves of once prominent American companies.

But whereas most of the rest of the American computer peripherals industry has largely vanished, the American HDD industry remained dominant in the face of competition from Asia and Europe.[4] Although U.S. firms such as Priam, Prairietek, Conner Peripherals, Ministor, and Hewlett Packard exited, so did firms from other countries. Mitsubishi, Matsushita, Rodime (the first firm to introduce the 3.5-inch disk drive), Olivetti, BASF, Sony, Philips, and Siemens are among the formidable foreign companies unable to remain in the industry. Asian and European PC makers bought HDDs from U.S. firms. South Korea, for example, depended almost entirely on American companies to meet the HDD requirements of its major PC exporters (MR, 1991). In Europe PC companies such as Amstrad also purchased American disk drives. Compared with the computer industry, the American HDD industry held a roughly steady 75 percent of the global market throughout the 1980s and then increased its share to more than 80 percent by 1992 (Figure 2). By 1995 U.S. global market share reached 85 percent, where it had been in the early 1970s.

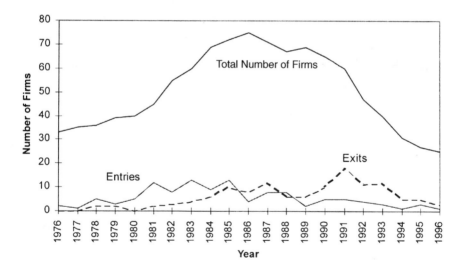

FIGURE 1 Number of firms in the HDD industry, 1976-1996.

---

[4]An important exception is the printer industry. Although the United States lost the impact printer market, it has a huge lead in laser printers.

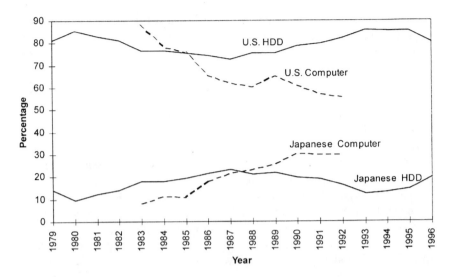

FIGURE 2 HDD and computer market shares, 1979-1996.
Source: The Data Storage Industry Globalization Project Worldwide; Disk/Trend, Inc.

Although the market share for American computer manufacturers fell throughout the 1980s, the American floppy drive industry practically disappeared, and the world increasingly turned to non-American suppliers of other computer components and peripherals, American firms continued to be the overwhelming source for HDDs. It is clear that the disk drive industry owed its birth to the American computer industry. But given the fate of other computer peripherals, a path-dependence argument is incomplete. The United States has been the inventor of other promising technologies that it relinquished to the Japanese. Why could it hold onto HDDs but not floppy disk drives, monitors, or optical storage devices? One possible factor is innovation.

INNOVATIVE CAPACITY

Many argue that Japanese and American firms have distinctly different methods of innovation and innovative capabilities compared with their Western counterparts. Japanese firms are said to possess several organizational, incentive, and communication advantages that are conducive to innovation (Aoki and Rosenberg, 1987; Aoki, 1990). The Japanese system has its strengths and weaknesses relative to the stylized facts about American innovative capabilities. New product introductions tend to be faster; strengths in incremental product modifications based on careful engineering make Japanese firms better at innovations along a predictable technological trajectory (Imai et al., 1985; Odagiri and Goto, 1993;

Mansfield, 1988). However, the links between scientific research and invention are weaker in Japan than they are in the United States. Although Japanese firms are adept at "the better known, closer-at-hand technologies," they are less suited to choose "bolder, riskier, and more visionary technologies" that lead to pivotal new products or process technologies (Okimoto and Nishi, 1994).

Given these characteristics, Japanese firms would be expected to perform better in situations where continuous incremental improvements, tight engineering tolerances, and manufacturing strength are the bases of competitive success and less well in segments incorporating radical or unproven technologies that rely on more fundamental technical research.

Innovations in Disk Drives

Rapid product or process innovation is a necessary condition for competitive success in all high-technology industries. This is especially true for disk drives. Although technological advances in semiconductors have generally been credited for most of the price and performance improvements in computers, fewer people are aware that progress in disk drive speed and capacity kept pace. The amount of data that can be stored on a square inch of a disk grew almost 30 percent a year between 1957 and 1990; since then it has increased about 60 percent a year. Data-transfer rates have increased while average access times have fallen. Between 1980 and 1995, the price per megabyte of storage fell at an annual rate of 40 percent (CRN, 1997). All of these advances were accomplished on increasingly smaller disk drives. Since the 1970s the disk drive's size, called form factor, has decreased, from 14 inches to 5.25 and 3.5 inches in the 1980s. These "architectural innovations" (Henderson and Clark, 1990) challenged the competencies of incumbent disk drive companies, and the inability of firms to make the transition to smaller form factors has been cited as a central reason behind firm failure (Christensen and Bower, 1996). Can U.S. success in the industry be explained by the greater innovative capabilities of its firms?

One American company, IBM, served as the technological fountainhead for the industry and continues to demonstrate remarkable technological leadership.[5] As Table 1 shows, IBM established the industry and introduced many key innovations—the first removable disk pack drive, the Winchester standard, the first drive with ferrite, thin film, and magneto-resistive heads, and the first 8-inch disk drive, which proliferated with the development of minicomputers. More than any other institution, IBM displayed engineering brilliance in overcoming critical technical constraints. The 1301 disk drive in particular pioneered in areas that led to follow-on improvements in storage density and access times. Nonetheless, the co-evolution of technology and competition in the HDD industry confounds the

---

[5]For a technical history of IBM's first 25 years of innovation in the industry, see Harker et al. (1981) and Stevens (1981).

TABLE 1 IBM "Firsts" in the HDD Industry

Firsts in HDD	Model	Year	Megabytes	Number of Disks	Disk Diameter (inches)
First disk drive	IBM RAMAC	1956	5	50	24
First disk drive with air-bearing heads	IBM 1301	1962	28	25	24
First disk drive with removable disk pack	IBM 1311	1963	2.68	6	14
First disk cartridge drive	IBM 2310	1965	1.024	1	14
First disk pack drive	IBM 2311	1965	7.25	6	14
First disk drive with ferrite core heads	IBM 2314	1966	29.2	11	14
First track following servo system.	IBM 3330-1	1971	100	12	14
First disk drive with low mass heads, lubricated disks, sealed	IBM 3340 Winchester	1973	70	4	14
First disk drive with thin film heads	IBM 3370	1979	571.4	7	14
First 8-inch HDD	IBM 3310	1979	64.5	6	8
First disk drive to use MR heads and PRML	IBM 681	1990	857	12	5.25

Source: Quantum Corporation web page based on information in Disk/Trend, Inc.

conventional wisdom about differences in Japanese and American innovative capability in two ways.

First, Japanese firms have been stronger than theory would predict in technologically advanced new products. One measure of this strength is the shift to different form factors, each representing architecturally distinct product generations. IBM introduced the 14-inch and 8-inch form factors, but since then young entrepreneurial firms, rather than older incumbents, have pioneered most architectural innovations (Christensen and Rosenbloom, 1995).

Theory suggests that Japanese firms would lag behind their American competitors in the shift to new form factors. At first glance this seems to be the case. Eight of the first ten companies to introduce 5.25-inch drives were American, led by Seagate in July 1980. The other two were European—Rodime and Olivetti. Three other firms shipped 5.25-inch drives by the end of 1981, but only one of those, Nippon Peripherals, was a Japanese firm. By the end of 1982, 13 more firms had begun shipping 5.25-inch drives, 7 of these were Japanese firms, including Fujitsu and Hitachi. In 1983, 14 more firms, 5 of which were Japanese, made the shift to 5.25-inch drives. Thus, from 1980 to 1993 only 13 of 41 HDD firms that shipped 5.25-inch drives were Japanese, and these firms were a year or more behind in introducing their drives. Among firms that *still* made disk drives at the end of 1996, however, the Japanese were quicker than most of their U.S. counterparts in moving to 5.25-inch drives. Fujitsu, Hitachi, and NEC all shipped 5.25-inch drives before or concurrently with Quantum, Maxtor, Micropolis, and IBM. Western Digital, another leader in 1996, did not make disk drives until 1988 when it acquired Tandon's HDD operations.

A similar story can be told regarding the shift to the 3.5-inch form factor. The European company Rodime was the first to ship 3.5-inch drives, in September 1983, and the next three were American—Microcomputer Memories, Microscience International, and MiniScribe—all of which shipped in 1984. The first Japanese firm to ship a 3.5-inch drive was Nippon Peripherals in February 1985. All five of these early innovators have since exited the industry. Of the HDD firms surviving at the end of 1996, the first to ship 3.5-inch drives was Hewlett Packard in March of 1985; but Hitachi, Fuji Electric, NEC, and Fujitsu followed close behind. More important, as occurred with the transition to 5.25-inch drives, these Japanese firms were quicker to make the shift to 3.5-inch drives than *every* U.S. firm that is a leader today. IBM did not introduce 3.5-inch drives until May 1986. Seagate first shipped 3.5-inch drives during the third quarter of 1987, the same date the Japanese firm Toshiba began shipping and one year after the Japanese firm Seiko Epson had begun shipping 3.5-inch drives. Quantum and Maxtor did not make the move into 3.5-inch drives until 1988, and Micropolis waited until 1991. In the shift to the 2.5-inch form factor the Japanese firm JVC was among the first movers. Other Japanese firms were no slower at adopting the new form factor than surviving American firms.

Japanese firms have also been among the leaders in incorporating advanced technology in their disk drives, specifically the new, thin film magneto-resistive (MR) recording heads. MR heads are designed to read media with very high recording densities and are the reason that growth in areal density—the amount of data that can be squeezed onto a given space of a disk—had jumped up to a 60 percent annual rate since 1990. Unlike previous head technologies that function like small electromagnets, MR heads use a thin strip of magneto-resistive material deposited on the head that senses the strength of the magnetic patterns on the disk and creates corresponding electrical pulses. The MR strip cannot write data, however, and so a traditional thin film component must be placed on the head next to the MR strip (Quantum, 1997; EBN, 1996; EET, 1996). Because the switch to MR heads requires corresponding changes in media and electronics technologies and because they are very difficult to make, many companies have been slow to commit resources to the new technology, choosing instead to try to increase capacity through conventional technologies.

Stylized notions of American and Japanese innovative capabilities suggest that U.S. firms would be more likely to move first into smaller market segments with more sophisticated technology while abandoning to firms from other countries the market segments dominated by older technology. American firms would thus be expected to lead the way into MR technology. Similarly, some would argue that Japanese drive designers would push technological improvements using the inductive thin-film technology with which they are familiar rather than make the complex shift to MR heads. In one sense these suppositions are true; IBM invented MR technology and entered the market with it almost three years before the nearest competitor. Yet, three of the next six companies to introduce disk drives with MR heads were Japanese companies.

Moreover, Japanese technological strength is further revealed by looking at the areal density of a disk drive. Areal density encapsulates in one picture a company's ability to bring together head and media technologies and is a major feature of the technology race in HDDs. As Table 2 shows, the Japanese are also among the leaders in areal density. The table ranks firms according to the disk drive with the highest areal density each offers as of 1997 (Disk/Trend, 1997). Once again, though IBM is clearly far ahead, three of the top five are Japanese. This ranking changes frequently, as the newest product to the market seems to embody the highest areal density, but the illustration nonetheless demonstrates Japanese innovativeness.[6]

A second exception to the conventional wisdom is that Japanese firms have also been *weaker* than theory would predict. Within a given form factor, technology has evolved in ways that should have given the Japanese an advantage. All companies have technology roadmaps, and technological progress has moved along well-known paths, especially in the technological development of the current generation of disk drives employing inductive thin film heads and disks. IBM was the first company to ship disk drives with thin film inductive heads in 1979; drives with thin film media appeared four years later. Innovations in areal

TABLE 2 Highest Areal Density, as of May 1997

Company	Areal density (megabits per square inch)
IBM	2638.0
Hitachi	2013.0
Quantum	1646.0
Toshiba	1308.0
Fujitsu	1300.0
Maxtor	1193.0
Seagate	1108.0
JTS	1008.0
Micropolis	959.2
Samsung	884.0

Source: Disk/Trend Report, 1997.

---

[6]Firms also competed in the desktop market in terms of "volumetric" density or how much capacity one could cram into the slot allotted to the disk drive. One trick in mechanical design was the introduction of "half-high" disk drives in which more disks were stacked closer together. A company might be a leader in areal density (data on a disk) but a laggard in volumetric density. Some say that IBM did not understand this distinction. Unfortunately, systematic data to test this notion are unavailable. I thank Frank Mayadas for bringing this to my attention.

density during the next decade involved improvements to these two increasingly understood technologies. Japanese firms would thus have been expected to advance more quickly along this technological trajectory while simultaneously obtaining cost advantages through more efficient manufacturing, but the reverse is in fact true. American firms have dominated this largest segment of the disk drive market and are making interesting adaptations to the basic technology.[7] In this way, American firms have been most responsible for extending the life of inductive head technology, which innovation theory would not predict.

Overall, there is little evidence that the Japanese are less innovative than successful American companies according to these key measures. They have not been far behind their U.S. competitors on the technological frontier, and they have even introduced advanced new products before leading U.S. companies. Although innovation has been necessary for all companies to stay in the game, it has not been a sufficient condition.[8]

FORM OF INDUSTRIAL ORGANIZATION

Many scholars argue that the Japanese form of industrial organization, with its complex interfirm relations, may have distinct advantages (Aoki, 1988; Gerlach, 1992; Teece, 1992). Although the evidence comes almost entirely from the automobile industry, the general claim is that Japanese firms are less vertically integrated than their American counterparts and maintain closer relationships with suppliers, often through some equity holdings (Aoki, 1990; Hill, 1995; Dertouzos et al., 1989). By combining market incentives with relational contracting, Japanese companies are reportedly more cost effective, flexible, and faster in coordinating operations than their more vertically integrated competitors. Was there, ironically, something about the *American* form of industrial organization that sustained U.S. advantage in the HDD industry?

Backward Integration: Components and HDD Assembly

The basic issue is whether Japanese and American disk drive firms practiced different methods of organizing production and delivery. I focus on four of the most important disk drive components—the recording heads that read and write the data, the disk to which data are written and stored, the motor used to rotate the

---

[7]Improvements to inductive technology include "proximity" or virtual-contact heads. These involve significant enhancements to etched air-bearing and transducer technologies.

[8]It is important to note that I have not addressed the ability of firms to introduce successive generations of products. When product cycles are so short, firms face intense pressures to stay competitive in terms of capacity, performance, and interfaces. Keeping design teams together in such a pressurized environment is difficult. It is possible that American firms have been better at this than those from Japan and Europe. I hope to explore this possibility in a later paper.

disk, and the semiconductors that control the drive and manage the flow of information between it and the computer. I also consider the extent of contract assembly of disk drives.

In contrast to the microcomputer industry (see, for example, Langlois, 1992), vertical integration has been an important, although not universally implemented, strategy for HDD firms. I compared the degree of backward integration into components for a sample of 28 firms, which included both surviving firms and firms that exited the industry. Of the 28 firms, 16 were still producing HDDs in 1995. Ten of the firms are Japanese, one is Canadian, and the remainder are from the United States. The Canadian firm, Northern Telecom, is listed with U.S. firms because its disk drive operations were in the U.S. as a result of acquisitions of two American companies. The degree of backward integration across these firms was compared at four different points—1983, 1987, 1991, and 1995. The combined global market share of the 28 firms was 85 percent in 1983, 91 percent in 1987, 98 percent in 1991, and 99 percent in 1995. The data show that backward integration has clearly been an important strategy in the industry and one that has become more prevalent over time. In 1983, 75 percent of the HDD firms in our sample were vertically integrated in one or more key components. This number increased to 91 percent in 1991 and 94 percent in 1995.

In-house assembly of HDDs has also been the dominant model in the industry, regardless of nationality. Contract assembly relationships have been common, but they have not accounted for a large share of total production. Of the more than 100 firms that shipped disk drives under their brand names since 1976, only 20 used contract assemblers. The majority of firms that engaged contract assemblers did so because they were small and had limited resources or competed in niche segments. The important exceptions to the general model are Quantum and IBM. Probably 30 percent of all disk drives shipped in 1996 were done on a contract basis for these two firms. *All* of Quantum's disk drives are assembled by Matsushita-Kotobuki-Electronics. IBM has used its former English disk drive subsidiary, spun off in 1994 and now called Xyratex, to assemble drives. In 1997 Xyratex assembled about 2 million drives for IBM. IBM has also used a Thai subcontractor, Saha Union, to assemble 2.5-inch and 3.5-inch drives designed by its Japanese disk drive operation in Fujisawa.

With the exceptions of Quantum and IBM, the vast majority of units shipped have come from HDD companies' own factories. All leading firms except Quantum have maintained a strong manufacturing capability. IBM, for example, no longer sources drives from Xyratex and is building a plant in Thailand to make its own. Moreover, most of those performing contract work have themselves been HDD firms rather than specialist assemblers. Several of those who contracted work to others also engaged in contract assembly themselves, including IBM. Interestingly, many American disk drive firms performed contract assembly. Of the companies that have contracted *all* assembly to others, only Quantum and tiny Nomai (France) survive.

Despite this dominant theme of in-house assembly and the general trend toward backward integration into components, the level of integration varies among firms. There does not appear to be any systematic difference in backward integration between leading Japanese and American firms, however. Highly integrated firms exist on both sides of the Pacific. Seagate, IBM, Fujitsu, and Hitachi make virtually all of the key HDD components in-house, although they are not completely self-sufficient. Some Japanese and American HDD firms for a time only assembled HDDs and did not integrate upstream into components—namely Kyocera and JTS. Those that never integrated into one of the four components were typically the smallest firms in the industry. Most firms in both countries lie somewhere between the two extremes, producing one or two key HDD components. If one focuses only on integration into heads and media, the story is much the same.

Moreover, the closest observable interfirm relationships appear to be *between* American and Japanese firms—Quantum and MKE, and Integral Peripherals and Fuji Electric, which owns a small share of Integral and at one time assembled Integral drives intended for the Japanese market. Samsung also provided most of the initial $1 million in start-up financing for the American firm Comport and assembled all of Comport's drives. More recently, IBM has announced a contract assembly relationship with NEC, which will manufacture IBM drives later in their life cycles and use IBM components. At the component level, the top three independent media companies are Komag (U.S.), Fuji Electric (Japan), and Mitsubishi Chemical (Japan). Only Fuji Electric has an equity relationship with an HDD firm, holding a small percent of Fujitsu. Yet, Fuji Electric's largest customers are MKE/Quantum and Seagate Technology, not Fujitsu (TrendFocus, 1996a). For recording heads, Read-Rite (U.S.), TDK (Japan) and its Hong Kong subsidiary SAE Magnetics, and Yamaha (Japan) are the three largest independent producers (TrendFocus, 1996b), and none has an equity relationship with any HDD producer.

Forward Integration: Computer Systems

Is there a systematic difference between Japanese and American firms in forward integration into computer assembly? It is true that virtually all of the surviving Japanese HDD firms make computers, and that none of the American firms, save IBM, do. This fact misses much of the evolution of the industry, however. Throughout the 1970s and 1980s many of the largest American HDD manufactures were computer makers. Conversely, numerous Japanese companies began to make HDDs during the 1980s but did not make computers.

Nonetheless, conventional wisdom in the industry is that the Japanese HDD industry has been at a competitive disadvantage because of its heavy reliance on captive sales. This dependence, it has been argued, tends to slow the speed and degree of innovation because captive drive makers may not be subject to the same competitive pressures as noncaptive firms. George Scalise, former Maxtor president and chief executive, has expressed an additional disadvantage facing captive makers: "The history of computer systems manufacturers in most instances has been that building peripheral products like disk drives is not a core business that can generate volume and economies of scale to be cost competitive" (EN, 8/21/89). Finally, captive HDD manufacturers are said to find it difficult to sell drives to outside computer firms. According to Micropolis founder, Stuart Mabon, "To enter into a large OEM relationship with a disk drive company means disclosing future computer plans. Most large OEM computer companies would prefer not to disclose those plans with a competitor" (EN, 8/21/89).

The captive market has indeed been whittled away, but the perception of Japanese companies as over-reliant on internal sales can be challenged on at least two fronts. First, as already discussed, captive sales have not made Japanese firms notably slower to innovate than successful American HDD firms. Second, the Japanese HDD industry has not been more reliant on captive sales than the American industry. If anything, the opposite has been true. Between 1983 and 1993 captive sales accounted for a slightly higher percentage of total HDD revenue among American firms than Japanese (Figure 3). Only for the periods before 1983 and after 1994 is it possible to argue that the Japanese have been more reliant on captive sales than the Americans. The differences between Japanese and American firms in their reliance on captive sales are too small to provide a compelling account of competitive advantage.

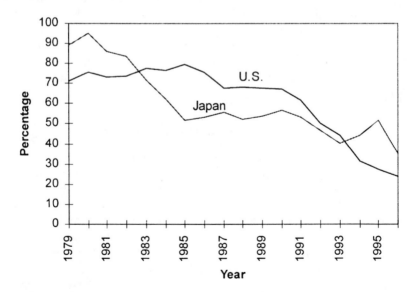

FIGURE 3 Captive revenue as a percentage of total revenue.
Source: Disk/Trend, Inc.

In summary, it is unlikely that variation in mode of organizing among U.S. and Japanese firms is responsible for the sustained U.S. dominance in the HDD industry. Until very recently, the industry has been characterized by a great amount of diversity in organizing, with specialists coexisting alongside vertically integrated firms. During the last decade, both Japanese and American firms have increasingly relied on backward vertical integration while reducing their reliance on captive sales to an in-house computer business. In fact, their integration strategies have been remarkably similar.

GLOBALIZATION OF ASSEMBLY

If the degrees of vertical integration, innovation, and path dependence in the disk drive industry explain only a part of the competitive advantage of U.S. firms, what else accounts for the American dominance?[9] One important ingredient has been the globalization of assembly. Innovation is critical, but companies have to be equally effective at transferring new products quickly into volume production while keeping costs down in the face of rapid price erosion. The president of Seagate, the world's largest disk drive company, says that his company is happy to be a follower rather than an innovator but to *outproduce* its competitors (SoS, 1996). The centerpiece of this production strategy has been overseas assembly.

In general, American firms have not been known for their manufacturing prowess. Yet U.S. disk drive companies have demonstrated that this generalization does not hold for all industries. American disk drive companies competed squarely in and came to dominate the low-margin, high-volume segments—the price and capacity points most in demand by users of personal computers. Judged by what scholars have had to say about the manufacturing failures of American firms in other industries, this is an extraordinary accomplishment. American industry achieved it primarily by being the first as a group to shift assembly offshore to lower-cost locations, where it quickly constituted an entire value chain of activities. If Silicon Valley is the geographical synonym for innovation, then non-Japan East Asia has come to signify low-cost assembly and logistics management. A look at the movement of the industry overseas and how it furthered U.S. competitiveness is instructive.

Home-Based Assembly: 1956–1982

In the 1960s and 1970s, before the introduction of the 5.25-inch disk drive, assembly of disk drives by American firms occurred primarily in Silicon Valley, the Los Angeles area, Minneapolis, Oklahoma City, and the region around Boston. Some HDD firms that were vertically integrated into computers also as-

---

[9]This section draws heavily on McKendrick and Hicken (1997).

sembled disk drives in Europe. IBM manufactured drives in Germany, England and Italy; Control Data manufactured in Portugal and had a joint venture in Romania; and Burroughs had operations in Scotland. Burroughs also had assembly operations in Brazil and Canada. Japanese and European companies, with the exception of Germany's BASF, which had an operation in Silicon Valley, all assembled in their home countries. The principal rationale among the firms that located assembly outside their home market was proximity to customers: placing assembly in those markets where governments, banks, and insurance companies—the primary customers for their computer systems—were likely to look favorably upon firms committed to local assembly of systems and peripherals.

In 1982 and 1983 Seagate, Computer Memories, Ampex, and Tandon, all independent producers, became the first companies to move HDD assembly to locations for reasons other than access to host country markets. These firms began to assemble drives in what they saw as the best location from a cost standpoint, selecting low-wage areas in Asia, particularly Singapore. By the end of 1983 assembly was scattered geographically, but still overwhelmingly in the countries where the firm had its headquarters. Virtually all of the production of HDDs in 1983 was concentrated in two countries, the United States (72.3 percent of shipments) and Japan (12 percent of shipments). With almost 5 percent of global shipments, Europe produced more disk drives than all of Asia outside of Japan. In 1983 U.S. firms produced some 93 percent of their drives in the U.S., while Japanese firms produced all of theirs in Japan.

A Shift in the Center of Gravity: 1983–1990

The experiences of Seagate, Tandon, and Computer Memories in Southeast Asia began to influence other American HDD firms. The perceived success of Seagate's Singapore facility, in particular, spurred several other HDD producers to adopt a similar cost-based siting strategy. Table 3 shows the movement of overseas disk drive assembly among firms headquartered in America, Japan, and elsewhere. Many American firms followed Seagate's lead and chose Singapore as their first overseas manufacturing site. American HDD companies also opened overseas facilities in other low-cost Asian locations such as Taiwan and Hong Kong.

In the span of just seven years, a dramatic change in the locus of assembly occurred. By 1990 Singapore was the world's largest producer of HDDs, accounting for 55 percent of global output, measured in shipments, with the rest of Southeast Asia accounting for only a percentage point more. As more firms located in Southeast Asia, supporting industries emerged in the region so that by 1990 three-fourths of the parts needed to produce a disk drive could be purchased there (LAT, 1990; BT, 1993). The revealed global strategies of American and Japanese firms could not have been more different. By 1990, eight years after the first HDD was produced in Singapore, American firms assembled two-thirds of

TABLE 3 Timing and Direction of Overseas HDD assembly

Year	American firms	Location	Japanese firms	Location	Other firms	Location
As of 1982	Burroughs	Scotland, Brazil, Canada			BASF	U.S.
	Control Data	Portugal				
	IBM	Germany, England				
1982	Seagate Technology	Singapore				
1983	Ampex	Hong Kong				
	Computer Memories	Singapore				
	Hewlett-Packard	England				
	Tandon	India				
	Tandon	Singapore				
1984	IBM	Japan			Rodime	U.S.
	Maxtor	Singapore				
	MiniScribe	Singapore				
1985	Microscience International	Singapore				
1986	Micropolis	Singapore	Fujitsu	U.S.		
	Tandon	Korea				
1987	Conner Peripherals	Singapore	NEC	U.S.	Rodime	Singapore
	Control Data	Singapore				
	Cybernex	Singapore				
	Microscience International	Taiwan				
	Priam	Taiwan				
	Seagate Technology	Thailand				
1988	IBM	Brazil				
	Unisys	Singapore				
	Western Digital	Singapore				
1989	Conner Peripherals	Malaysia				
	Kalok	Philippines				
	Micropolis	Thailand				
	Syquest	Singapore				

continues

TABLE 3 Continued

Year	American firms	Location	Japanese firms	Location	Other firms	Location
1990	Conner Peripherals	Scotland				
	Microscience International	China				
1991	DEC	Germany	Fujitsu	Thailand		
	Prairietek	Singapore	Xebec Co. Ltd.	Philippines		
1992	Integral Peripherals	Singapore	Toshiba	U.S.		
	MiniStor	Singapore				
1993	Conner	China				
1994	DEC	Malaysia	TEAC	Philippines		
	Hewlett-Packard	Malaysia				
	JTS	India				
	Quantum	Malaysia				
	Western Digital	Malaysia				
1995	Avatar	Thailand	Hitachi	Philippines		
	IBM	Singapore	NEC	Philippines		
	IBM	Hungary				
	Maxtor	Thailand				
	Seagate Technology	Ireland				
1996	Iomega	Malaysia	Toshiba	Philippines		
	IBM	Thailand	Fujitsu	Philippines		
1997	Maxtor	China			Tae Il Media	China

Note: Table excludes subcontracting arrangements.
Source: The Data Storage Industry Globalization Project, U.C., San Diego.

their disk drives in Southeast Asia. What began as a variation from the norm became a collective phenomenon. In contrast, Japanese companies assembled almost none in Southeast Asia, and only 2 percent in the rest of Asia. Japanese companies instead continued to manufacture predominantly in Japan, where they produced 95 percent of their disk drives.

As a group, Japanese firms were clearly hesitant to abandon a strategy that, up to the mid-1980s, appeared to be working—namely, exporting from Japan. In 1984, for example, TEAC Corp. was shipping almost 60 percent of its output to the United States. Even as late as 1989, both Matsushita and Hitachi invested in Japanese manufacturing capability for 3.5-inch drives, judging that applying more automation to drive assembly would enable them to overcome the otherwise higher costs of manufacture in Japan. As the yen strengthened against the dollar and they turned their attention abroad, the United States, not Asia, was the site of their first overseas manufacturing investments. Fujitsu opened a U.S. plant in 1986, NEC followed in 1987, and Toshiba entered in 1992. At one point, Fujitsu reportedly intended to manufacture nearly all of its disk drives in the United States (CW, 1985). Toshiba explained that its strategy in HDDs was proximity to the market—to respond to market needs more effectively by designing and building products closer to the markets where they were sold (LAT, 1991). Nor was Southeast Asia the chosen strategy for new Japanese entrants. After they entered in 1985, Fuji Electric, JVC, Seiko Epson, and Alps Electric all confined their manufacturing to Japan.

Strategic Convergence: 1990–1996

For high-volume, low-priced, and low-to-medium capacity drives, where cutting costs was paramount, Southeast Asia was clearly the location of choice for American companies, and their strategy increasingly confined the Japanese to niches in the high-capacity segments. This was a surprising switch because high-volume, low-cost manufacturing is an area where the Japanese traditionally excel. Eventually the success of the American firms impelled the Japanese to follow with investments in Southeast Asia. Between 1991, when Fujitsu began production in Thailand, and 1996, all the principal Japanese HDD firms gradually shifted manufacturing to Southeast Asia, principally the Philippines.

Fujitsu's move to Thailand was motivated when one of its major production facilities in Japan reached maximum capacity (CI, 1/3/92). In addition to expanding the Japanese facilities and investing in the United States, as it had done in the past, Fujitsu decided to manufacture drives in Thailand and retooled an existing recording heads facility for production of low-capacity 3.5-inch drives (IDC, 2/28/91). Production stayed at low levels until 1993 when the appreciation of the yen forced Fujitsu to move a large share of its manufacturing to Thailand. By the end of 1995 Fujitsu was doing nearly all of its volume manufacturing at the Thailand facility and a new facility in the Philippines. The president and

CEO of Fujitsu Computer Products of America cited the move to Southeast Asia as one of the prime factors behind the company's rapid growth in 1996. Fujitsu doubled its worldwide hard-drive revenues for 1996 and experienced a 123 percent growth in shipments compared with overall 1996 market growth of 17 percent (BWI, 6/17/97).

NEC, Hitachi, and Toshiba soon joined Fujitsu overseas. NEC completed its own HDD facility in the Philippines in 1995 and increased its off-shore production to 75 percent of total HDD output (COM, 10/9/95). Hitachi also made its first HDD investment in the Philippines in 1995 and had 90 percent of its 2.5-inch disk drive production there in 1997; it planned to soon make all 3.5-inch drives there as well.

By 1995 more than 64 percent of the world's disk drives were produced in Southeast Asia, generating nearly 61 percent of the industry's revenue (Table 4). HDD production in the United States fell to below 5 percent of world shipments, generating less than 9 percent of world revenues, while production in Japan fell to 15.7 percent of shipments and 13.3 percent of revenue. By 1995 the U.S. industry produced two-thirds of its total assembly in Southeast Asia, and Japanese firms had greatly increased their presence in Southeast Asia, producing nearly 55 percent of their HDDs in the region. Virtually all of the remaining drive production for Japanese firms was still located in Japan—45 percent, compared with 18 percent still located in the United States or Japan for U.S. firms.[10] By the mid-1990s, then, the geographic distribution of Japanese assembly had begun to resemble that of their American competitors.

The Value Chain Follows

Through continued investment in the region, nearly every part of the HDD value chain is now produced in Southeast Asia in some quantity, reinforcing its

TABLE 4 Distribution of 1995 Production (% Based on Unit Shipments)

| Type of firm | Locations of production | | | | | |
	U.S.	Japan	Southeast Asia	Other Asia	Europe	Total
All firms	4.5	15.7	64.2	5.7	9.9	100
U.S. firms	5.1	13.0	66.8	3.9	11.2	100
Japanese firms	0.0	45.2	54.8	0.0	0.0	100

Sources: Disk/Trend, Inc., and Globalization Project Database.

---

[10]American firms also extended the global assembly strategy to low-cost areas of Europe, Ireland, and Hungary between 1990 and 1995.

preeminence as the center of HDD and components production. Seagate offers a good illustration. In almost every year since its initial investment in 1982, Seagate has reinvested in Singapore—upgrading existing facilities or building new ones. The largest investments include a $56 million investment in 1988, a $100 million investment in 1992, and a $200 million investment in 1994. Seagate has also invested heavily in Thailand and Malaysia in upstream activities such as motors, heads, and printed circuit board assemblies and has recently opened plants in Indonesia (circuit boards), China (HDDs), and the Philippines (labor-intensive head assembly). Today it is the largest private employer in both Singapore and Thailand.

Independent manufacturers of media and heads have also moved into the region, further reinforcing it as the focus of the industry's global strategy. The first head maker to invest in the region was Applied Magnetics, which opened a plant in Singapore in 1983. Read-Rite, another head company, opened or acquired facilities in Thailand and Malaysia in 1991. The first media maker to locate production outside the United States or Japan was Domain Technology, Inc., which began volume production in Singapore in 1988. Komag and Stor-Media invested in Malaysia and Singapore, respectively, in 1993 and 1995. The first investment in Southeast Asia by a Japanese media company was Hoya Media's Singapore plant in 1996. As of 1995 nearly 70 percent of the firms that make heads or head assemblies had plants in Asia (excluding Japan), while 36 percent of the firms had plants in Southeast Asia. Among media producers, 81 percent had plants in Asia in 1995, while 38 percent were producing in Southeast Asia.

American HDD assemblers initiated the move to Southeast Asia and much of the value chain followed. By 1995, more than 60 percent of global employment in the HDD industry, including upstream activities, was in Asia outside of Japan (Gourevitch et al., 1997).[11] After a decade of investment by both multinationals and local supplier firms, low-cost Asia has become the region of choice for the HDD industry. The technical imperatives of the industry ultimately led to a convergence of American and Japanese strategic posture.

Global Strategy and National Advantage

Despite recent Japanese movements into Southeast Asia, American industry was able to sustain its advantage by being the first to implement this global strategy. How, exactly, did this strategy confer an advantage on American firms? According to industry participants, American industry's early move into Southeast Asia gave it the time to establish regional manufacturing, secure comple-

---

[11]This percentage is actually understated because it does not capture employment associated with a few of the least expensive components going into a disk drive—base-plates, condensers, capacitors, screws, and so forth. These are sourced almost entirely from vendors in Southeast Asia.

mentary assets, and move down the learning curve at a time the competitive logic of the industry dramatically shifted to low-cost, high-volume manufacturing. An industry consultant who set up several overseas production facilities explained the benefits of making disk drives in Southeast Asia in this way: "While loaded labor cost is typically one-quarter of the U.S. equivalent, material cost can be 30 to 45 per cent less than in the U.S. But since the cost of material can account for as much as 80 per cent for some peripheral products, the real savings are achieved through local sourcing for materials rather than in savings of labor costs. Overall, cost savings of 30 to 40 per cent can be achieved in making peripherals in Southeast Asia versus the U.S." (EN, October 1984). Other benefits of assembly in Southeast Asia included lower overhead costs, government incentives, faster investment approvals, and a less costly but developed infrastructure such as precision machinery, die casting, and a pool of skilled personnel in process engineering. Disk drive firms reportedly cut costs by 30 percent by moving production to Singapore (FW, February 24, 1987).

These cost savings coincided with the emergence of the 3.5-inch disk drive to generate enormous advantages in that critical high-volume market. As is shown in Figure 4, the demand for 3.5-inch disk drives exploded in the late 1980s; as Table 5 indicates, the largest part of the 3.5-inch market was for noncaptive sales. It was America's success in this market that extended its advantage. During this period, America's dominance was led by its independent HDD firms—Seagate, Conner, Quantum, Maxtor, and Western Digital—and was dependent on their ability to ramp up low-cost, high-volume production in Southeast Asia, where, with the exception of Quantum, they assembled the overwhelming majority of their 3.5-inch drives.

It took three years for the U.S. industry to claim more than 50 percent of the noncaptive market. Although each of these firms trailed Hitachi, Fujitsu, and NEC in the introduction of this form factor, this global strategy allowed American industry to claim 90 percent of the noncaptive market by 1991. Operating from Southeast Asia, U.S. companies put tremendous pressure on higher cost Japanese manufacturers even in the Japanese market. The retail price in Japan of a 20-megabyte drive made by Japanese producers cost about 200,000 yen, whereas drives made by Singapore-based U.S. firms were beginning to be sold at a retail price of under 70,000 yen (COMLINE Daily News Computers, April 15, 1988).

This is a surprising twist: Japanese and European firms were early to market with an innovation but were ultimately squeezed by the price competition brought to bear by American firms. American firms developed the ability to evolve quickly to new products, with fast, smoothly executed production ramps in overseas locations. Effective execution of production ramps also required U.S. manufacturers to become highly skilled in managing the stream of component parts coming from a diverse supplier infrastructure, both vertically integrated and independent, located in many countries. In contrast, according to the business press, the cost structure of Japanese manufacturing was simply not competitive with

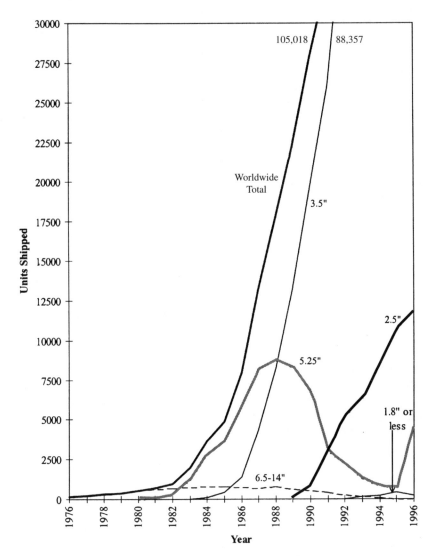

FIGURE 4 The market for disk drives by size (thousands of units).
Source: Disk/Trend, Inc.

that of American industry in the product segments most in demand. The timing, direction, and scope of globalization thus extended the leadership of America's disk drive industry by enabling it to move down the learning curve in overseas assembly while accumulating effective capabilities in managing internal and external international linkages in the value chain.

TABLE 5 The Market for 3.5-Inch Disk Drives (Thousands of Units)

Year	Captive shipments	Non-captive shipments	Total 3.5-inch shipments	Non-captive shipments (% of total)	U.S. market share of non-captive (% of total
1983	0	2	2	100	0
1984	0	67	67	100	11
1985	23	339	361	94	45
1986	250	1,108	1,358	82	56
1987	1,565	2,703	4,268	63	61
1988	2,310	5,899	8,209	72	71
1989	3,620	10,692	14,311	75	74
1990	3,564	16,336	19,900	82	82
1991	3,866	22,170	26,036	85	90
1992	3,972	32,342	36,314	89	92
1993	3,904	39,368	43,272	91	94
1994	4,364	55,980	60,343	93	95
1995	4,623	73,153	77,776	94	93
1996	4,679	83,678	88,357	95	84

Source: Disk/Trend, Inc.

FUTURE PROSPECTS: THE LOCUS OF R&D AND GLOBAL KNOWLEDGE MANAGEMENT

Now that locational strategies are similar across companies, what will continue to sustain America's advantage in disk drives? The industry faces two
critical challenges. The first is generating and exploiting innovations in a context
that has changed considerably since the industry's beginnings 40 years ago.
Although IBM remains a powerful presence, the sources of knowledge for the
industry have broadened considerably, with universities taking on a more important role than at any time in the past. The second challenge remains implementation. Now that all the major disk drive firms have physically separated assembly
from product development, competitive advantage will continue to hinge on the
management of that long-distance relationship. Overall, the United States remains well positioned with regard to both challenges. But there is some concern
that U.S. disk drive and component companies are underinvesting in R&D at a
time when Japanese and Korean companies have evolved into much stronger
competitors.

Trends in the Locus of R&D

The context for innovation in the industry has shifted during the last 15
years.[12] Until the mid-1980s, corporations were the source of almost all research

---

[12]This section has benefited from conversations with Barry Schechtman, director, National Storage
Industry Consortium; Ami Berkowitz, professor, UCSD; Albert Hoagland, professor, Santa Clara
University; Dawn Talbot, librarian, UCSD; and researchers at IBM.

leading to innovations in disk drives, and technical advance was very much driven by engineering within the firm, principally by IBM. Today, companies can no longer engineer their ways into higher densities. For example, the industry is concerned that it is approaching a fundamental physical limit to the density of magnetic disk recording in its present form—the superparamagnetic limit. At the same time, the research underlying industrial applications is becoming more institutionally and geographically dispersed. Since the mid-1980s, more research underlying industrial applications is being conducted in more locations by a variety of research organizations.

It is hard to determine cause and effect in changes in the locus of R&D. One method is to look at patents, which have been hugely important in the history of the industry. One experienced manager claims that Conner Peripherals, which became a $1 billion business in the shortest time in history, owed more for royalties than its entire historical profits for the company! Unfortunately, according to industry and university sources, a meaningful count for important patents related to magnetic recording is nearly impossible because one cannot distinguish the critical patents from the minor ones.

As a convenient measure of the changes in the sources of R&D in the industry, this paper counts technical papers in the most relevant journals. Although research on magnetic recording traverses several journals, university and industry researchers suggested two in particular. One is a peer-reviewed journal, the IEEE Transactions on Magnetics (IEEE), which I cover in five-year increments since 1970. The second is The Magnetic Recording Conference (TMRC), which invites scholars to submit their papers. TMRC was used as a proxy for research quality because only those perceived to be doing the most interesting or advanced research are invited. Unfortunately, TMRC has published only since 1990, so it provides a picture of only current research. The institutional sources of research were also tracked: firms, universities, or "other," principally government or quasi-government research labs.

Table 6 summarizes the geographic trends in the sources of R&D, three of which I note here. First, while the relative share of publications by United States authors has declined since 1970 in IEEE, research by the United States is still the most influential as measured by invited publications in TMRC. Second, and somewhat surprisingly, the share of publications by authors from Japan and Europe has remained fairly steady, with Europe showing considerably more strength in magnetics than one would think from its market presence in HDD and components. Of course, magnetics research is applicable to several branches of industry, so this count is an imperfect measure of capability in data storage. Third, however, when measured by TMRC, which is more specific to magnetic recording research, the positions reverse, with Japan having a higher representation than Europe.

Research publications were also classified according to the institutional affiliation of their authors. The percentages in Table 7 refer to the share of a given

TABLE 6 Location of Authors Publishing in Academic Journals on Magnetics, 1970-1995

Year	Total number of articles	United States	% of total	Japan	% of total	Europe	% of total	Other	% of total	Cross-national Authorship	% of total
IEEE Transactions on Magnetics											
1970	37	19	51	4	11	9	24	5	14	0	0
1975	47	25	53	6	13	13	28	3	6	0	0
1980	68	30	44	12	18	20	29	6	9	0	0
1985	64	24	38	14	22	16	25	5	8	5	8
1990	81	27	33	10	12	23	28	16	20	5	6
1995	83	23	28	15	18	22	27	15	18	8	10
TMRC											
1990	19	12	63	4	21	2	11	0	0	1	5
1991	23	19	83	3	13	1	4	0	0	0	0
1992	25	22	88	1	4	0	0	0	0	2	8
1993	28	24	86	2	7	2	7	0	0	0	0
1994	30	24	80	3	10	2	7	0	0	1	3
1995	28	23	82	4	14	1	4	0	0	0	0

region's publications—for example, the share of all American papers that were contributed by American firms or universities. The most obvious fact about this research is the extent to which corporations are still involved in fundamental research in the United States and Japan but not elsewhere. Among firms, IBM remains the strongest company in magnetic recording technologies for hard disk drives and still has an enormous reservoir of technical talent. But its share of articles in IEEE declined during the period covered, from 40 percent of all articles contributed by U.S. firms in 1970, to 22 percent in 1985, and to zero in 1995. Quantum was the only disk drive firm that contributed articles that year. This last figure for IBM is surely an outlier, however, since IBM was affiliated with between 20 and 40 percent of all American papers to appear in TMRC between 1990 and 1995.

There is some similarity between the United States and Japan in the institutional affiliation of authors. As expected, university research has featured more prominently in both countries during the last two decades. Of all U.S.-authored papers appearing in IEEE in 1990, almost 60 percent were by researchers affiliated with universities, and 25 to 45 percent of American papers appearing in TMRC were contributed by universities. Magnetics research in Japan is slightly more concentrated in Japanese firms than in universities or other institutions. Japanese firms accounted for more than two-thirds of Japanese papers appearing in IEEE in 1970 or 1975, still more than half in 1995, and virtually all of the high-quality research on magnetic recording as measured by TMRC. Japanese univer-

TABLE 7 Affiliation of Authors Publishing in Academic Journals on Magnetics, 1995 (Percent)

Year	United States			Japan			Europe			Other			Cross-national authorship		
	Firms	Universities	Other	Firms	Universities	Other	Firms	Universities	Other	Firms	Universities	Other	Firms	Universities	Other
IEEE Transactions on Magnetics															
1970	60	30	11	83	17	0	56	33	11	0	40	60	0	0	0
1975	72	22	6	67	17	17	19	58	23	33	33	33	0	0	0
1980	67	30	3	35	56	8	15	42	43	0	78	22	0	0	0
1985	75	25	0	57	43	0	19	75	6	0	80	20	0	59	41
1990	46	49	5	43	45	13	17	72	11	2	98	0	7	80	13
1995	20	59	22	54	38	8	5	74	21	16	80	4	0	96	4
TMRC															
1990	75	25	0	100	0	0	100	0	0	0	0	0	100	0	0
1991	60	40	0	67	33	0	100	0	0	0	0	0	0	0	0
1992	55	44	2	100	0	0	0	0	0	0	0	0	29	54	17
1993	69	31	0	100	0	0	100	0	0	0	0	0	0	0	0
1994	54	46	0	67	33	0	100	0	0	0	0	0	67	33	0
1995	74	26	0	100	0	0	100	0	0	0	0	0	0	0	0

sities are stronger in magnetic recording than they were in the 1970s, however, contributing anywhere from 40 to 55 percent of the publications in IEEE since 1980.

Overall, the United States remains strong in research related to HDDs, with American universities becoming a more important source. The Japanese also have research depth in magnetic recording, some of which may be concealed by these measures. In particular, it is likely that not all relevant Japanese research appears in these journals.[13] But it is also true that research with commercial usefulness can come from anywhere, and companies need to have the capability to evaluate and incorporate new knowledge into industrial applications. In 1988, for instance, two scientists from France and Germany working independently published papers that directly influenced the development of so-called "giant" magneto-resistive heads, which IBM planned to incorporate into its disk drives in 1998. Following is a description of the efforts underlying these trends.

Basic and Applied Research in the United States: University and Collaborative R&D

Magnetics research is the basic foundation for the disk drive industry, and Europe was the pioneer in magnetism. BASF (Germany) and Philips (The Netherlands) had strong labs, and Philips is still strong in magnetics related to optical storage. In the United States, Bell Telephone Laboratories, IBM, General Electric, and Westinghouse all had magnetic recording research programs during the 1940s and 1950s. The material basis was a continual refinement on earlier developments in Europe. By 1980 only two large groups in the United States were still at work on applied magnetics in any systematic way, one at Bell Telephone Laboratories, and the other at IBM. IBM is still the primary American source of industrial research with a direct impact on the HDD industry; Seagate, Quantum, and Western Digital do not have the R&D resources of IBM. But IBM is no longer the only locus of important research relevant to magnetic recording.

Government Support. Compared with many other important industries, government support for data storage in general, and HDD in particular, has been

---

[13]For a time, the Center for Magnetic Recording Research (CMRR) at the University of California, San Diego, hired translators in Japan and the United States to monitor technical developments in magnetics appearing in Japanese technical journals. CMRR then sent out abstracts to its member companies, who might request the full text. Because of industry interest, this cost quickly escalated and CMRR dispensed with the practice. But the experience indicates that interesting Japanese research does not always appear in English language journals, even if they are often the most desirable outlet for a Japanese researcher. Between 1985 and 1994, the IEEE published the IEEE Translation Journal on Magnetics in Japan, which translated important research from Japanese-language journals. After some lag, these articles were cited 111 times in English language publications. I am grateful to Dawn Talbot for this information.

barely measurable. During the 1950s the Naval Research Lab did its own magnetic recording research; Oak Ridge Laboratories worked on neutron fractions, a subject fundamental to magnetism; while the Office of Naval Research sponsored some research. In the 1950s the National Bureau of Standards also did some research developing the world's first rotating disk storage device, which came to IBM's attention (Bashe et al., 1986: 280; Rabinow, 1952). But government influence on industry is hard to trace and appears diffuse or very indirect at best. The government's most explicit commitment to the storage industry gained expression only in the 1990s. The Defense Department's Advanced Research Projects Agency (DARPA) and the National Institute of Standards and Technology (NIST) have begun to consider data storage a critical industry and have helped to underwrite some recent research through the National Storage Industry Consortium (See below).

In addition to providing limited funding, NIST itself engages in magnetics research at its Electronics and Electrical Engineering Laboratory. The laboratory's Electromagnetic Technology Division collaborates with the magnetic recording industry in developing metrology to support future recording heads and media with their ever-increasing data density. The division also provides new measurement methods, instrumentation, imaging and characterization tools, and standards in support of the magnetics industry. Compared with private and university research, its role has been quite small, but NIST has announced its interest in enlarging its magnetic and storage programs (Rhyne, 1996).

University Research. In the 1950s and 1960s the theoretical basis of research in magnetism was fairly primitive, and no American university had an academic center doing work in magnetic materials. In fact, the most eminent people in magnetics were in or came from industry. Although important theoretical advances may have come from academe, progress in magnetism was very much a story of individuals. That is, basic research affecting the HDD industry from the 1950s through the early 1980s was excited and generated by individuals and industrial interests rather than organized university or government programs.

In the late 1970s, the HDD industry recognized that the technological trajectory had gone as far as it could without more fundamental research. The magnetic recording industry (tape, optical, and floppies as well as disks) in the United States brought its concerns to the National Science Foundation (NSF), arguing that unless academia got involved, the United States risked losing its international lead in the industry. But the research foundations for magnetic storage are so interdisciplinary that universities typically were not organized to address the industry's problems. Given that physics, chemistry, materials science, mechanical engineering, and electrical engineering are all relevant to storage, it was difficult to find a home department.

In recognition of this need, the NSF and the industry funded a Magnetics Research Center at Carnegie Mellon University in 1982, which was subsumed into the Data Storage Systems Center, an NSF Engineering Research Center since 1990. The Magnetics Research Center opened with mostly federal and some industry funding. Initially, its researchers attacked a broad spectrum of problems in materials development. The center's first major effort was to understand the basic chemical and physical processes that hindered the development of better recording materials and equipment, including the development of a theoretical model of the recording process. Firms did not yet understand many of these fundamental issues. The center's main focus today includes the integration of storage systems into high performance computing and research on high density magnetic and magneto-optic disk and tape recording including the electronic and mechanical subsystems.

In 1983, industry took the lead in establishing the Center for Magnetic Recording Research (CMRR) at the University of California, San Diego, a location more convenient to Silicon Valley than was Carnegie Mellon.[14] CMRR was established by seven sponsoring firms, with UCSD contributing land and faculty; there was no federal assistance. The number of industrial sponsors has since more than doubled, and they provide part of the research expenses, which are leveraged by grants from federal agencies. To ensure academic independence, no restrictions are imposed on the public release of research results, and faculty and staff agreed not to patent any of their work. CMRR has more focus on long-term fundamental research than CMU—tribology, signal processing, mechanics, information theory, micromagnetic modeling, and research on new kinds of magnetic particles.

These are the two most prominent centers for magnetic recording research, but other universities have received industry support and are strong in fields related to disk drives, including U.C. Berkeley, Stanford, Minnesota, Alabama, Washington University, and George Washington University. Santa Clara University became an important source of engineering graduates for the industry following the establishment of its Institute for Information Storage Technology in 1984. In fact, university graduates with training in magnetic storage have had a profound impact on the industry. Historically HDD firms had to train graduates for a couple of years because many new engineers knew only one piece of the industry or had little knowledge about magnetic recording. In the near future, the industry will be entirely headed by a generation of graduates of these centers, people who deeply understand all aspects of the industry.

National Storage Industry Consortium (NSIC). Formal collaboration among U.S. firms in R&D is a new phenomenon. The big donors to the university centers gradually observed that everybody was working on similar issues, and they wanted more diverse and complementary research. At the same time, the HDD

---

[14]One might wonder why U.C. Berkeley or Stanford did not become the industry's choice. In fact, both were, but neither institution was ready to commit to building a center.

industry thought that it should get more federal monies and questioned why semi-conductors should get such favorable treatment. These concerns spawned the industry consortium idea. It was useful to use NSIC as a forum to govern university research underwritten by industry donors instead of meeting with each academic group independently to try to guide research.

Incorporated in 1991, NSIC is a nonprofit mutual benefit corporation whose membership consists of about 60 corporations, universities, and national laboratories with common interests in the field of digital information storage, not only HDDs but also optical and tape storage. During its first five years, NSIC started only one project a year, and DARPA or NIST supported 50 percent of the costs. The primary emphasis of these projects was to advance less established storage technologies. In 1996 that focus changed to support projects of more immediate interest to NSIC's members but that were still in the precompetitive stages of research. About 40 percent of the $100 million that has flowed through NSIC has come from federal tax dollars, mostly from NIST and DARPA. To date, the HDD industry has received support for three projects, with much of the work being done by universities. Each project has focused on increasing the perceived limits on areal density.

University and consortial research are helping U.S. HDD firms focus on long-term and fundamental issues that present technology barriers. Although the industry has come to value the role of NSIC and university research generally, there is a perception among many managers and academics that among U.S. firms only IBM is doing sufficient internal applied technology aimed at proprietary solutions. These observers are concerned that U.S. disk drive firms, as well as independent component companies, are too focused on near-term needs and thus not well positioned to absorb new developments that will arise from external sources such as universities. Although fear of foreign competition has always been overblown in this industry, the changing competitive context may make the U.S. industry more vulnerable than it has ever been.

Foreign Sources of R&D

Japan is the source for the most significant R&D outside the United States, but R&D efforts in South Korea, Singapore, and Europe are worth noting.

Japanese Firms. Among Japanese firms, Fujitsu and Hitachi have the greatest depth in technology. In some areas their technical base rivals IBM's and completely dwarfs that of other U.S. manufacturers. Both companies have strong Japanese research labs. To complement its domestic research related to disk drives, in 1987 Fujitsu acquired Intellistor, a data storage and subsystems design company in Longmont, Colorado. In 1991 it was made part of a new subsidiary, Fujitsu Computer Products of America, and became Fujitsu's North American research arm. Longmont is Fujitsu's only magnetic storage R&D operation outside of Japan. In 1995, although it held only a small percentage of the market, Fujitsu announced its plan to capture 20 percent of the worldwide market for hard disk drives by the end of 1996. One part of the plan was a commitment to increase manufacturing capacity in the Philippines and Thailand, consistent with the convergence in global strategy described in the previous section. The other component was the development of second generation MR head technology and its incorporation into a larger range of drives. Fujitsu has strong internal magnetic recording research and development capabilities, skills surpassed only by IBM's. In fact, the company experienced more than 100 percent unit growth in 1996 and offers drives with among the highest areal density of any manufacturer as well as a broadened product line that includes high-end drives for servers.

Although it has a smaller presence in disk drives than Fujitsu, Hitachi has greater research depth according to some observers. It is strong in science and has contributed more articles to TMRC than any other Japanese disk drive company. Hitachi is also the most likely firm to be first after IBM in introducing giant magnetoresistive (GMR) heads. According to one manager of a U.S. corporate R&D lab, Hitachi lags behind Fujitsu and American firms primarily because it has less effective technology transfer.

In 1995, the four largest Japanese HDD firms—Fujitsu, Hitachi, Toshiba, and NEC—agreed to form a consortium to research and develop HDDs and data storage devices. Like NSIC, the Storage Research Consortium intends to sponsor work in which industry and universities collaborate. With an expenditure of $2.2 million, its resources are currently smaller than are NSIC's. But the companies hope to attract as many as 30 others, expand the consortium's resource base, and involve other parts of the value chain including makers of disks, heads, circuit boards, and test equipment. Membership in the consortium is open to foreign companies that have Japanese development or production facilities. Thus, Seagate is effectively excluded, while IBM, which develops drives in Fujisawa, and Komag, the world's largest independent maker of disks and with a Japanese joint venture, presumably are eligible. It is too early to judge the consortium's impact.

Korean Firms. South Korean firms have been at the edges of the disk drive industry for more than a decade, but only in the last few years have they made the financial investments necessary to compete in this high-tech commodity business—not only for volume manufacturing but also for R&D.

Hyundai entered the industry in 1995 through its acquisition of Maxtor, one of the most successful American start-ups of the early 1980s. The parent firm has announced its goal is to become the world's second largest HDD supplier by 2000 and the world's leader by 2005. It intends to invest $1 billion by 2000 and set up a global network linking production bases in China and Thailand, component suppliers in Singapore and Hong Kong, an R&D base in the United States, and headquarters in Korea (CDSN, 1996).

Maxtor's U.S. R&D team is the center of Hyundai's disk drive operations. Maxtor will concentrate on R&D, marketing, and production of high-end HDDs. Singapore will be the headquarters for procurement and a production plant for high-end drives. Thailand and China will focus on assembly of low-end units. Hyundai will also maintain production in Korea, where the firm's storage division will be headquartered. Although Hyundai will certainly attempt to strengthen the technical capabilities of its Korean-based operation, it is unlikely that its U.S. R&D work can be overtaken in the near term. In fact, Maxtor recently announced plans to establish a new engineering center in California headquarters, joining its California-based Advanced Technology engineering group, which focuses on heads/media integration. The new California engineering center will complement Maxtor's existing engineering operations in Colorado, where it develops high-end drives for the desktop PC market (BWI, 9/12/97).

For its part, Samsung has announced that it intends to become the *fourth* largest HDD company in the world by 2001 (KEW, 1996). It entered the disk drive industry initially as an investor in and contract manufacturer for a small U.S. start-up. After the American company failed in early 1990, Samsung continued to make its drives and then began to develop its own. But the company limped along for several years with products just behind the market and scouted around for acquisitions in recording heads and another disk drive company while reportedly losing money in the business. It renewed its commitment in 1996 when it completed a $370 million investment in a new HDD plant. Besides adding to its manufacturing muscle, Samsung continued to build up its development center in San Jose, California, where it had been developing disk drives since the early 1990s. The center is responsible for advanced engineering, product development and qualification, marketing, product planning, and technical support. With its new manufacturing and R&D resources, some observers think Samsung has the potential to be a force. But given its earlier problems in product development, R&D is likely to stay centered in California for some time.

Singapore's Data Storage Institute. Singapore has organized an intensive effort to move up the technology ladder in the HDD industry and has established a Data Storage Institute. Started in 1992 as the Magnetics Technology Centre, its role expanded in 1996 when it became one of three new research institutes at the National University of Singapore. It will receive government support to the tune of an initial S$30 million for the building and S$55 million over three years. The new institute will do research and participate in joint programs with multinational corporations involved in HDD, opto-electronic, and disk media technologies. It has more than 160 researchers and is expected to train about 40 engineers every year for employment in the HDD industry.

One interesting aspect of the institute is the help it is receiving from leading American HDD firms and researchers. IBM and Carnegie Mellon University are on its advisory panel, more than half its corporate members come from the American industry, and its first project in 1992 came from Seagate. The institute will not focus on fundamental research as much as U.S. universities do. Instead, it will leverage Singapore's strength in HDD manufacturing to improve process technologies, including testing, as well as offer more direct and immediate support to industry than do American research universities. It is also exploring innovations at the component level, such as a collaborative effort with IBM, Motorola, Fujitsu, Hitachi, and a local Singaporean company to "push the benchmarks" on channel chips that do the read/write function on hard disks.

Early in 1997 the government hived off the Data Storage Institute and other research facilities into separate and independent companies. Their primary roles will be to support industry in technology development through "greater responsiveness" (BT, 1997). Given Singapore's place in the global HDD network, and the government's unique commitment to the industry, the Data Storage Institute will likely evolve into a center of excellence. At this stage, the institute complements rather than competes with American university research.

The European "Scotsman" Project. Despite its considerable research base in magnetics, Europe has only two indigenous disk drive development companies remaining, Calluna Technologies (Scotland) and Nomai (France), and very few suppliers of primary components. The "Scotsman" (Strategic Components, Technologies and Systems in Magnetic Storage) project is a collaboration initiated in February 1996 under the Esprit research program of the European Union to work on head technology. In addition to Calluna and Nomai, the other members are Myrica (U.K.), which is Nomai's development subsidiary; Silmag S.A. (France); and Xyratex, Ltd. (U.K.), the IBM spin-off. Half the $5 million in funding is provided by the European Commission and half is provided by the partners. The primary technology is expected to come from Silmag, which has developed what some say is a leading recording head technology. Although no one expects Europe to obtain a leading position as a result of the project, it is intended to maintain European expertise in magnetic storage and in the removable disk drive niches in which Calluna and Nomai operate.

The Relationship between Product Development and Volume Manufacture

In the course of the industry's evolution, pockets of technical sophistication developed in the United States (the Los Angeles area, Silicon Valley, Minnesota, Colorado, and, to a lesser extent, the Boston region), Japan, and Europe (the United Kingdom, the Netherlands, France, and Germany). Through industry consolidations, surviving firms have found themselves in possession of R&D assets in more than one location. At the same time, the shift of assembly away from a firm's home base means that the management of technical knowledge between geographically dispersed facilities has become a critical organizational task. Table 8 lists all HDD firms in operation as of mid-1997 and the location of their

TABLE 8 Location of HDD Product Development and Assembly, 1997[a]

Company	Product development	Assembly	Company	Product development	Assembly
Seagate	California, U.S.	Singapore	Hitachi	Japan	Japan Philippines
	Colorado, U.S.	Thailand			
	Singapore	Ireland	Samsung	S. Korea California, U.S.	S. Korea
	Oklahoma, U.S.	Malaysia			
	Minnesota, U.S.	Oklahoma, U.S. China	Micropolis	California, U.S.	Singapore
IBM	California, U.S.	Singapore	Iomega	Utah, U.S.	Malaysia
	New York, U.S.	Thailand[b]	JTS	California, U.S.	India
	Japan	Hungary England[c]	SyQuest	California, U.S.	Malaysia
Quantum	California, U.S.	Japan[d] Singapore[d] Ireland[d]	Avatar Systems	California, U.S.	Thailand
Western Digital	California, U.S.	Singapore	Calluna Technologies	Scotland, UK	England[c] Scotland, UK
	Minnesota, U.S.	Malaysia	Gigastorage	California, U.S.	China[e]
Toshiba	Japan California, U.S.	Japan Philippines	Integral Peripherals	Colorado, U.S.	Singapore
			Sequel	California, U.S.	California, U.S.
Fujitsu	Japan Colorado, U.S.	Japan Philippines Thailand	Nomai	France Scotland, UK	England[c]
Maxtor	Colorado, U.S.	Singapore	Raymond Engineering[f]	Connecticut, U.S.	Connecticut, U.S.
NEC	Japan	Japan Philippines	Sagem[f]	France	France

[a]Firms listed in rough order of HDD revenue.
[b]Contract manufacture by Saha Union.
[c]Contract manufacture by Xyratex.
[d]Contract manufacture by Matsushita-Kotobuki Electronics.
[e]Not yet shipping: in negotiations with the Chinese government.
[f]Makes small numbers of "ruggedized" drives.
Sources: Author's data.

product development and volume manufacturing facilities. Locations for component development and manufacturing in vertically integrated firms are omitted. One of the most remarkable characteristics of the HDD industry is that, with very few exceptions, product development is geographically separated from volume manufacturing and, among the leaders, by great distances.[15]

An important question is whether the trend witnessed over the last 15 years in the internationalization of assembly will extend to other core organizational tasks. What are the implications of the growing importance of university research for how globally dispersed a firm's R&D can be? Specifically, will R&D follow manufacturing offshore, or do other forces act as countermagnets? Is the industry defying those who argue that remote manufacturing can cause quality and service problems that outweigh any apparent savings? Three outcomes are possible: R&D could follow assembly abroad; assembly could return to be closer to product development; or the industry could reach some manageable equilibrium.

The likely scenario is that the current organization of the industry will persist. That is, most disk drive design and development, along with pilot production lines, will remain concentrated in the United States and Japan, and volume manufacturing will continue to be physically separated and situated in countries where assembly is cheaper. Firms offer a number of interrelated reasons for why this kind of organizational arrangement is effective and durable. The short explanation is that product transfer has become straightforward, and any costs associated with transfer and coordination are paid for with just one day of high volume manufacturing in a lower cost location.

Typically, companies conduct pilot production proximate to product development because of the greater risk with design during initial assembly. Companies also form product transfer teams consisting of product developers and process engineers from both the home product development facility and the volume manufacturing facility. Today, a product transfer team might be as big as 40 people, and the team stays with the product from pilot through ramp-up overseas. Then the manufacturing team takes over responsibility for volume assembly.

Seagate's Malaysian facility, for example, has had a good experience with product transfer and can ramp up quickly.[16] Six to eight engineers in the United States write the code, do mechanical design and testing, and work with the process people in the domestic facility to stabilize yields during pilot production, which might involve as many as 10,000 drives if the product is especially advanced. A few weeks before the transfer to Malaysia, quality, operations, and lead operators in Malaysia go the United States to prepare for transfer. Then six to ten people from the U.S. team go to Malaysia for 3 to 5 weeks to ensure a good

---

[15]The exceptions are for the very smallest disk drive firms. The last ten companies listed in the right-hand side of Table 8 had less than one half of one percent of the global market in 1996.

[16]This information was provided by the plant manager during the author's visit.

start. The product release date is typically met because, according to the plant manager, "transfer is almost routine now, very smooth." Transfer is also effective to Seagate's Thai facility, which makes less sophisticated products and hosts fewer U.S. engineers during transfer but sends more staff to the United States before transfer because its workforce is less skilled.

Western Digital uses a formal new product introduction process that allows it to achieve 92 percent yields in overseas assembly.[17] The firm has a special group to manage the process which it claims gives the company an advantage in time to market, time to volume, and time to high yields. In product design, the company uses a typical "gating" process common to well-managed high technology firms. The product concept and then the product itself need to pass certain gates on its way to manufacturing. These are milestones that have to be met at each step. Design, engineering-level build, test, and tooling buildup take place in the United States because there are still bugs that need fixing. Concurrent engineering is occurring in Asia where production level equipment is introduced and components are chosen. Transfer teams to and from Asia then ensure a smooth product transition. The transfer team from Asia visits San Jose to work on the pilot line to "wring out" the process, while the U.S. team stays with the product all the way to ensure manufacturability. The transfer process involves 40 people for 30 days.

An important reason for the success of this model is that both the technology and the assembly process are better understood than they were 15 years ago, new products are increasingly designed so that they do not disrupt existing manufacturing processes, and computer information systems lower the costs of long distance management. Although firms vary somewhat in their ability to minimize changes to products that might otherwise require substantial changes in tooling for the assembly process, companies try to maintain substantial commonality in components across products. Western Digital, generally thought of as the leader in this regard, has 70 to 80 percent commonality between products.

At the same time that companies have accumulated skills in design, transfer, and ramp-up, the quality of the infrastructure in Singapore and Malaysia has facilitated technology transfer and rapid ramp-up to volume manufacture. The disk drive industry has developed a large base of skilled professionals in the region with specialized industry knowledge, and the Malaysian and Singaporean governments have been aggressive in offering complementary services, such as rapid investment approvals, access to land, and labor training programs. As a consequence, there appear to be considerable cost savings with little lost in product yields or volume output. As product cycles shorten, ramping up has become even *faster* in Asia. In 1995 Western Digital ramped up production from zero to 750,000 units within three months (CRN, 1995). In 1996 Quantum/MKE went from zero to 7 million disk drives in nine months (NST, 1996). Moreover, over-

---

[17]This information was provided by a Western Digital vice president during the author's visit to the firm's Malaysian facility.

seas assembly does not appear to cost companies in terms of yield. When IBM shifted production from California to Singapore, it not only ramped up quickly (from October to December), but it did so with no loss of yield. Like other drive companies, IBM could find all the engineering and managerial skills it needed in Singapore, and the Singaporean government facilitated the move by approving the investment quickly and even leasing IBM a plant that the government had specifically prepared for disk drive assembly. According to one IBM manager, the company could ramp to volume manufacturing in Singapore faster than anywhere else in the world, including the United States.

Coordinating technical activities with volume manufacturing across national boundaries has become standard practice for the industry. The model for American firms is design and pilot production in the United States, fast ramp-up in Singapore, and matured products and process transferred out to Malaysia, Thailand, or China. The system has become routinized, and American firms excel at it. Other than niche players, firms that did not adopt this organizational model, or executed it poorly, have all exited the industry.

Summary

Some industry managers and academics involved in magnetic recording see the current era as a watershed in industrial applications in magnetics, with universities contributing to industry in fundamental ways. Although differences remain in the priorities and interests of academe and industry, they have become more and more aligned. America's university centers have also had profound influences on the industry in another way; their graduates populate the data storage industry and in the near future will be its leaders. Some in industry, however, express mild concern that some U.S. companies, plus the independent U.S. media and recording head manufacturers, are not investing enough in in-house applied technology to enable them to absorb and commercialize university research quickly enough. By contrast, Fujitsu and Hitachi are much more heavily involved in applied technology work, and some American observers fear that the growing importance of more fundamental technological research will play into their strengths, especially now that they have also adopted low-cost manufacturing strategies.

In the near term the United States and Japan are unlikely to be displaced as the centers of research and product development. University research in the United States has become a more significant factor in the industry's technical evolution; the research labs of the big Japanese firms continue to make important advances in data storage; and Korean producers depend primarily on U.S. R&D. The only major impetus to shift the locus of R&D out of these countries comes from Singapore, which assembles more disk drives than any other country in the world. Yet even though Seagate and others may elect to do more product development in their Singapore subsidiaries, and the Data Storage Institute is making progress, the technical resources and depth in the United States and Japan continue to attract investment in R&D.

It also seems clear that American HDD firms know how to manage international operations and coordination between home-based product development and foreign assembly, including its international supply chain. Barring a discovery that product yields would in fact be greater if assembly were brought home to be closer to product development, there is little indication that the physical separation of development and manufacturing cannot be sustained.

CONCLUSION

Path dependence, industrial organization, and innovation all contributed to American success in the disk drive industry. The American HDD industry was built by successful computer firms, which enabled the industry to achieve an early lead over European and Japanese drive manufacturers. In addition, product development capabilities and some degree of vertical integration have been necessary conditions for industrial performance. Each of these contributed to America's initial industrial advantage, yet taken together they are insufficient in explaining the ability of American firms to *sustain* their dominance. Differences between American and Japanese firms along these dimensions do not appear strong enough to explain the persistence of American leadership in the industry. This chapter suggests that a potentially important yet overlooked variable in studies of national industrial advantage may be the scope, timing, and direction of an industry's overseas manufacturing operations.

By being the first to shift assembly offshore, American firms were able to learn the organizational technology of international coordination and production. Although their activities were dispersed, they were at the same time concentrated in key regions—research and development in the United States, labor-intensive assembly in low-cost Asia, and somewhat more skilled assembly activities in Singapore (Gourevitch et al., 1997). American firms combined the benefits of low-cost, high-volume assembly with sophisticated management of these value networks. Innovative firms that failed to shift assembly abroad exited the industry or else claimed imperceptible shares of the market.

The history of the disk drive industry differs from other high-technology industries in additional ways. First, the American HDD industry excels at manufacturing. This is contrary to what researchers have observed in other industries, where Japanese firms are leaders in manufacturing. The business press initially expected that pattern to hold for disk drives as well: "Once in production, a disk drive is basically a commodity product that must be assembled as quickly and as cheaply as possible—something that the Japanese are expert at doing" (BW, 1984). If anything, they lagged behind American firms in their ability to ramp to volume manufacturing. Especially interesting is that the vast majority of assembly was conducted in-house. Although companies have frequently resorted to contract manufacturing, such arrangements have played a small role quantitatively.

Second, the American industry was largely ignored by the federal government and university departments during its first two and a half decades. Certainly, disk drive programs in private firms benefited at least indirectly from federal monies earmarked for computers and semiconductors. Yet technical progress in disk drives went largely unnoticed by those outside the industry and was achieved through heroic mechanical and materials engineering efforts in firms, especially in IBM, rather than through publicly funded research. Moreover, unlike software (Mowery, 1996), where the federal government played a prominent role in developing computer science as an academic field, in data storage the private sector initiated the establishment of academic programs specifically for magnetic recording, although the federal government then stepped forward with critical funding. These programs also emerged much later than those targeted at the computer, semiconductor, and software industries.

REFERENCES

Anderson, P. (1995). "Microcomputer manufacturers" in *Organizations in Industry: Strategy, Structure and Selection*, Glenn R. Carroll and Michael T. Hannan, eds. New York: Oxford University Press.

Aoki, M. (1988). *Information, Incentives and Bargaining in the Japanese Economy*. New York: Cambridge University Press.

Aoki, M. (1990). "Toward an economic model of the Japanese firm." *Journal of Economic Literature* 28(March):1-27.

Aoki, M., and N. Rosenberg. (1987). "The Japanese firm as an innovating institution." CEPR Discussion Paper No. 106. Stanford University.

Bashe, C., L. Johnson, J. Palmer, and E. Pugh. (1986). "Disk storage," in *IBM's Early Computers*. Cambridge, MA: MIT Press.

BT *(Business Times)*. (May 3, 1993). "Plugging into and Asian gold mine."

BT *(Business Times)*. (April 1, 1997). "NSTB spins off seven research institutes." 2.

BW *(Business Week)*. (February 6, 1984). "The disk-drive boom has suppliers spinning." 68.

BWI *(Business Wire)*. (September 12 , 1997). "Maxtor to enter server disk drive market."

BWI *(Business Wire)*. (June 17, 1997). "Fujitsu ranked the fastest growing hard drive manufacturer."

CDSN *(Computer Data Storage Newsletter)*. (1996). Vol. 9, no 10. Mike Cannon, President & CEO of Maxtor: "We'll be profitable in 1997." 1.

Chandler, A. (1990). *Scale and Scope: The Dynamics of Industrial Capitalism*, Boston: Belknap Press.

Christensen, C.M., and J.L. Bower. (1996). "Customer power, strategic investment, and the failure of leading firms," *Strategic Management Journal* 17:197-218.

Christensen, C.M., and R.S. Rosenbloom. (1995). "Explaining the attacker's advantage: Technological paradigms, organizational dynamics, and the value network," *Research Policy* 24:147-162.

CI *(Computergram International)*. (January 13, 1992). "Thailand: Fujitsu slashed local production of small hard drives as prices collapse."

COM *(COMLINE Daily News Computers)*. (October 9, 1995). NEC hard disk drive plant in Luzon now operational.

COM *(COMLINE Daily News Computers)*. (April 15, 1988). "Competition Increasing Between Japanese and U.S. Magnetic Disk Drive Makers."

CRN (*Computer Reseller News*). (December 4 , 1995). "Western Digital's Burger speaks out on state of storage industry." 133.

CRN (*Computer Reseller News*). (February 24 , 1997). "Will emerging new digital technologies ever stop?"

CW (*Computerworld*). (December 9, 1985). "'Made in Japan' tag penetrating components market: U.S.-Japan pacts create added equipment sales." 64.

Dertouzos, M. L., R. K. Lester, and R. M. Solow. (1989). *Made in America: Regaining the Productive Edge*, Cambridge, MA: MIT Press.

Disk/Trend. (Various years). *Disk/Trend Report: Rigid Disk Drives.* Mountain View, CA: Disk/Trend, Inc.

EBN (*Electronic Buyer's News*). (April 29, 1996). "MR technology may finally have its year."

EET (*Electronic Engineering Times*). (August 26 , 1996). "MR heads inch toward mainstream."

EN (*Electronic News*). (October 22 , 1984). "Peripheral manufacturers making move to SE Asia; cheaper labor, materials draw U.S. firms seeking to be more price-competitive." 53.

EN (*Electronic News*). (August 21, 1989). "Seagate/Imprimis deal forces industry shift." 1.

FT (*Financial Times*). (June 1, 1988). "Seagate sets sail for Singapore." 16.

FW (*Financial World*). (February 24, 1987). "Disk drive technology is changing so quickly that only a few U.S. companies can keep up with it. But those survivors should prosper—and soon." 124.

Florida, R., and M. Kenney. (1990). *The Breakthrough Illusion: Corporate America's Failure to Move from Innovation to Mass Production*, New York: Basic Books.

Gerlach, M.L. (1992). "The Japanese corporate network: A blockmodel analysis," *Administrative Science Quarterly* 37:105-139.

Gourevitch, P., R. Bohn, and D. McKendrick. (1997). "Who Is Us? The Nationality of Production in the Hard Disk Drive Industry," *The Data Storage Industry Globalization Project Report* 97-01. San Diego, CA: Graduate School of International Relations and Pacific Studies, U.C. San Diego.

Harker, J.M., D.W. Brede, R.E. Pattison, G.R. Santana, and L.G. Taft. (1981). "A Quarter Century of Disk File Innovation," *IBM Journal of Research and Development* 255:677-689.

Henderson, R., and K. Clark. (1990). "Architectural innovation: The reconfiguration of existing product technologies and the failure of established firms," *Administrative Science Quarterly* 35:9-30.

Hill, C.W.L. (1995). "National institutional structures, transaction cost economizing and competitive advantage: The case of Japan," *Organization Science* 6:119-131.

IDC (*IDC Japan Report*). (February 28, 1991). "Fujitsu will produce small hard disk drives in Thailand starting this summer."

Imai, K., I. Nonaka, and H. Takeuchi. (1985). "Managing the product development process: How Japanese companies learn and unlearn," in *The Uneasy Alliance: Managing the Productivity-Technology Dilemma*, Kim Clark et al., eds. Boston: Harvard Business School Press.

KEW (*The Korea Economic Weekly*). (October 31, 1996). Samsung Electronics opens a HDD plant.

Kogut, B., ed. (1993). *Country Competitiveness: Technology and the Organizing of Work*, New York: Oxford University Press.

Langlois, R. N. (1992). "External economies and economic progress: The case of the microcomputer industry," *Business History Review* 66:1-50.

LAT (*Los Angeles Times*). (June 25, 1990). "Why American high-tech firm recruits in Asian ricefields." D3.

LAT (*Los Angeles Times*). (August 6, 1991). "Toshiba develops disk drive of high capacity." D2.

Mansfield, E. (1988). "The speed and cost of industrial innovation in Japan and the United States: External vs. internal technology," *Management Science* 34:1157-1168.

McKendrick, D., and A. Hicken. (1997). "Global strategy and population level learning in the hard disk drive industry," *The Data Storage Industry Globalization Project Report* 97-05. San Diego, CA: Graduate School of International Relations and Pacific Studies, U.C. San Diego.

Mowery, D. (1996). "Spinning off and spinning on?: The federal government role in the development of the US computer software industry," *Research Policy* 25:947-966.

MR (*Market Reports*). (July 19, 1991). "Korea - disk storage devices."

NST (*New Straits Times*). (April 25, 1996). "Quantum makes new move in drive business."

Odagiri, H., and A. Goto. (1993). "The Japanese system of innovation: Past, present and future," in *National Innovation Systems*, R. R. Nelson, ed. New York: Oxford University Press.

Okimoto, D., and Y. Nishi. (1994). "R&D organization in Japanese and American semiconductor firms," in *The Japanese Firm: Sources of Competitive Strength*, M. Aoki and R. Dore, eds. Oxford: Oxford University Press.

Porter, M.E. (1990). *The Competitive Advantage of Nations*. New York: Free Press.

Quantum. (1997). "Storage resource center on the web." Available: www.quantum.com/src/storage_ basics.

Rabinow, J. (1952). "The notched-disk memory," *Electrical Engineering* (August):745-749.

Rhyne, V.T. (1996). Congressional testimony on behalf of the National Research Council's Panel for Electronics and Electrical Engineering, the Board on Assessment of National Institute of Standards and Technology NIST Programs. June 25, 1996.

SoS (*Scotland on Sunday*). (September 29, 1996). "Patents pay-off pending." 1.

Stevens, L.D. (1981). "The evolution of magnetic storage," *IBM Journal of Research and Development* 255:663-675.

Teece, D.J. (1992). "Competition, cooperation, and innovation: Organizational arrangements for regimes of rapid technological progress," *Journal of Economic Behavior and Organization* 18:1-25.

TrendFocus. (1996a). Rigid Media Information Service. Annual Study. Palo Alto, CA: TrendFocus, Inc.

TrendFocus. (1996b). HDD Recording Head Information Service. Annual Study. Palo Alto, CA: TrendFocus, Inc.

Van de Ven, A., and R. Garud. (1989). "A framework for understanding the emergence of new industries," in *Research on Technological Innovation, Management and Policy*, R. Rosenbloom and R. Burgelman, eds. Greenwich, CT: JAI Press.

Apparel[1]

PETER DOERINGER
AUDREY WATSON
Boston University

Economists have traditionally argued that technological change and improvements in human capital are the key determinants of productivity growth. Business historians favor broader explanations that include changes in managerial organization and conduct. Alfred Chandler (1977), for example, characterizes the period between 1870 and 1920 as a second industrial revolution because improvements in technology interacted with new management systems that could tap economies of scale, scope, and organizational learning. Some analysts argue that the world is now experiencing a third industrial revolution, based on a combination of new information technologies and modern manufacturing techniques (Greenwood, 1997; Best, 1990; Milgrom and Roberts, 1990).

The apparel industry in the United States participated modestly in the first two revolutions. The sewing machine was the technological innovation that paved the way for factory production of clothing beginning in the 1850s and 1960s. During the second industrial revolution, economies of scale and scope in apparel manufacturing resulted in a rate of productivity growth that actually exceeded the average for all manufacturing between 1860 and 1910 (Kuznets, 1952).

---

[1] This research is sponsored by the Alfred P. Sloan Foundation, through a grant to the Harvard University Center For Textile and Apparel Research. The paper draws heavily upon materials developed by our colleagues Frederick Abernathy, John T. Dunlop, Janice Hammond, and David Weil. We are also grateful to Bruno Courault, Lynn Oxborrow, and Elisabeth Parat, whose work on a counterpart Sloan Foundation study in France and the United Kingdom has added to our understanding of the U.S. experience. David Mowery, members of the New England Economy Study Group, and participants at a conference on small-scale enterprise organized by the French Ministry of Labor and the Center for Employment Studies provided helpful comments on earlier drafts of these materials. Kara Bunting contributed excellent research assistance.

The apparel industry, however, became the antithesis of most modern industries in the postwar economy. It is dominated by small and medium-sized firms, technological change has been modest, education requirements are relatively low, and the industry remains labor intensive. The predictable result has been a loss of market share to imports and a substantial decline in employment. With two exceptions—commodity products, such as socks and men's underwear that can be mass-produced at low cost using capital-intensive technologies and high-fashion products that are not sensitive to price—the prognosis for the apparel industry through the early 1980s was one of continuing decline in market share and jobs.

That prediction is now being reassessed (Abernathy et al., 1995; *New York Times,* 1998). The apparel industry is adopting modern information technologies, domestic suppliers are serving new just-in-time replenishment markets, and labor productivity has been rising at rates comparable to those in all manufacturing. The *sustainability* of these trends, however, is less certain. Domestic production may have speed advantages over offshore production, but speed and cost are substitutes in the sourcing decision, and there is no domestic monopoly on production speed. Either increases in domestic production costs or faster production and delivery speeds among offshore suppliers could threaten the revival of production inside the United States.

This chapter examines recent developments in apparel production channels in the United States. It focuses on the growth of new domestic markets for just-in-time apparel supply and on the prospects for U.S. apparel manufacturers to develop the rapid response production capabilities needed to secure these markets against foreign competition.

APPAREL'S PLACE IN AMERICAN INDUSTRY

In 1995 U.S. apparel manufacturers shipped $62.9 billion in 1992 dollars of product, representing slightly under 2 percent of U.S. manufacturing output (Figure 1; Table 1). In addition to clothing products, the industry also includes home furnishings and industrial products, such as automobile upholstery, and the share of output accounted for by these non-clothing sectors has been growing in recent years.

The apparel industry has long been one of the nation's larger employers. Although employment has fallen by 38 percent since 1970 (Figure 2; Table 2), apparel still accounts for 4.6 percent of manufacturing employment. Earnings in apparel, which were once close to the manufacturing average, are now only 55 percent of average earnings in manufacturing, and real earnings have fallen more than 13 percent since 1970 (Table 3).

Skill and education levels are low in apparel. Ninety percent of production workers are unskilled or semi-skilled (Mittelhauser, 1997), and the percentage of the apparel workforce with less than a ninth-grade education is about double that of the average for manufacturing (Arpan et al., 1982). The apparel industry is

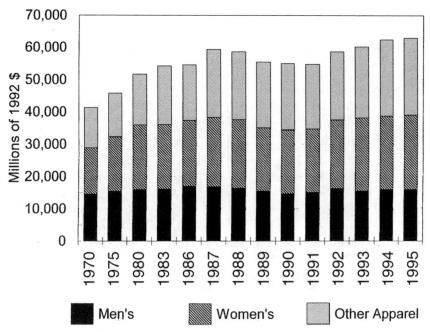

FIGURE 1 Real value of shipments (millions of 1992 dollars), 1970-1995.
Source: Bureau of the Census, *Annual Survey of Manufactures* (various years).

TABLE 1 Real Value of Shipments (millions of 1982 dollars), 1970-1995

Year	All apparel	Men's and Boys' apparel	% of total	Women's and girls' apparel	% of total
1970	41,231.2	14,511.9	35.2	14,351.3	34.8
1975	45,816.6	15,416.7	33.6	16,786.7	36.6
1980	51,614.2	16,008.5	31.0	19,914.7	38.6
1985	54,279.1	16,086.2	29.6	19,904.1	36.7
1986	54,486.0	17,025.6	31.2	20,269.8	37.2
1987	59,264.4	16,854.5	28.4	21,335.0	36.0
1988	58,220.2	16,339.1	28.1	21,235.6	36.5
1989	55,370.0	15,416.6	27.8	19,423.3	35.1
1990	54,866.8	14,555.2	26.5	19,606.4	35.7
1991	54,636.3	15,040.5	27.5	19,606.4	35.9
1992	58,548.8	16,158.7	27.6	21,192.9	36.2
1993	60,062.6	15,471.8	25.8	22,539.8	37.5
1994	62,331.6	16,014.9	25.7	22,538.5	36.2
1995	62,879.9	15,944.7	25.4	22,947.6	36.5

Note: Men's and boys' apparel is the sum of SIC codes 231 and 232; women's and girls' apparel is
the sum of SIC codes 233 and 234. Values of shipments are deflated by the producer price
indices for all apparel products, men's apparel, and women's apparel, respectively.
Source: U.S. Bureau of the Census, *Annual Survey of Manufactures* (various years).

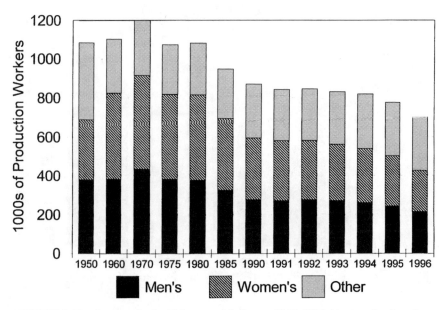

FIGURE 2 Employment in the U.S. apparel industry, 1950-1996 (thousands of production workers).
Source: Bureau of Labor Statistics, *Employment and Earnings* (various years).

also a major employer of women and minorities. Female workers make up about 75 percent of the workforce, accounting for almost 11 percent of all females in manufacturing. Around 15 percent of the apparel workforce is African-American, 24 percent is of Hispanic origin, and a substantial percentage of the remainder is Asian (U.S. Department of Labor, 1994).

Apparel is an industry of small firms (Figures 3 and 4; Tables 4 and 5). Average number of emloyees is 38, and two-thirds of all establishments employ fewer than 20 workers. Average establishment size, however, varies considerably across product sectors. With 109 employees, the average men's wear establishment is more than three times the size of the average women's wear establishment (Figure 3; Table 4). Establishment size had been growing until the early 1980s, when this trend reversed across all product categories. Firms with fewer than 20 employees account for less than 10 percent of the industry's workforce, however, while 37 percent of the workforce is employed in establishments with 250 or more employees (Figure 5; Table 6).

With many small plants, relatively limited economies of scale, and little vertical integration, apparel manufacturing is the quintessential example of a competitive industry. The four largest dress manufacturers, for example, account for only 6 percent of their market, and the eight largest have only a 10 percent market share.

From the beginning of mass production, however, apparel manufacturing has been characterized by elaborate contracting networks in which production is divided among "inside" shops (or "manufacturers"), "outside" shops, (or "contractors"), and "jobbers." Clothing jobbers are intermediaries, but they often play a much more extensive role than that of a mere middleman between suppliers and retailers. They design clothing, purchase and often cut material, deliver the fabric parts to contractors for assembly, and market the completed garments. Inside shops resemble manufacturers in other industries. They design their products, buy raw materials, produce their goods in company-owned facilities, and then

TABLE 2 Employment in the Apparel Industry, 1899-1996 (thousands of Production Workers)

Year	Total	Men's clothing	Women's clothing
1899	338	158	84
1914	548	226	169
1925	515	224	127
1935	631	277	221
1940	819	324	226
1950	1080	380	428
1960	1098	383	439
1970	1196	436	478
1975	1067	382	433
1980	1079	377	436
1985	944	325	367
1986	926	315	348
1987	922	312	344
1988	912	306	336
1989	907	295	333
1990	869	277	317
1991	841	271	309
1992	844	276	304
1993	829	272	289
1994	815	262	278
1995	772	244	259
1996	695	215	227

Note: Data from 1899 to 1935 are not directly comparable to later data. 1899 to 1935 data on men's clothing represent employment in outerwear, work clothing, shirts, and nightwear. 1940 to 1996 total employment figures are all production workers for SIC 23, apparel and related products. Men's clothing is the sum of production workers in SIC codes 231 and 232; women's clothing is the sum of production workers in SIC codes 233 and 234, except for 1940, for which SIC 234 is not available.

Sources: 1914-1935 total production workers, U.S. Department of Commerce, Bureau of the Census, *Census of Manufactures*; employment in men's and women's clothing, 1914-1935, Drake and Glasser, *Trends in the New York City Clothing Industry*; 1940-1996, Bureau of Labor Statistics, *Employment and Earnings and Supplement to Employment and Earnings* (various years).

TABLE 3 Relative and Real Average Weekly Earnings in the Apparel
Industry, 1909-1996

Year	Weekly apparel earnings as a percentage of weekly manufacturing	Real weekly apparel earnings in 1982-1984 dollars
1909	92.8	N/A
1914	86.2	N/A
1925	94.3	N/A
1935	88.6	N/A
1950	76.5	185.1
1960	62.9	190.7
1970	63.3	217.4
1975	58.5	207.4
1980	55.9	195.9
1985	54.0	193.8
1986	54.1	195.6
1987	54.1	193.5
1988	54.1	191.4
1989	54.5	189.0
1990	54.1	183.0
1991	55.0	183.9
1992	55.0	184.3
1993	54.3	182.5
1994	54.3	185.7
1995	54.9	185.5
1996	55.3	187.5

Note: Average weekly earnings for 1909 to 1935 are estimated by dividing total payrolls in men's and
    women's clothing by total wage earners in both sectors, and dividing the result by 52. Men's
    clothing includes outerwear, work clothing, shirts, and nightwear. Real earnings are deflated
    by the consumer price index, 1982-1984 = 100.
Sources: 1909 to 1935 apparel payrolls and wage earners, Drake and Glasser, *Trends in the New York
    Clothing Industry*; 1940 to 1996 apparel wages and all manufacturing wages, Bureau of
    Labor Statistics, *Employment and Earnings* and *Supplement to Employment and Earnings*
    (various years).

market the output. Outside shops serve as contractors for jobbers and inside
shops (Teper, 1937).

The apparel industry is labor intensive. Assets per employee were only 14
percent of the manufacturing average as late as the mid-1980s (Murray, 1995;
Rothstein, 1989), and new capital expenditures per worker average less than 15
percent of the manufacturing average (Table 7).

The pace of technological change in apparel has also been relatively modest
(Murray, 1995). The dimensional instability of fabric has made the actual sewing
process difficult to automate (Dunlop and Weil, 1996). More manufacturing innovation has occurred in the preproduction stages. Computer-aided design (CAD)
systems have reduced fabric waste and speeded the size-grading of patterns, while

marker-making and fabric cutting can now be performed by computer-guided lasers. The high cost of these systems, however, makes their adoption prohibitive for all but the largest firms (Murray, 1995; Rothstein, 1989). More recently, new information technologies are being adopted to link manufacturers to retailers. These systems allow producers to receive electronic point of sale data from stores and to track orders from production through delivery (Abernathy et al., 1995).

The single most important factor affecting the apparel industry has been the globalization of the supply chain. Prior to the 1970s, imports accounted for only about 10 percent of the domestic sales. Although both domestic shipments and domestic value added have continued to grow in real terms, domestic production has steadily lost market share to imports and over half the U.S. market is supplied by foreign producers (AAMA, 1997).

The rate of import penetration has been controlled by the Multi-Fiber Agreement (MFA), a complex system of tariffs and quotas which allowed Asian countries to be the dominant source of U.S. imports. Special trade privileges have been granted to Mexico under the North American Free Trade Agreement (NAFTA) and to Caribbean basin countries. These changes in trade policy are shifting production to the western hemisphere as domestic firms are outsourcing assembly to these regions to take advantage of preferential trade arrangements.

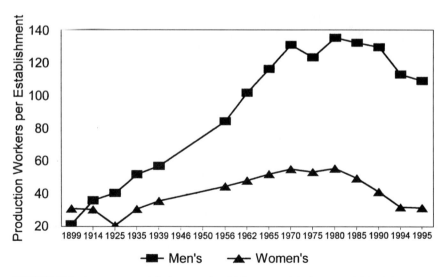

FIGURE 3 Average establishment size U.S. apparel industry, 1899-1995.
Sources: 1899-1939, industry averages, *Historical Statistics of the United States*; men's and women's, Seidman, *The Needle Trades*; Bureau of the Census, *County Business Patterns*.

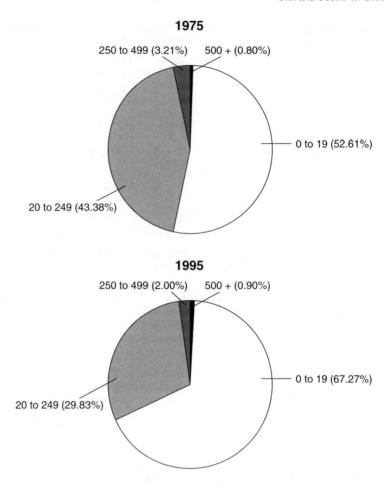

FIGURE 4 Establishments by size class, 1975 and 1995.
Source: Bureau of the Census, *County Business Patterns* (various years).

Even greater import pressure can be expected as the MFA is phased out early in the next century.

The challenge for the domestic clothing industry is to find ways to shift the basis of global competition from manufacturing cost, where the United States is at a comparative disadvantage in all but the most capital-intensive products, to the speed of supply, where proximity to markets gives the U.S. industry an edge. While employment continues to fall in the industry, these declines are concentrated in those sectors where foreign competition matters most, and the industry is now reporting employment gains in markets where just-in-time supply is important (*New York Times*, 1998).

THE INFLEXIBLE PRODUCTION SYSTEM

Since World War II, apparel production has been dominated by a relatively inflexible form of mass production known as the progressive bundle system (PBS). Under PBS, bundles of cut and partly sewn clothing parts are sequentially assembled into complete garments as they pass from work station to work station. Labor is highly specialized, many tasks take only seconds to perform, and the total labor content of a garment is measured in minutes. Workers can become very proficient at these specialized tasks, but learning curves can be as long as six months on the more skilled sewing jobs.

PBS assembly lines are difficult and costly to balance because of speed and quality problems inherent in working with soft fabric. Frequent style changes further raise line-balancing costs, making PBS most efficient when there are long production runs of each style. Individual employee differences in proficiency and level of fatigue also raise line-balancing costs.

For these reasons, large buffer stocks between work stations are used to prevent bottlenecks, resulting in long throughput times for individual garments. For

TABLE 4 Average Establishment Size in the Apparel Industry, 1899-1995

Year	Industry average	Men's clothing	Women's clothing
1899	27	21	31
1914	30	36	30
1925	28	41	21
1935	33	52	31
1939	37	57	36
1946	38	N/A	N/A
1950	42	N/A	N/A
1956	42	84	45
1962	46	102	48
1965	51	116	52
1970	57	131	55
1975	52	123	53
1980	58	136	55
1985	48	132	49
1990	43	130	41
1994	38	113	31
1995	38	109	31

Note: Industry averages 1899-1914 are number of production workers per establishment. 1899 to 1914 men's are for "men's, youths' and boys'" and women's are for "women's and children's". For 1946 to 1995, industry figures are for SIC code 23; men's for SIC codes 231 and 232, and women's for SIC codes 233 and 234.

Sources: Industry averages, 1914-1939, *Historical Statistics of the United States*; men's and women's, Seidman, *The Needle Trades*; 1946-1995, U.S. Bureau of the Census, *County Business Patterns* (various years).

TABLE 5 Percent of Establishments in Each Size Category, 1921 and 1946-1995

	No wage earners	0 to 5 wage earners	6 to 20 wage earners	21 to 50 wage earners	51 to 100 wage earners	101 to 500 wage earners	501 and over wage earners
1921	2.4	24.0	43.1	20.3	6.5	2.8	0.7

	0 to 19 employees	20 to 49 employees	50 to 99 employees	100 to 499 employees	Over 500 employees
1946	56.8	25.1	10.4	7.3	0.4

	0 to 3 employees	4 to 7 employees	8 to 19 employees	20 to 49 employees	50 to 99 employees	100 to 249 employees	250 to 499 employees	Over 500 employees
1950	20.7	14.0	22.1	24.1	10.8	6.2	1.5	0.5
1956	19.7	13.1	21.2	24.1	12.1	7.3	1.8	0.6
1962	18.9	12.7	20.2	24.1	12.9	8.1	2.3	0.7
1965	17.8	11.8	20.4	24.1	13.5	8.8	2.7	0.9
1970	15.7	11.5	20.0	24.4	14.1	9.9	3.4	1.0

	0 to 4 employees	5 to 9 employees	10 to 19 employees	20 to 49 employees	50 to 99 employees	100 to 249 employees	250 to 499 employees	Over 500 employees
1975	27.1	11.4	14.0	21.7	12.2	9.4	3.2	0.8
1980	22.8	11.9	15.3	22.1	13.0	10.2	3.7	1.1
1985	29.1	13.3	15.1	19.3	10.7	8.6	2.9	1.0
1990	32.5	14.9	14.4	17.5	9.7	7.6	2.6	0.8
1994	37.8	15.0	14.2	15.5	8.2	6.3	2.1	0.9
1995	37.6	15.0	14.6	15.6	8.0	6.2	2.0	0.9

Sources: 1921, U.S. Department of Commerce, Bureau of the Census, *Census of Manufactures*; 1946-1995, U.S. Department of Commerce, Bureau of the Census, *County Business Patterns* (various years).

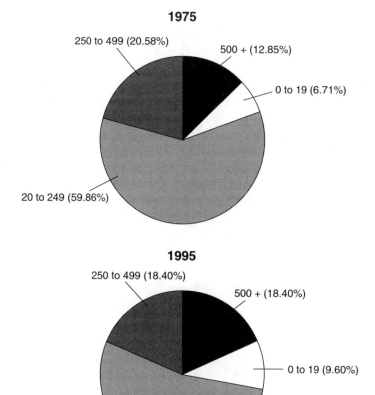

FIGURE 5 Employment by establishment size class, 1975 and 1995.
Source: Bureau of the Census, *County Business Patterns* (various years).

example, standard industry practice is to have one day's buffer stock between operations. This implies that a given pair of pants requiring 40 operations will take 40 days to move through the line, even though the average direct labor time for a pair of pants is only about 24 minutes (Dunlop and Weil, 1996).

This inflexible system had its origins in the development of mass markets in the 1920s and 1930s, but it became widespread in the manufacturing and retailing environment of World War II. The mass demand for military garments, War Production Board regulations limiting the variety of civilian styles that could be offered, and the preferences of consumers for "quality over variety" created opportunities for long production runs of identical products (Disher, 1947). Long

TABLE 6 Percent of Employment in Each Size Category, 1975-1995

	0 to 4 employees	5 to 9 employees	10 to 19 employees	20 to 49 employees	50 to 99 employees	100 to 249 employees	250 to 500 employees	Over 500 employees
All apparel								
1975	0.9	1.7	4.2	14.4	16.8	28.6	20.6	12.9
1980	0.8	1.6	4.2	13.2	16.5	27.8	21.3	14.6
1985	0.9	1.8	4.4	12.8	15.5	27.6	20.5	16.5
1990	1.2	2.3	4.6	12.9	15.8	27.2	20.5	15.4
1994	1.4	2.7	5.1	12.8	15.0	25.5	19.0	18.5
1995	1.5	2.7	5.4	13.1	15.0	25.5	18.4	18.4
Men's and boys' furnishings								
1975	0.2	0.4	1.2	4.6	10.6	33.2	32.9	16.8
1980	0.2	0.3	1.2	4.4	9.6	30.9	32.8	20.6
1985	0.2	0.4	0.9	4.1	9.9	33.3	29.5	21.7
1990	0.2	0.4	0.9	4.2	9.5	35.3	30.0	19.6
1994	0.2	0.5	1.1	4.9	10.5	30.3	26.7	25.8
1995	0.2	0.5	1.2	5.3	11.2	31.3	26.4	23.9
Women's and misses' outerwear								
1975	0.6	1.5	5.2	23.2	24.5	28.0	13.1	3.9
1980	0.6	1.7	5.8	22.4	25.1	27.0	12.4	5.0
1990	1.1	2.7	6.3	20.1	21.8	27.5	12.7	7.7
1994	1.6	39	8.1	21.3	20.0	23.5	10.2	11.4
1995	1.5	3.8	8.7	21.4	20.1	22.2	9.7	12.7

Note: Men's and boys' furnishings is SIC code 232; women's and misses' outerwear is SIC code 233.
Source: U.S. Department of Commerce, Bureau of the Census, *County Business Patterns* (various years).

TABLE 7 New Capital Expenditures per Production Worker (1982 dollars), 1960-1995

Year	All industries	All apparel	Men's and boys' apparel	Women's and girls' apparel
1960	2476.7	240.6	191.9	195.5
1970	4085.8	637.9	539.3	429.2
1975	5093.9	617.8	410.6	613.0
1980	5878.8	627.2	638.0	495.2
1985	6346.4	717.3	583.5	534.1
1986	5915.9	722.6	627.5	502.1
1987	5751.2	723.3	663.2	549.5
1988	5682.8	657.3	618.1	578.1
1989	6628.4	806.7	669.3	677.0
1990	6839.8	769.2	540.4	535.4
1991	6780.9	700.5	585.3	391.5
1992	6863.3	885.4	832.6	628.4
1993	6629.8	886.7	759.2	490.0
1994	7039.7	1018.2	612.0	562.4
1995	7655.4	1100.2	635.8	559.3

Note: Men's and boys' apparel is the sum of SIC codes 231 and 232; women's and girls' apparel is the sum of SIC codes 233 and 234. New capital expenditures are deflated by the producer price index for capital equipment.

Source: U.S. Bureau of the Census, *Annual Survey of Manufacturers* (various years).

runs offered substantial scale economies, particularly through the adoption of Tayloristic PBS manufacturing systems and the substantial division and specialization of labor. The result was the growth of relatively large, highly efficient firms that specialized in the mass production of garments.

The war also changed the apparel industry from a "buyer's market" to a "seller's market." Retailers, "anxious for all available supplies," could no longer demand rapid response production as they had in earlier decades. As a result, they were forced to accept the longer order times and production smoothing long sought by manufacturers (Disher, 1947).

After the war, clothing styles were no longer regulated, and style change again became important (Frank, 1953). Instead of returning to the prewar system of relatively flexible production, however, the demand for mass fashion at reasonable prices and the low unit costs made possible by PBS allowed the large manufacturing sector to continue to dominate the production channel. Manufacturers defined the styles of garments to be produced at different price points, set production and delivery schedules, and consequently had considerable influence over the level of inventories held by retailers (Disher, 1947). Many large manufacturers also developed "branded" products that gave them a marketing advantage over "store" brands.

The efficiency of this system kept clothing costs low at both the wholesale and retail levels. During the 1950s, before imports became significant, retail apparel prices rose less than 7.5 percent (less than one-third the rate of all consumer prices) and wholesale apparel prices were comparably stable (Tables 8 and 9). Low prices, coupled with the control of style and marketing by manufacturers, presumably lowered the resistance of mass retailers to the large inventories and long supply times inherent in the inflexible mass production system.

The accommodation that retailers made to the PBS system, however, laid the foundation for the subsequent shift from domestic to offshore supply chains. When lower-cost foreign clothing became available in the 1960s and 1970s, the adaptations that retailers had already made to inflexible domestic mass production predisposed them to view foreign suppliers as relatively easy substitutes for domestic suppliers. The even longer lead times and greater inflexibility of foreign supply channels were only an extension of the inflexible characteristics of the domestic supply chain, and the costs of added inflexibility were more than compensated for by the labor cost advantages of imports.

THE FLEXIBLE PRODUCTION CHANNEL

Currently, new demands from downstream retailers and inflexibilities in fabric supply, as well as pressures from foreign competition, are transforming the inflexible manufacturing-driven domestic production system. Unlike apparel, the retail and textile sectors are characterized by large firms and rising levels of concentration. For example, the four largest apparel retailers held 17.9 percent of the market in 1992, compared with 6.4 percent in 1972. The corresponding figures are 27.6 percent and 11.2 percent for women's specialty shops and 53.1 percent and 38.8 percent for department stores (Bureau of the Census, Census of Retail Trade, various years). Increased concentration, along with the availability

TABLE 8 Percentage Change in Consumer Price Indices, 1950-1996

	All items	Apparel less footwear	Men's and boys' apparel	Women's and girls' apparel
1950–1960	22.82	7.42	11.06	5.39
1960–1970	31.08	27.22	31.78	26.63
1970–1980	181.67	44.19	43.73	33.70
1980–1990	58.62	32.04	34.68	27.60
1990–1996	20.05	4.48	6.06	1.80
1950–1996	551.04	171.82	200.47	131.78

Source: Bureau of Labor Statistics.

TABLE 9 Percentage Change in Producer Price Indices, 1950-1996

	All finished goods	All apparel	Men's and boys' apparel	Women's and girls' apparel	Children's and infants'apparel
1950–1960	18.44	4.72	7.03	1.43	16.10
1960–1970	17.66	16.80	24.57	10.37	23.53
1970–1980	123.92	55.61	78.32	38.38	48.13
1980–1990	35.45	32.47	31.65	33.60	32.38
1990–1996	10.15	6.47	9.90	3.27	6.16
1950–1996	365.60	168.45	244.01	113.73	198.54

Source: Bureau of Labor Statistics.

of offshore suppliers, has shifted decision-making power within the production channel from clothing manufacturers to mass retailers.

Increased cost pressures following the wave of leveraged buyouts and mergers in retailing in the 1980s encouraged retailers to reduce the costs of inventories by adopting new information technologies, such as electronic point of sale (EPOS) data and computerized ordering and stock management programs. These cost-cutting practices are known as "lean retailing" (Berg et al., 1996). The proliferation of clothing styles, colors, and sizes as well as the shortening of product life cycles in the 1980s further intensified the incentives for adopting lean retailing practices. In one industry sample, the number of stock-keeping units (SKUs)—the codes that define individual products by style, color, and size—increased by 63 percent between 1988 and 1992. At the same time, the year-to-year turnover in styles rose. Between 1984 and 1992, the number of selling seasons in a year grew from 2.8 to 3.2 for basic fashion products and from 2.9 to 3.7 for fashion products (Abernathy et al., 1995). Within selling seasons, the importance of "short-cycle" products—those with a planned sales period of only 5 to 10 weeks, compared with 12 to 20 weeks for traditional "seasonal" products—has also risen, accounting for more than one-third of apparel consumption in 1988 (Rothstein, 1989).

More products and more rapid style change tend to raise inventory and markdown costs and to increase the possibility of lost sales. They also raise uncertainty about consumer demand because there are fewer products with a market history and less time in a season to adapt to demand fluctuations.

In the face of these pressures, mass retailers sought further reductions in inventories and markdowns through the extension of lean retailing to include just-in-time supplies. Leadership for this change came across a wide spectrum of retailing—mass merchandisers such as Wal-Mart, large department store chains such as Dillards, national variety chains such as J.C. Penney, and national specialty chains such as The Limited (Abernathy et al., 1995). These extended lean retailing efforts have focused on "basic fashion" products positioned between commodity products and more fashionable women's wear products.

In principle, just-in-time supply allows retailers to place smaller initial orders because replenishment supplies can be obtained throughout the selling season in response to actual sales. As a result, inventories, stockouts, and markdowns would be reduced. Just-in-time delivery is not consistent with the supply capabilities of the inflexible domestic PBS system, however, and is beyond the reach of distant offshore supply chains. Introducing a quick supply capability into domestic PBS supply channels has involved both a willingness among retailers to pay a cost premium for quick and accurate fulfillment of replenishment orders and to provide the information systems needed to link domestic clothing manufacturing to retail sales data.

The main instrument for building just-in-time supply chains has been the transfer of new information technologies from lean retailers to apparel manufacturers. Examples include electronic data interchange (EDI) of point-of-sale data between retailers and clothing manufactures and the use of EPOS computer programs to trigger quick-response shipments and initiate new production.

The spread of new computer and information technologies within the domestic supply chain has been well documented by the Harvard Center for Textile and Apparel Research (Abernathy et al., 1995). Between 1988 and 1992 the use of bar coding at the detailed product level grew by 2.75 times; EDI links between clothing retailers and manufacturers grew by more than 7 times; the receipt of EPOS data on either an individual store or a company-wide basis rose from 13 percent of sales volume to 32 percent; and the use of programmed ordering models also increased (Abernathy et al., 1995).

The effective use of EPOS and EDI by clothing manufacturers, however, requires considerable knowledge about the information systems and lean retailing practices of specific retailers. Much of this knowledge is proprietary and cannot be readily acquired through arms-length market relationships. Lean retailers, therefore, have reinforced information linkages by forming privileged business relationships, or "partnerships," with their just-in-time suppliers to facilitate the transfer of "match-specific" knowledge. These partnerships often involve technical assistance in improving operations planning, management, and logistics between retailers and manufacturers.

INDICATORS OF PERFORMANCE UNDER
JUST-IN-TIME PRODUCTION

Rapid replenishment capabilities have spread through the supply chain. For example, 73 percent of sales volume for national chains and 66 percent of sales volume for mass merchants in 1992 were replenished on a daily or weekly basis (Abernathy et al., 1995). These data understate the growing importance of the practice, however, because the retail sectors that use rapid replenishment most intensively have also been increasing their overall share of domestic apparel production. Between 1988 and 1992 the percentage of domestic shipments going to national chains rose by 64 percent and that going to mass merchants rose by 12 percent, compared with a 7 percent decline for department stores and a constant share for specialty stores (Abernathy et al., 1995).

Adopting information technologies, forming retailer-supplier partnerships, and achieving higher replenishment speeds has brought corresponding improvements in clothing manufacturing practices. For example, work-in-process inventories fell by one-third, from three weeks of supply to two weeks, between 1988 and 1992, while manufacturing throughput times have been reduced (Abernathy et al., 1995; Dunlop and Weil, 1996; Berg et al., 1996).

Firms that adopt the new information technologies and other innovations associated with rapid replenishment capabilities perform better than those that do not. Comparisons between "high innovation" and "traditional" apparel firms show large differences in the time from fabric purchase through completion of manufacturing (79.2 days compared with 128.9 days), replenishment supply speeds (9.2 days compared with 26.3 days), and operating profits (10.5 percent compared with 5 percent) (Abernathy et al., 1995). Even firms that adopt the minimal level of innovation consistent with supplying lean retailers perform substantially better than traditional firms.

Aggregate Productivity Growth

Given these sweeping changes in the apparel supply channel, it is appropriate to look for wider evidence of their impact on business performance. One aggregate indicator of performance is the growth in productivity. To establish a baseline from which to compare productivity growth during the lean retailing period, we calculated the trend rate of growth in real value-added per worker hour between 1950 and 1995. This trend rate for the entire industry is 3.3 percent a year, a record that is slightly better than that for all manufacturing (Table 10; Figure 6). We also looked at a second baseline period, 1950-1970, when the PBS manufacturing system dominated the apparel industry and when imports were relatively unimportant. Trend annual productivity growth in this period was very close to that for the entire postwar period—3.4 percent between 1950 and 1960 and 3.2 percent between 1960 and 1970 (Table 10). Trend productivity growth was somewhat higher for women's and girls apparel and somewhat lower for men's and boy's apparel (Figures 7 and 8).

These aggregate trends, however, are the net result of a far more complicated set of forces than the organization of production in the industry. These forces include growth of imports since the early 1970s and shifts in the importance of mass manufacturers as well as technological change. The magnitudes of some of these changes, such as those associated with imports and increased offshore contracting, are presumably large relative to those associated with manufacturing technology. Although productivity growth cannot be partitioned into its compo-

TABLE 10 Value-Added Per Worker Hour (in 1982 dollars), 1950-1995

	All manufacturing	All apparel	Men's and boys' apparel	Women's and girls' apparel
1950	13.4	4.9	5.1	4.4
1960	20.3	7.0	7.3	6.4
1970	28.6	9.6	9.8	8.9
1975	31.6	11.4	11.0	10.6
1980	32.9	13.2	12.8	12.6
1985	40.3	16.7	16.6	15.7
1986	43.3	17.4	17.1	17.1
1987	45.5	18.6	17.9	18.3
1988	47.2	18.4	17.0	18.9
1989	46.7	18.3	16.9	17.7
1990	45.8	18.3	16.3	17.9
1991	46.5	18.7	17.5	18.5
1992	49.1	19.7	18.2	20.2
1993	50.0	19.9	17.1	20.0
1994	52.1	21.2	19.4	20.6
1995	53.3	21.8	20.8	20.5
Annual Growth Rates:				
1950–1995 (%)	3.1	3.3	3.1	3.4
1950–1960 (%)	4.1	3.4	3.5	3.8
1960–1970 (%)	3.4	3.2	2.9	3.4
1970–1980 (%)	1.4	3.2	2.7	3.4
1980–1990 (%)	3.3	3.3	2.4	3.5
1970–1987 (%)	2.7	3.9	3.5	4.2
1987–1995 (%)	2.0	2.0	1.9	1.4
1990–1995 (%)	3.1	3.4	4.9	2.7

Note: Men's and boys' apparel is the sum of SIC codes 231 and 232; women's and girls' apparel is the sum of SIC codes 233 and 234. Manufacturing value-added is deflated by the producer price index for all finished goods; apparel value-added figures are deflated by the producer price indices for all apparel products, men's apparel, and women's apparel respectively.

Source: U.S. Bureau of the Census, *Annual Survey of Manufactures* (various years).

nent parts with any precision, the direction and timing of these different influences can be examined.

Import Penetration

The period 1970-1987 largely predates lean retailing and approximates a time when import growth was the dominant influence on the industry. Productivity growth during this period was 3.9 percent a year, substantially above the long-term trend, and it was particularly high in women's and girls' apparel. We suspect that the conventional wisdom of imports displacing low value-added domestic production is the primary explanation for this superior productivity performance.

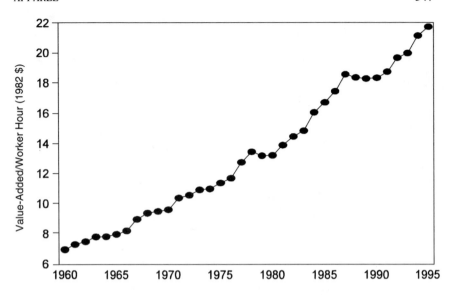

FIGURE 6 Real value added per worker hour, all apparel, 1960-1995.
Source: U.S. Bureau of the Census, *Annual Survey of Manufactures* (various years).

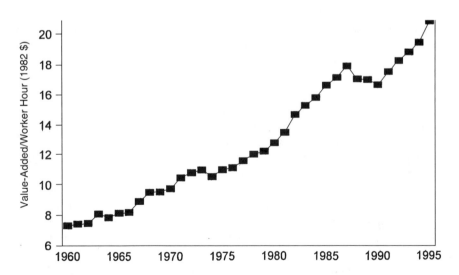

FIGURE 7 Real value added per worker hour, men's apparel, 1960-1995.
Source: U.S. Bureau of the Census, *Annual Survey of Manufactures* (various years).

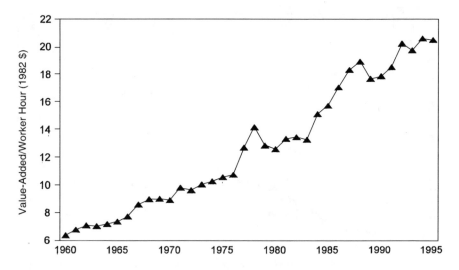

FIGURE 8 Real value added per worker hour, women's apparel, 1960-1995.
Source: U.S. Bureau of the Census, *Annual Survey of Manufactures* (various years).

The Effects of Lean Retailing

Productivity growth after 1987 and extending to the present is subject to a more complex set of influences than earlier periods. This was the time when lean retailing became important in creating new markets for rapid replenishment domestic production in certain product lines. But it was also a period that witnessed a shift in the origin of imports from Asia to the Caribbean basin and Mexico as well as continued growth in imports.

Lean retailing can affect productivity in several different ways, some of which operate in contradictory directions, and its net effect is difficult to predict. For example, it can raise productivity growth by promoting lean manufacturing practices that make labor more efficient. Conversely, if production batches are smaller or permit less specialization of labor as would be the case with modular production, the growth in labor productivity will be reduced.

A second effect occurs through the mix of products produced domestically. Lean retailing has focused on basic fashion products. To the extent that the value-added of basic fashion products is below the average for domestic production, an increase in replenishment production will also slow the observed rate of productivity growth.

A third consequence of lean retailing has been to favor large suppliers over smaller ones. If there are differences in value-added per worker by size of firm, then the changing size composition of firms will affect productivity growth.

Somewhat surprisingly, there is roughly a U-shaped distribution of real value-added per worker hour by size of establishment in the apparel industry with the smallest firms (fewer than 20 employees) often having the highest productivity. On balance, the growth in importance of large firms under the influence of lean retailing has probably had a relatively minor effect on aggregate productivity in the industry.

A final aspect of lean retailing is to shift responsibility for various "value-added services" required to make garments floor-ready from retailing to manufacturing. For example, the volume of floor-ready merchandise shipped rose from 40.6 percent to 50.3 percent between 1988 and 1992, and increases were also registered in the labeling and pricing of garments by clothing manufacturers (Abernathy et al., 1995). Business units with the highest replenishment pressures were the most likely to provide these services. Although data are lacking on the actual value-added per worker hour of these services, tasks like price tagging and floor-ready packing of garments presumably contribute less to value-added than the average assembly job and probably have had a negative effect on the growth in labor productivity.

Changing Trade Regimes

The shift in the national origin of U.S. apparel imports toward the Caribbean basin and Mexico began in 1985 as the result of preferential trading arrangements that eased restrictions on imports from these regions. The Caribbean basin trade preferences are conditional upon imports being assembled from U.S.-made parts, while NAFTA requires only that the fabric originates in North America. This requirement, along with geographic proximity, has encouraged domestic clothing manufacturers to retain high value-added design, cutting of U.S.-made fabric, and marketing functions while transferring lower value-added assembly work to foreign countries (Mittelhauser, 1997). This "contracting" effect contributes to raising average labor productivity in the U.S. apparel industry.

Net Effects on Productivity Change

Although the precise contribution of each of these factors to changes in productivity cannot be determined, their net effect can be observed. Productivity growth has decreased for the industry as a whole during the period 1987-1995, with the decline being sharpest for women's garments (Table 10). This picture, however, is misleading because the timing and pattern of productivity change vary by product.

For example, basic fashion has been the product sector where lean retailing should have the greatest effect on domestic productivity. We would expect to see the lean retailing directly raising productivity through its effects on lean manufacturing and its bias toward large firms Conversely, product lines such as fashionable dresses or blouses are least likely to be affected by either lean retailing or

TABLE 11 Annual Growth Rates of Real Value Added per Worker Hour, Selected Products

	Men's and boys' suits and coats (SIC 231)	Men's and boys' shirts (SIC 2321)	Women's and misses' dresses (SIC 2335)	Women's and misses' suits and coats (SIC 2337)
1960 to 1970	2.6	3.0	3.5	3.2
1970 to 1980	2.0	2.1	1.9	4.7
1980 to 1990	2.2	4.2	5.2	4.6
1970 to 1987	3.1	3.7	4.4	4.9
1987 to 1995	0.8	0.7	0.5	0.4
1990 to 1995	3.5	1.1	1.7	−1.3

Note: Men's and boys' shirts includes nightwear before 1987.
Source: U.S. Bureau of the Census, *Annual Survey of Manufactures* (various years).

import penetration and fashion products are also more likely to be produced by small firms where productivity growth has been relatively slow in recent years. These effects are likely to slow productivity growth in product lines such as fashionable dresses and blouses.

Examining productivity change between 1987 and 1995 in four of the largest product lines for which consistent data are available reveals a slowdown in productivity growth below the long-term trend. This slowdown, however, was concentrated in the late 1980s. After 1990, as lean retailing gained strength, there was a widespread rebound in most of the product lines (Table 11). Moreover, productivity growth in men's and boys' clothing, where style tends to be more "basic" than "fashion," is at an historical high, while the lowest productivity growth was recorded for product lines such as women's suits and coats and women's dresses, where we would expect to see the weakest influences of trade and lean retailing (Tables 10 and 11). These patterns lend support to the thesis that the effects of lean retailing on productivity are positive.

The Reinvigoration of the Large Firm Sector

A second way to assess the effects of lean retailing on the apparel supply chain is to examine growth patterns by size of firm. Between the 1950s and 1970, when the inflexible PBS production channel was expanding and before foreign imports and lean retailing were significant, the fraction of large establishments (250 or more employees) in the apparel industry more than doubled (Table 5). Thereafter, imports led to widespread employment declines, particularly among large PBS manufacturers whose markets were particularly vulnerable to import penetration.

One would expect, however, that the growth of rapid replenishment markets would limit or reverse this decline because of the preferred position of large firms in lean retailing partnerships. This effect is confirmed by the data. The employment share of the very largest firms (500+ employees) grew from 12.9 percent in 1975 to 15.4 percent in 1990 and to 18.4 percent in 1995 (Table 6).

This growth in large firms also appears to be associated with an increase in the importance of "inside" manufacturers. The share of total value-added held by inside manufacturers grew across a broad range of product lines throughout the period 1972 to 1992 (Table 12). A similar trend throughout this period can be seen in the share of employment accounted for by manufacturers in all of the product lines except women's dresses (Table 13). This growth in the share of value-added by manufacturers does not necessarily imply a reduction in contracting, since manufacturers also took on more value added services under lean retailing and had the option of foreign contracting as well. Comparable data for domestic contractors, however, confirm a reduction in the amount of domestic contracting activity by manufacturers.

The overall trend away from domestic contracting and toward consolidating production within manufacturing began when import penetration was the main influence on the apparel industry. During 1987-1992, the part of the "lean retailing" period for which there are data, this trend is limited to a more narrow range of products. The share of both value added and employment held by manufacturers continued to increase for the men's wear products examined (except for the value-added in men's and boys' shirts) and for women's blouses. Contrary to the long-term trend, however, contractors gained a larger share of value-added in all of the women's wear products and in men's and boys' shirts. The share of em-

TABLE 12 Percent of Value Added, by Type of Firm: Selected Products, 1972, 1987, and 1992

	1972	1987	1992
Manufacturers			
Men's and Boys' Suits and Coats (SIC 231)	55.5	69.0	77.4
Men's and Boys' Trousers and Slacks (SIC 2325)	45.9	54.0	60.0
Men's and Boys' Shirts (SIC 2321)	46.1	75.5	71.5
Women's and Misses' Blouses (SIC 2331)	25.2	38.8	44.9
Women's and Misses' Dresses (SIC 2335)	30.8	48.4	40.8
Women's and Misses' Suits and Coats (SIC 2337)	28.5	45.1	44.4
Contractors			
Men's and Boys' Suits and Coats (SIC 231)	31.1	22.2	18.5
Men's and Boys' Trousers and Slacks (SIC 2325)	34.5	20.8	13.8
Men's and Boys' Shirts (SIC 2321)	29.9	9.9	17.1
Women's and Misses' Blouses (SIC 2331)	40.5	31.1	32.6
Women's and Misses' Dresses (SIC 2335)	38.1	28.5	34.3
Women's and Misses' Suits and Coats (SIC 2337)	34.0	24.0	25.1

Note: Men's and boys' shirts include nightwear in 1972.
Source: U.S. Bureau of the Census, *Census of Manufactures* (various years).

TABLE 13 Percent of Total Production Worker Employment, by Type of Firm: Selected Products, 1972, 1987, and 1992

	1972	1987	1992
Manufacturers			
Men's and Boys' Suits and Coats (SIC 231)	46.9	5.7	67.2
Men's and Boys' Trousers and Slacks (SIC 2325)	37.6	51.5	61.4
Men's and Boys' Shirts (SIC 2321)	40.9	56.6	71.1
Women's and Misses' Blouses (SIC 2331)	25.7	26.6	32.4
Women's and Misses' Dresses (SIC 2335)	32.4	27.2	24.4
Women's and Misses' Suits and Coats (SIC 2337)	27.9	32.7	30.0
Contractors			
Men's and Boys' Suits and Coats (SIC 231)	48.5	39.8	31.2
Men's and Boys' Trousers and Slacks (SIC 2325)	54.8	45.6	34.1
Men's and Boys' Shirts (SIC 2321)	51.8	41.9	27.9
Women's and Misses' Blouses (SIC 2331)	69.0	68.5	63.6
Women's and Misses' Dresses (SIC 2335)	61.1	69.5	72.4
Women's and Misses' Suits and Coats (SIC 2337)	63.8	60.5	62.4

Note: Men's and boys' shirts include nightwear in 1972.
Source: U.S. Bureau of the Census, *Census of Manufactures* (various years).

ployment held by contractors also increased for women's dresses and suits and coats (Tables 12 and 13). By and large, we read these data as confirming the conclusion that lean retailing has shifted domestic production toward manufacturers in product lines, such as men's wear, that have a relatively lower fashion content, while at least some higher fashion products, such as dresses, are becoming more contractor-intensive.

THE LIMITS TO GROWTH OF REPLENISHMENT MARKETS

By 1992, almost half of all sales by domestic producers were being shipped on a weekly or shorter replenishment basis. The substantial progress that has been made in serving rapid replenishment markets and in improving the performance of the U.S. apparel industry is cause for optimism. If past trends can be extended, the apparel industry faces a brighter future than would have been predicted a decade ago. Some signs, however, point to limits on future improvements in rapid replenishment speeds and business performance.

One set of limiting factors may be the difficulty of managing rapid response production. Despite new information technologies and various other changes in supply chain management, rapid replenishment appears to be accomplished, in part, by manufacturers' holding larger inventories of finished goods from which they can provide frequent shipments. This result can be seen in data on "innovative" clothing firms—the types of suppliers that are most likely to be serving rapid replenishment markets—that hold larger inventories of finished goods and have fewer turns per year of finished goods inventories than do less innovative firms (Abernathy et al., 1995). Our surveys in the United Kingdom show that reliance on inventories of finished goods is an even more common device for meeting rapid replenishment pressures than in the United States (Doeringer et al., 1998). What appears to be happening, in part, is that lean retailing is shifting inventories of finished goods from retailers to clothing manufacturers.

Moreover, the flexible production channel being constructed to serve rapid replenishment markets is biased toward large firms (Abernathy et al., 1995). A cluster of factors that link large suppliers to lean retailing explain this bias. First, large firms are better able than small firms to finance and manage the costly information technologies required by lean retailers. Second, there are likely to be substantial coordination efficiencies within partnerships between mass retailers and large suppliers compared with those with a fragmented supply chain of numerous smaller firms. The link between lean retailing and large suppliers is further reinforced by the focus of rapid replenishment on basic fashion products. Domestic suppliers of basic fashion products tend to be larger than suppliers of fashion products (Abernathy et al., 1995).

Mass retailers in the United Kingdom show a similar tendency to form partnerships with large suppliers. Some large retailers in the United Kingdom are even providing performance incentives to their partners by promising to increase the volume of orders to suppliers who meet quick response targets. These retailers eventually plan to phase out their smaller suppliers.

The rationale of building just-in-time supply chains around large firms and basic fashion products may prove short-sighted. Although large firm production channels have made substantial strides in speeding production and delivery to replenishment markets, they may never be able to achieve the response speeds and small order sizes that are reported historically for small firms (Magee, 1930) or that we routinely find in our interviews in the small firm sector.

The tendency of large firms to accomplish rapid replenishment by increasing finished goods inventory may also indicate that diminishing returns are occurring in the gains to lean manufacturing in large firms. If so, reliance on large firms for rapid replenishment may restrict the future growth of lean retailing markets for basic fashion products and is likely to inhibit the extension of lean retailing to fashion products.

PROSPECTS FOR THE SMALL FIRM SECTOR

Not only have the largest firms been growing in importance in the apparel industry, but there has also been substantial growth in the relative importance of firms with fewer than 20 employees (Tables 5 and 6). One explanation of small firm growth is that rising imports forced firms in the medium-size range (between 20 and 249 employees) to downsize. This downsizing thesis is consistent with the timing of the decline of medium-size firms and the growth of small firms beginning after 1970 and accelerating since the late 1980s.

According to this interpretation, the industry is "hollowing out" and becoming more dualistic (Palpacuer, 1996). In the future the industry will consist of a core of large, highly efficient, information-intensive manufacturers that will serve the relatively secure markets for commodity apparel products and basic fashion replenishment items. A periphery of small, marginal firms will be left with highly uncertain residual markets.

A second hypothesis is that the hollowing-out of the middle of the industry and the relative growth of small firms is related to changing patterns of outsourcing. For example, the downsizing of medium-sized firms to become small firms can be seen as a response to new opportunities for international specialization within the production channel as well as to declining demand for domestic production. Such "internationally specialized" small firms will continue to buy fabric, make patterns, cut fabric, and market the final product, but their garments will be assembled in the Caribbean basin and Mexico. This specialization will allow the United States to retain the more highly skilled workers and the higher value-added tasks in which it holds a comparative advantage.

Indirect evidence of such specialization is found in the changing occupational mix in the apparel industry. Between 1983 and 1994, employment in apparel manufacturing fell by 16 percent, but this decline was concentrated in sewing and unskilled jobs. Employment in other occupations such as designers, technicians, and marketing that are involved in onshore pre- and post-assembly activities either grew or remained unchanged during this period (Mittelhauser, 1997).

A third possibility is that the growth in small firms reflects a set of production advantages that are important in fashion markets. Compared with basic fashion, fashion products are likely to have smaller initial orders, more uncertain markets, and shorter product seasons. Small firms have a proven capacity to produce small lots of fashion garments with short throughput times, whereas large firms are not sufficiently flexible to serve such markets. These advantages of speed and flexibility may also position small firms for a role in serving rapid replenishment markets and may be a foundation for extending rapid replenishment to fashion.

History provides ample precedent for this possibility. Today's rapid replenishment pressures from lean retailers echo those of the 1920s and 1930s. This was a period when the new and rapidly growing demand for mass fashion led to product proliferation. "In the manufacturing industry there developed a period of the wildest experimentation in design. This in turn led to great confusion, not only among manufacturers as to what to make, but also among retailers as to what to buy," according to the National Retail Dry Goods Associations (1936).

Because of this "great confusion," retailers were generally reluctant to order goods ahead of actual demand, preferring instead to order a few items at the start of the season followed by rapid replacement of the styles that sold (Teper, 1937). Uncertainty about fashion trends was further aggravated by the "piracy" of cloth- ing designs. Lower-priced copies of garments appeared with startling rapidity, sometimes before the originals reached the stores, lessening the value of the origi- nal designs. This problem of piracy increased the rate at which the better-priced manufacturers came up with new styles and created further incentives for retail- ers to bring new styles to market as quickly as possible (Teper, 1937).

Problems of fashion uncertainty and piracy of styles were particularly preva- lent in women's garment production and aggravated the seasonal volatility of that sector (Grieg, 1949). Women's wear production was traditionally scheduled around two selling seasons, spring and fall, with each having a mid-season "fill- in" period. Styles were set and orders placed relatively close to the selling season, and production and delivery followed orders with a relatively short lag. Spring styles for women, for example, were shown in early December, with the first orders delivered in January and the fill-in orders occurring in March (Carpenter, 1972). By 1939 an estimated 125,000 different dress styles were produced in New York; slightly fewer than half of these were moderately priced garments with an average production run of 997. The remaining styles represented better- priced dresses with an average production run of only 267 (Hochman, 1941).

Unlike current lean retailing partnerships, lean retailing efforts in this period relied on the rapid response capability of small firms. Pressures from retailers for ever more rapid supply response and lower costs led to the development of quick response mass production systems (Bryner, 1916). A "frantic insistence upon immediate deliveries when orders are finally placed" (Teper, 1937) increased the pressures for rapid response manufacturing. Most orders placed after the begin- ning of the season were "for immediate delivery, that is, a week or ten days" (Magee, 1930). Retailers also frequently returned merchandise that had been ordered but went unsold (Teper, 1937).

The quickest of these rapid response systems was the "piece" or "complete garment" method, which was widely used in small and medium-sized shops (Bryner, 1916). Under this system a single operator performed the basic assem- bly of the garment, with an additional worker completing finishing operations such as felling linings and making buttonholes.

Similar systems and similar production speeds and flexibility still appear to be common, but today's small firms are often characterized as inefficient because they lack advanced technologies and sophisticated management practices. Other factors may offset these inefficiencies, however; the data on productivity by size of firm shows that firms with fewer than 20 employees often outperform all but the largest firms.

Although these various pieces of evidence point to the potential for positive efficiency contributions from rapid replenishment suppliers in the small-scale sector, it is premature to conclude that this sector can be integrated into lean retailing production channels in the foreseeable future. The weaknesses in information technology and modern management practices are a barrier to small firms supplying lean retailers. Bringing small firms into lean retailing production channels and extending rapid replenishment to fashion products will require new institutions that can link mass retailers to the small firm sector.

INSTITUTIONAL OBSTACLES TO RESTRUCTURING APPAREL PRODUCTION CHANNELS

Lean retailing is currently providing the motivation and the leadership necessary for efficient restructuring of domestic apparel production channels based upon rapid supply chain response. Progress toward this goal, however, is impeded by incompatibilities in relationships between sectors, missing institutional links, and institutional biases in channel reforms.

One obstacle to production channel reform is the self-reinforcing character of efficient relationships among the different sectors of the channel. Once production and distribution arrangements become compatible with channel-wide efficiency, any one sector in the channel will find it difficult to respond to changing markets or technology unless the other sectors also adapt accordingly.

Efficient transformation of the production channel is also slowed by the adaptation biases inherent in the established institutions of the channel, or what economists call "path dependency." Established production arrangements mean that production channels evolve in directions that extend, rather than radically depart from, existing systems. This "systemic" aspect of production channels biases the restructuring process toward incremental, rather than radical, changes in channels. One example, as noted earlier, is the successful adaptation of mass retailers to the efficiencies of the inflexible domestic supply system of the postwar period. Once this adaptation was made, it became more likely that retailers would next choose to lower costs further by turning to even slower and lower-cost supplies from Asia, instead of by developing rapid replenishment capabilities within the domestic supply chain.

A related obstacle is the trend toward growth of the large-firm sector. Present reforms in production channels revolve around the relationship between mass retailers and large clothing suppliers. Mass retailers that are promoting restructuring through organizational relationships with their suppliers find it easier to coordinate production and logistics with a relatively small number of partners. As a result, retailers are deliberately reducing the number and increasing the size of their suppliers.

As large firms gain a bigger market share, they become less dependent upon contractors and jobbers. The decline in contractors was documented earlier; the decline in jobbers has been even faster. In 1967 there was one jobber for every 3.2 contractors in the women's clothing sector, but that number fell to one jobber for every 6.9 contractors by 1992 (Bureau of the Census, Census of Manufacturers, various years). Similar changes can be seen in specific product lines (Table

TABLE 14 Number of Contractors per Jobber, by Sector, 1972, 1987, and 1992

	1972	1987	1992
Manufacturers			
Men's and Boys' Suits and Coats (SIC 231)	2.66	5.28	5.47
Men's and Boys' Trousers and Slacks (SIC 2325)	2.86	6.39	5.41
Men's and Boys' Shirts (SIC 2321)	2.53	5.00	5.19
Women's and Misses' Blouses (SIC 2331)	3.47	6.64	8.65
Women's and Misses' Dresses (SIC 2335)	3.50	12.47	6.60
Women's and Misses' Suits and Coats (SIC 2337)	2.62	4.06	5.23

Note: Men's and boys' shirts include nightwear in 1972.
Source: U.S. Bureau of the Census, *Census of Manufactures* (various years).

14). Much of this decline in jobbers occurred before lean retailing was important, but it has been sustained or extended during the lean retailing period in four of the six product lines we examined.

As production networks between large and small suppliers and between small suppliers and large retailers become more tenuous, the potential for lean retailing and just-in-time production to create new jobs for the domestic supply chain may be threatened. Domestic job growth depends on continually increasing replenishment speeds.

Filling Institutional Gaps in Apparel Production Channels

Reversing the exclusion of flexible, small-firm production from lean retailing systems requires a reconception of the role of intermediaries in apparel production channels. Historically, the problems in the small-firm sector—lack of scale economies, limited managerial capacity, and fragmentation—have been offset by various intermediaries between small producers and retailers. The traditional intermediaries have been jobbers, and sometimes "inside" manufacturers. Lean retailers, however, have been eliminating such intermediaries from their production channels by choosing to develop direct relationships with large manufacturers and by performing more of the design function once controlled by manufacturers. At the same time, manufacturers are relying less on domestic contracting.

One alternative to jobbers as intermediaries would be for small firms to form multifirm partnerships along the model of trade associations. Historical experience has shown, however, that competitive market pressures have limited the success of trade associations in coordinating business decisions among their members. Furthermore, many small firms lack the managerial sophistication and foresight needed to participate effectively in such collective efforts.

Another alternative is for trade unions to serve as intermediaries, either on their own or in conjunction with trade associations or jobbers. Unions once provided managerial assistance to improve the efficiency of the fragmented supply chain, and the most successful examples of coordination through trade associations occurred within a framework of collective bargaining with unions. For example, the Amalgamated Clothing Workers contributed to improved industry performance during the 1920s by conducting efficiency studies aimed at lowering assembly costs, providing management assistance to troubled companies, and eliminating inefficient work rules (Fink, 1977). The International Ladies Garment Workers Union (ILGWU) played a similar role through its industrial engineering department, established in 1930 (Disher, 1947). The ILGWU also encouraged "larger production units, closer relationships between jobbers and their contractors, training for management, planning, cost accounting, and fair trade practices to govern dealing with retailers" (Seidman, 1942). It is unlikely that unions, however, could play the same intermediary role today, largely because union membership in the industry has declined substantially.

A third possibility is for governmental or not-for-profit organizations to serve as intermediaries. Examples of such not-for-profit organizations are $(TC)^2$, an organization sponsored by apparel and textile companies, unions, and the U.S. government which provides technology-based R&D and technical assistance to the U.S. apparel industry, and the Garment Industry Development Corporation which offers both technical assistance and marketing services to the small-scale fashion sector in New York City. Similar organizations have been set up in France and the United Kingdom (Doeringer et al., 1998). None of these organizations, however, is providing the full range of intermediation services needed to link small-scale flexible producers with lean retailers.

Another institutional obstacle is the lack of quick response relationships between apparel firms and their textile suppliers. Despite the enormous changes that have occurred in clothing production channels in recent years, minimum orders for fashion fabric averaged about 3500 square yards in 1992, and minimum delivery speeds averaged more than two and one-half months (Abernathy et al., 1995). These numbers are almost unchanged from 1988.

Clothing manufacturers have been unable to change textile supply practices through market forces, while mass retailers lack the direct economic relationships with the textile sector through which they might initiate reforms. Increasing capital intensity in textile production may be one explanation for these rigidities, but examples can be found in the United Kingdom of large retailers coordinating fabric supplies through direct partnerships with textile manufacturers.

The Possibility of Global Rapid Response

Replenishment markets have also been protected because delivery lags and throughput times average about one-third longer for foreign than for domestic production (Abernathy et al., 1995). Recent developments in trade policy, however, may be undermining the long-term future of rapid replenishment production in the United States by laying the foundation for quick response production channels in Mexico and the Caribbean basin.

Special trade legislation exempts imports from the Caribbean from quantity restrictions, and tariffs are levied on only the value-added from assembly, provided the garments are sewn from U.S.-made parts. Imports from Mexico face a preferential tariff regime, with tariffs to be phased out over ten years, as long as the garments are of NAFTA origin, defined as being made of NAFTA-produced fabric (Trebilcock and Howse, 1995). These provisions encourage geographic specialization within the production channel whereby capital- and skill-intensive stages of production remain in the United States while labor-intensive clothing assembly is performed offshore.

A by-product of this arrangement is that it encourages new contracting networks in which nearby offshore contractors could become part of a rapid replenishment supply chain managed by U.S. manufacturers and jobbers. If such "partnerships" can somewhat narrow the disadvantages of longer throughput and slower delivery times, the cost advantages of offshore products might be sufficient for them to enter rapid replenishment markets in the United States. This possibility may explain a part of the increase in imports from the Caribbean basin and Mexico from 7 percent of total imports in 1984 to 29 percent in 1995 (AAMA, 1996; Mittelhauser, 1997). The potential for such global rapid response production channels makes it all the more imperative that domestic response times be shortened.

CONCLUSION

New information technologies, new forms of coordination between retailers and suppliers, and the emergence of quick replenishment clothing markets are improving the economic prospects of production channels in the United States. Sustaining this scenario in the future, however, depends upon retailers' being able to continue offsetting the higher cost of clothing produced in the United States by the cost savings of lean retailing. The faster, the more reliable, and the more accommodating the domestic supply chain, the larger the proportion of goods that can be produced in the United States.

The sustainability of the advantages of lean retailing production channels remains somewhat in doubt. Lean retailing has favored large firms because of their advantages in acquiring new technologies and developing more sophisticated planning, production, and logistics practices. In addition, lean retailers find it easier to form partnerships with large firms, rather than trying to coordinate supplies among a number of small firms. The large-firm sector, however, may encounter internal obstacles to continuing to raise delivery speeds.

Small firms have the potential for very fast throughput and are a potential source of high-speed replenishment production. Small firms have been excluded from the rapid response production channel in the United States, however, and the current institutional structure of that channel may preclude their integration with lean retailing. Meanwhile, international sourcing of rapid response production may become possible by using low-cost suppliers that are "closer to home" (Murray, 1995).

Further reforms to speed production and delivery are critical if domestic clothing suppliers are to retain their position in replenishment markets for basic fashion products and if they are to expand into replenishment markets for more fashionable products. Such efforts, however, must be differentiated by type of product, size of firm, and position in the apparel production channel. The current focus of policies to achieve these reforms often neglects these important distinctions.

For example, few innovations are equally applicable to all products and sizes of firm. Technologies that can be successfully adopted by large firms are likely to be different from those that small firms can use, and technologies for serving mass markets for basic fashion may differ from those appropriate to lower-volume fashion markets. The results of current R&D initiatives, such as those carried out by $(TC)^2$, are available to all firms in the apparel industry, but they are often tailored to the larger firms that have traditionally led in the adoption of innovation.

Policies for training and management assistance, because they are often organized on a local or regional level, tend to be differentiated by size of firm and product. These programs address issues of work organization, quality control, production skills, and commercial knowledge that are often a barrier to integrating small firms into mass retailing supply channels, but they do not tackle the larger problem of how to establish such linkages.

The challenge for policy is to introduce reforms that focus on the institutional processes that govern the relationships among the different sectors of the production channel. These may also involve developing new intermediaries that can provide access to information technologies, develop new production channel relationships, and stabilize production in fragmented markets. Lean retailers are currently positioned to lead these reforms, but other candidates include clothing unions, jobbers, employer associations, and government agencies.

REFERENCES

Abernathy, F., J.T. Dunlop, J.H. Hammond, and D. Weil. (1995). "The information-integrated channel: a study of the U.S. apparel industry in transition." *Brookings Papers on Economic Activity: Microeconomics 1995*. Washington, DC: Brookings Institution.
American Apparel Manufacturers Association. (Various years). "Focus: An Economic Profile of the Apparel Industry." Arlington, VA.

Arpan, J.S., J. de la Torre, and B. Toyne. (1982). *The U.S. Apparel Industry: International Challenge, Domestic Response.* College of Business Administration, Georgia State University, Atlanta.

Berg, P., E. Applebaum, T. Bailey, and A. Kalleberg. (1996). "The Performance Effects of Modular Production in the U.S. Apparel Industry." *Industrial Relations* 35(3, July):356-373.

Best, M. (1990). *The New Competition.* Cambridge, MA: Harvard University Press.

Braun, K. (1947). *Union-Management Co-Operation.* Washington, DC: The Brookings Institution.

Bryner, E. (1916). *The Garment Trades.* Cleveland: The Survey Committee of the Cleveland Foundation.

Carpenter, J.T. (1972). *Competition and Collective Bargaining in the Needle Trades, 1910-1967.* Ithaca: New York State School of Industrial and Labor Relations, Cornell University.

Chandler, A.D. Jr. (1977). *The Visible Hand: The Managerial Revolution In American Business.* Cambridge, MA: Harvard University Press.

Disher, M.L. (1947). *American Factory Production of Women's Clothing.* London: Devereaux Publications, Ltd.

Doeringer, P., B. Courault, L. Oxborrow, E. Parat, and A. Watson. (1998). "Apparel Production Channels: Recent Experience and Lessons for Policy from the U.S., UK, and France," Geneva: International Institute for Labour Studies, Business and Society Programme. Discussion Paper DP/97/1998.

Drake, L., and C. Glasser. (1942). *Trends in the New York Clothing Industry.* New York: Institute of Public Administration.

Dunlop, J.T., and D. Weil. (1996). "Diffusion and Performance of Modular Production in the U.S. Apparel Industry." *Industrial Relations* 35(3, July):334-55.

Dunlop, J.T., and D. Weil. (1993). "The Diffusion of Human Resource Innovations: Lessons from the Apparel Industry." Working Paper, Harvard Center for Textile and Apparel Research.

Dunlop, J.T., and D. Weil. (1995). "Human Resource Innovations in the U.S. Apparel Industry." Working Paper, Harvard Center for Textile and Apparel Research.

Fink, G., ed. (1977). *Labor Unions.* Westport, Conn.: Greenwood Press.

Frank, B. (1953). *Progressive Apparel Production.* New York: Fairchild Publications.

Galenson, W. (1960). *The CIO Challenge to the AFL.* Cambridge, MA: Harvard University Press.

Greenwood, J. (1997). *The Third Industrial Revolution.* Washington, DC: AEI Press.

Grieg, G.B. (1949). *Seasonal Fluctuations in Employment in the Women's Clothing Industry in New York.* New York: Columbia University Press.

Hochman, J. (1941). *Industry Planning Through Collective Bargaining: A Program for Modernizing the Dress Industry.* New York: International Ladies Garment Workers Union.

Hufbauer, G., and K.A. Elliott. (1994). *Measuring the Costs of Protection in the United States.* Washington, DC: Institute for International Economics.

Kuznets, S. (1952). *Changes in the National Incomes of the United States of America Since 1870.* London: Bowes and Bowes.

Magee, M. (1930). *Trends in Location of the Women's Clothing Industry.* University of Chicago Press.

McNair, M., and E. May. (1960). "The American Department Store, 1920-1960: A Performance Analysis Based on the Harvard Reports." *Bureau of Business Research Bulletin* 166. Harvard University Graduate School of Business, Cambridge, MA.

Milgrom, P., and J. Roberts. (1990). "The Economics of Modern Manufacturing: Technology, Strategy, and Organization. " *American Economic Review* 80(June):511-528.

Mittelhauser, M. (1997). "Employment Trends In Textiles and Apparel, 1973–2005." *Monthly Labor Review* 121(8, August):24-35

Murray, L.A. (1995). "Unraveling Employment Trends in Textiles and Apparel." *Monthly Labor Review* 118(8, August):62-72.

National Retail Dry Goods Association. (1936). *Twenty-five Years of Retailing, 1911-1936.* New York.

New York Times. (1998). "Apparel Making Regains Lost Ground" (January 13, 1998): A19.

Palpacuer, F. (1996). "Strategies Competitives, Gestion des Competences et Organisation en Reseaux: Etude de cas de l'Industrie New Yorkaise de l'Habillement." Unpublished doctoral thesis, Universite Montpelier I.

Rothstein, R. (1989). *Keeping Jobs in Fashion: Alternatives to the Euthanasia of the U.S. Apparel Industry*. Washington, DC: Economic Policy Institute.

Seidman, J. (1942). *The Needle Trades*. New York: Farrar & Rinehart.

Teper, L. (1937). *The Women's Garment Industry*. New York: International Ladies' Garment Workers' Union.

Trebilcock, M.J., and R. Howse. (1995). *The Regulation of International Trade*. London: Routledge.

U.S. Department of Commerce. Bureau of the Census. (Various years). *Annual Survey of Manufactures*. Washington, DC: U.S. Government Printing Office.

U.S. Department of Commerce. Bureau of the Census. (Various years). *Census of Manufactures*. Washington, DC: U.S. Government Printing Office.

U.S. Department of Commerce. Bureau of the Census. (Various years). *Census of Retail Trade*. Washington, DC: U.S. Government Printing Office.

U.S. Department of Commerce. Bureau of the Census. (Various years). *County Business Patterns*. Washington, DC: U.S. Government Printing Office.

U.S. Department of Commerce. Bureau of the Census. (1975). *Historical Statistics of the United States, Colonial Times to 1970*. Washington, DC: U.S. Government Printing Office.

U.S. Department of Labor. Bureau of Labor Statistics. (Various years). *Employment and Earnings*. Washington, DC: U.S. Government Printing Office.

U.S. Department of Labor. Bureau of Labor Statistics. (1994). *Employment, Hours, and Earnings: United States 1909-1994*. Washington, DC: U.S. Government Printing Office.

Pharmaceuticals and Biotechnology[1]

IAIN COCKBURN
University of British Columbia and
National Bureau of Economic Research
REBECCA HENDERSON
Massachusetts Institute of Technology and
National Bureau of Economic Research
LUIGI ORSENIGO
Università Commerciale Luigi Bocconi
GARY P. PISANO
Harvard Business School

The pharmaceutical industry has been by almost any measure outstandingly successful. It is one of the few high-technology industries that American firms have dominated almost since its inception, and it is one of the few in which American firms continue to have an indisputable lead. During the 1980s and 1990s, double-digit rates of growth in earnings and return on equity were the norm for most pharmaceutical companies, and the industry as a whole ranked among the most profitable in the United States.[2]

To what degree can this success be viewed as a triumph of U.S. public policy? This question cannot be answered definitively because the roots of the industry's success are complex, and causality cannot be attributed to any single factor with precision. A plausible case can be made, however, that in the case of the pharmaceutical industry public policy has played a particularly important role in contributing to the global success of American firms.

Public policy has always played an enormously important role in shaping the pharmaceutical industry in the United States. On the supply side, public funding for health-related research supplies both new knowledge and highly trained employees to pharmaceutical firms. New drugs can be sold only with the explicit

---

[1] This paper draws on an ongoing program of work exploring the determinants of research productivity in the pharmaceutical industry, funded by the Program on the Pharmaceutical Industry and the Center for Innovation in New Product Development under NSF Cooperative Agreement Number EEC-9529140. Their support is gratefully acknowledged.

[2] Note that these figures are based on accounting rates of return. Figures that are recalculated to account for heavy spending by the industry on advertising and research suggest that rates of return were actually somewhat lower than the accounting figures would suggest.

approval of the federal government, an approval that is typically granted only after potential candidates have passed a series of rigorous clinical tests. On the demand side, the federal government has an enormous impact on the market for new drugs, both by virtue of its role as a major consumer of drugs through its funding of Medicare and Medicaid and through its regulation of how pharmaceutical firms may advertise and market their products. Public policy toward the protection of intellectual property also has a very significant effect because the pharmaceutical industry is one of the few in which intellectual property protection plays a central role in product market competition. Public policy also plays a more indirect but nevertheless important role in shaping the industry through its effects on both the labor markets and the markets for new capital, particularly the market for venture capital.

Taken together these policy instruments have been instrumental in building an exceptionally strong industry. Before World War II public policy played little role in shaping the industry's evolution. In the postwar period, however, the federal government's heavy investment in basic research, its support of a strong intellectual property regime, and its imposition in 1962 of tight product approval criteria combined to help create an industry whose leading firms were not only increasingly able to translate scientific advances into effective therapies but also well positioned to exploit the new opportunities opened up by the revolution in molecular biology.

The molecular biology revolution made the role of public policy in shaping the industry even more important. The revolution was initially based in the universities, and the size and strength of the American commitment to health-related research ensured that U.S. universities were at the frontier of the new science. But public policy also proved very important in shaping the ways in which the new science affected the pharmaceutical industry. The industry used molecular biology in two forms—as a new process technology in making large molecular weight drugs and as a new research tool in searching for more conventional, small molecular weight drugs. The vast majority of drugs prescribed today are "small" molecular weight drugs—relatively small, simple molecules that can be synthesized in a test tube and that often can be taken orally. "Large" molecular weight drugs, are much, much larger. They usually cannot be directly synthesized but must be "grown" or "expressed" and cannot usually be taken orally.

The first trajectory was, at least initially, unambiguously competence destroying and was most effectively exploited by new entrants. In the United States an institutional environment that not only supported universities in making the fundamental breakthroughs necessary to exploit the new science but also supported their translation into small, flexible, aggressively funded new firms led to the birth of an entire industry segment, the biotechnology firms.

At the same time the second trajectory—the adoption of the tools of biotechnology as search tools—proved to be competence destroying for those pharmaceutical firms that had not fully made the transition to "science-based" or "rational" drug discovery. It thus reinforced the dominance of the large scientifically based firms, a large majority of which were located in the United States and which owed much of their success to the U.S. public policy regimes of the 1970s and 1980s.

The remainder of this chapter expands on this argument. We begin by discussing the evolution of drug discovery research technology and the role of public policy in shaping U.S. success prior to the molecular biology revolution. We suggest that a number of public policies played instrumental roles in building an American industry that was among the strongest in the world. The third section lays the foundation for a discussion of the impact of public policy on the industry in the wake of the revolution in molecular biology. We suggest that molecular biology as a process technology—"biotechnology"—was competence destroying for the vast majority of established firms, while molecular biology used as a research tool was competence enhancing for those firms that had already made a transition to science driven, or more "rational" drug discovery. Further, we describe the ways in which the revolution shaped the evolution of the industry across the world, focusing particularly on the ways in which response in the U.S. was very different, and in many ways much more effective, than responses in Europe and Japan. Finally, we discuss the role of public policy in shaping this differential response. We suggest (as have many before us) that public policy was instrumental in laying the foundations for the explosion of vibrant "new biotechnology firms" that characterized the American response to "biotechnology." We also suggest that the ability of many of the established American firms to respond effectively to the challenges of the new science was predicated on skills that they had developed during the previous era, skills developed partly in response to an environment largely shaped by American public policy.

THE PHARMACEUTICAL INDUSTRY BEFORE THE MOLECULAR BIOLOGY REVOLUTION

The history of the pharmaceutical industry can be usefully divided into three major epochs. The first, corresponding roughly to the period 1850-1945, was one in which little new drug development occurred and in which the minimal research that was conducted was based on relatively primitive methods. The large-scale development of penicillin during World War II marked the emergence of the second period of the industry's evolution. This period was characterized by the institution of formalized in-house R&D programs and relatively rapid rates of new drug introduction. During the early part of the period the industry relied largely on so-called "random" screening as a method for finding new drugs, but in the 1970s the industry began a transition to "guided" drug discovery or "drug development by design," a research methodology that drew heavily on advances in molecular biochemistry, pharmacology, and enzymology. The third epoch of the industry had its roots in the 1970s but did not begin to flower until quite recently as the use of the tools of genetic engineering in the production and discovery of new drugs has come to be more widely dispersed.

Understanding the evolution of the industry in the first two periods is important because their history illustrates the role public policy played in shaping the industry and because both the industrial and institutional structure of the industry and the organizational capabilities of individual firms were molded during these early periods.

Early History

By almost any measure pharmaceuticals is a classic high-technology or science-based, industry. Yet drugs are as old as antiquity. For example, the Ebers Papyrus lists 811 prescriptions used in Egypt in 550 B.C. Eighteenth century France and Germany had pharmacies where pharmacists working in well-equipped laboratories produced therapeutic ingredients of known identity and purity on a small scale. Mass production of drugs dates back to 1813, when J.B. Trommsdof opened the first specialized pharmaceutical plant in Germany. During the first half of the nineteenth century, however, standardized medicines for treating specific conditions were virtually nonexistent. A patient instead would be given a customized prescription that would be formulated at the local pharmacy by hand.

The birth of the modern pharmaceutical industry can be traced to the mid-nineteenth century with the emergence of Germany and Switzerland as leaders of the new synthetic dye industry. This was due in part to the strength of German universities in organic chemistry and in part to Basel's proximity to the leading silk and textile regions of Germany and France. During the 1880s dyestuffs and other organic chemicals were discovered to have medicinal effects, such as antiception. It was thus initially Swiss and German chemical companies such as Ciba and Sandoz, Bayer and Hoescht, leveraging their technical competencies in organic chemistry and dyestuffs, that began to manufacture drugs, usually based on synthetic dyes, later in the nineteenth century. For example, the German company Bayer was the first to produce salicylic acid (aspirin) in 1883.

Mass production of pharmaceuticals also began in the United States and the United Kingdom in the later part of the nineteenth century, but the pattern of development was quite different from that of Germany and Switzerland. Whereas Swiss and German pharmaceutical activities tended to emerge within larger chemical-producing enterprises, the United States and the United Kingdom witnessed the birth of specialized pharmaceutical producers such as Wyeth (later American Home Products), Eli Lilly, Pfizer, Warner-Lambert, and Burroughs-Wellcome. Up until World War I German companies dominated the industry, producing approximately 80 percent of the world's pharmaceutical output.

In the early years the pharmaceutical industry was not tightly linked to formal science. Until the 1930s, when sulfonamide was discovered, drug companies undertook little formal research. Most new drugs were based on existing organic chemicals or were derived from natural sources such as herbs, and little formal testing was done to ensure either safety or efficacy. Harold Clymer, who joined SmithKline in 1939, noted:

> [Y]ou can judge the magnitude of [SmithKline's] R&D at that time by the fact I was told I would have to consider the position temporary since they had already hired two people within the previous year for their laboratory and were not sure that the business would warrant the continued expenditure.

World War II and wartime needs for antibiotics marked the drug industry's transition to an R&D-intensive business. Alexander Fleming discovered penicillin and its antibiotic properties in 1928. Throughout the 1930s, however, it was produced only in laboratory-scale quantities and was used almost exclusively for experimental purposes. With the outbreak of World War II, the U.S. government organized a massive research and production effort that focused on commercial production techniques and chemical structure analysis. More than 20 companies, several universities, and the Department of Agriculture took part. Pfizer, which had production experience in fermentation, developed a deep-tank fermentation process for producing large quantities of penicillin. This system led to major gains in productivity and, more important, laid out an architecture for the process and created a framework in which future improvements could took place.

The commercialization of penicillin marked a watershed in the industry's development. Due partially to the technical experience and organizational capabilities accumulated through the intense wartime effort to develop penicillin, as well as to the recognition that drug development could be highly profitable, pharmaceutical companies embarked on a period of massive investment in R&D and built large-scale internal R&D capabilities. At the same time there was a very significant shift in the institutional structure surrounding the industry. Whereas before the war public support for health-related research had been quite modest, after the war it boomed to unprecedented levels, helping to set the stage for a period of great prosperity.

Golden Age for the Industry: 1950-1990

The period from 1950 to 1990 was a golden age for the pharmaceutical industry, as the industry in general, and particularly the major U.S. players, firms such as Merck, Eli Lilly, Bristol-Myers, and Pfizer, grew rapidly and profitably. R&D spending literally exploded and with them came a steady flow of new drugs. Drug innovation was a highly profitable activity during most of this period. Statman (1983), for example, estimated that accounting rates of return on new drugs introduced between 1954 and 1978 averaged 20.9 percent (compared to a cost of capital of 10.7 percent). Between 1982 and 1992, firms in the industry grew at an average annual rate of 18 percent.

Several factors supported the industry's high average level of innovation and economic performance. One was the sheer magnitude of both the research opportunities and the unmet needs. In the early postwar years, there were many physical ailments and diseases for which no drugs existed. In every major therapeutic category, from pain killers and anti-inflammatories to cardiovascular and central nervous system products, pharmaceutical companies faced an almost completely open field. Before the discovery of penicillin, very few drugs effectively *cured* diseases.

Faced with such a "target-rich" environment but very little detailed knowledge of the biological underpinnings of specific diseases, pharmaceutical companies invented an approach to research now referred to as "random screening." Under this approach, natural and chemically derived compounds are randomly screened in test tube experiments and laboratory animals for potential therapeutic activity. Pharmaceutical companies maintained enormous "libraries" of chemical compounds and added to their collections by searching for new compounds in places such as swamps, streams, and soil samples. Thousands, if not tens of thousands, of compounds might be subjected to multiple screens before researchers honed in on a promising substance. Serendipity played a key role because the "mechanism of action" of most drugs, the specific biochemical and molecular pathways that were responsible for their therapeutic effect, was generally not well understood. Typically, researchers had to rely on the use of animal models as screens. For example, researchers injected compounds into hypertensive rats or dogs to explore the degree to which they reduced blood pressure. Under this regime it was not uncommon for companies to discover a drug to treat one disease while searching for a treatment for another.

Although random screening may seem inefficient, it worked extremely well for many years and continues to be widely employed. Several hundred chemical entities were introduced in the 1950s and 1960s, and several important classes of drug were discovered in this way, including a number of important diuretics, all of the early vasodilators, and several centrally acting agents, including reserpine and guanethidine.

In the early 1970s, the industry also began to benefit more directly from the explosion in public funding for health-related research that followed the war. Between 1970 and 1995, for example, support for the National Institutes of Health (NIH), the agency through which the vast majority of federal support for health-related research is channeled, increased nearly 200 percent in real terms, to over $8.8 billion a year or 36 percent of the federal nondefense research budget, an amount roughly equal to the total research expenditure of all the U.S. pharmaceutical firms (Figure 1).

Before the 1970s publicly funded research was probably most important to the industry as a source of knowledge about the etiology of disease. From the middle 1970s on, however, substantial advances in physiology, pharmacology, enzymology, and cell biology—the vast majority stemming from publicly funded

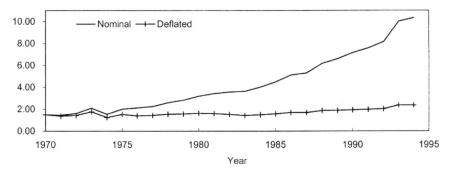

FIGURE 1 NIH total appropriations (billions of dollars).

research—led to enormous progress in understanding the mechanism of action of some existing drugs and the biochemical and molecular roots of many diseases. This new knowledge made it possible to design significantly more sophisticated screens. By 1972, for example, the structure of the renin angiotensive cascade, one of the systems within the body responsible for the regulation of blood pressure, had been clarified by publicly funded researchers, and by 1975 several companies had drawn on this research in designing screens for hypertensive drugs (Henderson and Cockburn, 1994). These firms could replace ranks of hypertensive rats with precisely defined chemical reactions. Instead of requesting "something that will lower blood pressure in rats," pharmacologists could request "something that inhibits the action of the angiotensin 2 converting enzyme."

In turn, the more sensitive screens made it possible to screen a wider range of compounds. Before the late 1970s, for example, it was difficult to screen the natural products of fermentation, a potent source of new antibiotics, in whole animal models. The compounds were available in such small quantities or triggered such complex mixtures of reactions in living animals that it was difficult to evaluate their effectiveness. The use of enzyme systems as screens made it much easier to evaluate these kinds of compounds. It also triggered a "virtuous cycle" in that the availability of drugs whose mechanisms of action were well known made possible significant advances in the medical understanding of the natural history of several key diseases, advances that in turn opened up new targets and opportunities for drug therapy (Gambardella, 1995; Maxwell and Eckhardt, 1990).

The industry's increasing reliance on advances in fundamental science dramatically increased the importance of public sector research in shaping industry productivity. Publicly funded research was important for several reasons. First, it provided the "raw knowledge" that undergirded many key discoveries. Table 1 illustrates the increasingly close relationship between the public and private sectors during the period. It summarizes detailed case histories of the discovery and development of 21 drugs identified by two leading industry experts as "having had the most impact upon therapeutic practice" between 1965 and 1992. Only 5

TABLE 1 History of the Development of the 21 Drugs with "Highest Therapeutic Impact" Introduced Between 1965 and 1992

Generic name	Trade name	Indication	Date of key enabling discovery	Public?	Date of synthesis of compound	Public?	Date of market intro	Lag from enabling discovery to market introduction
Old-fashioned or random drug discovery: screening of compounds in whole or partial animal screens								
Cyclosporine	Sandimmune	Immune suppression	NA		1972	N	1983	
Fluconazole	Diflucan	Antifungal	1978	N	1982	N	1985?	7
Foscarnet	Foscavir	CMV infection	1924	Y	1978	Y	1991	67
Gemfibrozil	Lopid	Hyperlipidemia	1962	N	1968	N	1981	19
Ketoconazole	Nizoral	Antifungal	1965	N	1977?	N	1981	16
Nifedipine	Procardia	Hypertension	1969	N	1971	N	1981	12
Tamoxifen	Nolvadex	Ovarian cancer	1971	Y	NA		1992	21
Mechanism driven research: screening of compounds against a very specific known or suspected mechanism								
AZT	Retrovir	HIV	Contentious	Y	1963	N	1987	16
Captopril	Capoten	Hypertension	1965	Y	1977	N	1981	29
Cimetidine	Tagamet	Peptic ulcers	1948	Y	1975	N	1977	18
Finasteride	Proscar	BPH	1974	Y	1986	N	1992	18
Fluoxetine	Prozac	Depression	1957	Y	1970	N	1987	30
Lovastatin	Mevacor	Hyperlipidemia	1959	Y	1980	N	1987	28
Omeprazole	Prilosec	Peptic ulcers	1978	N	1982		1989	11
Ondansetron	Zofran	Nausea	1957	Y	1983	N	1991	34
Propranolol	Inderol	Hypertension	1948	Y	1964	N	1967	19
Sumatriptan	Imitrex	Migraine	1957	Y	1988	N	1992	35
Drugs discovered through fundamental science								
Acyclovir	Zovirax	Herpes	?		?		1982	
Cisplatin	Platinol	Cancer	1965	Y	1967	Y	1978	13
Erythropoietin	Epogen	Anemia	1950	Y	1985	N	1989	39
Interferon beta	Betaseron	Cancer, others	1950	Y	Various	Y	Various	

Note: This sample was chosen in consultation with two experts: Professor Louis Lasagna (Tufts) and Professor Richard Wurtman (MIT). The sample includes five of the drugs (cyclosporine, nifedipine, captopril, cimetidine, and propranolol) that Maxwell and Eckhardt (1990), who used a similar criterion but focused on an earlier time period, included in their sample. Although our list is a selective and not necessarily representative sample of new drugs introduced in this period, and it is certainly the case that many potentially important drugs arising from more recent discoveries are still in development or have been just introduced, our qualitative work suggests that the general trend identifiable in these data is a plausible summary of recent trends in the relationship between the public and private sectors in the industry. For purposes of general comparison we list a "date of key enabling discovery" for each drug. The choice of any particular event as the "key enabling discovery" is bound to be contentious, since in pharmaceuticals, as in many fields, discovery usually rests on a complex chain of interrelated events. In the case of drugs discovered through screening we give the date of first indication of activity in a screen. In the case of mechanism based drugs, we give the date of the first clear description of the mechanism. Dates for the third class are only broadly indicative, and all should be used carefully.

Source: Maxwell and Eckhardt (1990), interview notes, detailed technical histories of each drug compiled from the public literature.

of these drugs, or 24 percent, were developed with essentially no input from the public sector. These data suggest that public sector research has become more important to the private sector over time.

Table 1 groups the drugs into three classes according to the research strategy by which they were discovered: those discovered by "random screening," those discovered by "mechanism-based screening," and those discovered through fundamental scientific advances. Broadly speaking, the degree of reliance on the public sector for the initial insight increases across the three groups, and as the industry has moved to a greater reliance on the second and third approaches, so too has the role of the public sector increased. The public sector was also important in providing highly trained employees for the private sector and in helping to sustain a "research ethos" within those private firms that aggressively embraced the new techniques and that was highly productive.

Efforts to measure the rate of return to public research have been very contentious and dogged by a variety of difficult practical and conceptual problems (Griliches, 1994; Ward and Dranove, 1995). However in a recent study Cockburn and Henderson (1998) suggest that differences in the effectiveness with which pharmaceutical firms access the upstream pool of knowledge created by public science correspond to differences in research productivity of as much as 30 percent. Zucker et al. (forthcoming) find very similar results in their study of the role of the public sector in supporting the growth of the newly founded biotechnology firms.

Although any estimate of this type must be treated with great caution, these results are consistent with the hypothesis that public sector research has been critically important to the industry's health. Most intriguingly from a public policy perspective, these authors found that a firm's connectedness to the public sector, measured by the coauthorship of scientific papers across institutional boundaries, is closely related to several other factors that enhance the productivity of privately funded pharmaceutical research. These include the number of "star scientists" employed by the firm and the degree to which the firm uses a researcher's reputation among his or her peers as a criterion for promotion.[3] These results are consistent with the hypothesis that the ability to take advantage of knowledge generated in the public sector requires investment in a complex set of activities that taken together change the nature of private sector research. Thus they raise the possibility that the ways in which public research is conducted may be as important as the level of public funding.

Despite their apparent importance, these new research techniques were *not* uniformly adopted across the industry. For any particular firm, the shift in the technology of drug research from "random screening" to one of "guided" discovery or "drug discovery by design" depended critically on the ability to take ad-

---

[3] The use of coauthoring behavior to measure connectedness to the public sector was pioneered by Zucker et al. (1997) in their study of the emergence of new biotechnology firms.

vantage of publicly generated knowledge (Cockburn and Henderson, 1996; Gambardella, 1995) and of economies of scope within the firm (Henderson and Cockburn, 1996). Smaller firms, those farther from the centers of public research, and those that were most successful with the older techniques of rational drug discovery appear to have been much slower to adopt the new techniques than were their rivals (Cockburn et al., 1998; Gambardella, 1995; Henderson and Cockburn, 1994). There was also significant geographical variation in adoption. The larger firms in the United States, the United Kingdom, and Switzerland were among the pioneers of the new technology, but Japanese and other European firms have been slow in responding to the opportunities afforded by the new science. In general, although the pharmaceutical industry is global in nature, companies from the United States, Switzerland, Germany, and the United Kingdom have dominated in the postwar period. French and Italian firms have not played major international roles. Japan is the second largest pharmaceutical market in the world and is dominated by local firms, largely for regulatory reasons; but Japanese firms have to date been conspicuously absent from the global industry. Only Takeda ranks among the top 20 pharmaceutical firms in the world, and until relatively recently the innovative performance of Japanese pharmaceutical firms has been weak compared with their U.S. and European competitors.

INSTITUTIONAL ENVIRONMENTS

Institutional forces have shaped the industry in the "pre-biotechnology world," providing powerful inducements to innovation. From its inception, the evolution of the pharmaceutical industry has been tightly linked to the structure of national institutions. The pharmaceutical industry emerged in Switzerland and Germany in part, because of strong university research and training in the relevant scientific areas. German universities in the nineteenth century were leaders in organic chemistry, and Basel, the center of the Swiss pharmaceutical industry, was the home of the country's oldest university, long a center for medicinal and chemical study. In the United States the government's massive wartime investment in the development of penicillin profoundly altered the evolution of American industry. In the postwar era, the institutional arrangements in four key areas, the public support of basic research, intellectual property protection, procedures for product testing and approval, and pricing and reimbursement policies, have strongly influenced both the process of innovation directly and the economic returns, and thus the incentives, for undertaking such innovation.

Public Support for Health-Related Research

Nearly every government in the developed world supports publicly funded health-related research, but countries vary significantly in both the level of support offered and in the ways in which it is spent. As reviewed earlier, public spending on health-related research in the United States is now the second largest item in the federal research budget after defense and is roughly equivalent to the research budget of the entire U.S. pharmaceutical industry. Both qualitative and quantitative evidence suggests that this spending has had a significant effect on the productivity of those large U.S. firms that were able to take advantage of it (Cockburn and Henderson, 1998; Maxwell and Eckhardt, 1990; Ward and Dranove, 1995).

Public funding of biomedical research also increased dramatically in Europe in the postwar period, although the United Kingdom spent considerably less than Germany or France, and total spending did not approach American levels (Table 2). Moreover, the institutional structure of biomedical research in continental Europe evolved quite differently from its evolution in the United States and the United Kingdom, creating an environment in which science is far less integrated with medical practice.

Science does not in general confer the same status within the medical profession in continental Europe as it does in the United Kingdom or the United States. Traditionally the medical profession has had less scientific preparation than is common in either the United States or the United Kingdom, and medical training and practice have focused less on scientific methods per se than on the ability to use the results of research. Moreover doctorates in the relevant scientific disciplines have been far less professionally oriented. Historically the incentives to engage in patient care at the expense of research have been very high. France and

TABLE 2 Breakdown of National Expenditures on Academic and Related Research by Main Field, 1987[a]

	Expenditure (1987 million dollars)						
	U.K.	FRG	France	Netherlands	U.S.	Japan	Average[b]
Engineering	436	505	359	112	1966	809	14.3%
	15.6%	12.5%	11.2%	11.7%	13.2%	21.6%	
Physical sciences	565	1015	955	208	2325	543	21.2%
	20.2%	25.1%	29.7%	21.7%	15.6%	14.5%	
Life sciences	864	1483	1116	313	7285	1261	36.3%
	30.9%	36.7%	34.7%	32.7%	48.9%	33.7%	
Social sciences	187	210	146	99	754	145	6.0%
	6.7%	5.2%	4.6%	10.4%	5.1%	3.9%	
Arts and humanities	184	251	218	83	411	358	6.8%
	6.6%	6.2%	6.8%	8.6%	2.8%	9.6%	
Other	562	573	418	143	2163	620	15.6%
	20.1%	14.2%	13.0%	14.9%	14.5%	16.6%	
Total	2,798	4,037	3,212	958	14,904	3,736	

[a]Expenditure data are based on OECD "purchasing power parities" for 1987 calculated in early 1989.
[b]This represents an unweighted average for the six countries (i.e., national figures have not been weighted to take into account the differing size of countries).
Source: Irvine et al. (1990, p. 219).

Germany have only recently implemented systems designed to free clinicians from their financial ties to patient-related activities, and, partly in consequence, within universities medically oriented research has played a marginal role compared with patient care.

The organizational structure of medical schools tends to reinforce these differences. In continental Europe medical schools and hospitals are part of a single organizational entity, whereas in the United States and the United Kingdom medical schools are generally independent of hospital administrations. This status allows them to give clear priority to their intrinsic goals of research and teaching. In principle, the European system should have some advantages. In practice, however, patient care has tended to absorb the largest fraction of time and financial resources.

The weakness of the research function within hospitals in continental Europe is one of the reasons that several governments have decided to concentrate biomedical research in national laboratories rather than in medical schools. However, the separation of the research from daily medical practice may have had negative effects on both the quality of the research and on the rate at which it diffuses into the medical community.

Protection of Intellectual Property

In many industries, successful new products quickly attract imitators. But rapid imitation of new drugs is difficult in pharmaceuticals. One reason is that pharmaceuticals has historically been one of the few industries where patents provide solid protection against imitation. Because small variants in a molecule's structure can drastically alter its pharmacological properties, potential imitators often find it hard to work around the patent. Although other firms might undertake research in the same therapeutic class as an innovator, the probability of their finding another compound with the same therapeutic properties that did not infringe on the original patent is usually quite small.[4]

The scope and efficacy of patent protection has varied significantly across countries, however. The United States and most European countries have provided relatively strong patent protection in pharmaceuticals. In contrast, until recently only *process* technologies could be patented in Japan and in Italy; not until 1976 in Japan and 1978 in Italy did patent law offer protection for pharmaceutical *products*. As a result, Japanese and Italian firms tended to avoid product R&D and to concentrate instead on finding novel processes for making existing molecules.

---

[4]This is not always the case. The history of the discovery of the ACE inhibitors provides a notable exception.

Procedures for Product Approval

Pharmaceuticals are regulated products. Procedures for approval have a profound impact on both the cost of innovating and on firms' ability to sustain market positions once their products have been approved. Since the early 1960s most countries have steadily increased the stringency of their approval processes. The United States and the United Kingdom have adopted by far the most stringent approval process of any industrial country, followed by the Netherlands, Switzerland, and the Scandinavian countries. Germany and especially France, Japan, and Italy have historically been much less demanding.

In the United States, the 1962 Kefauver-Harris Amendments were passed after the thalidomide disaster. This law introduced a proof-of-efficacy requirement for approval of new drugs and established regulatory controls over the clinical testing of new drug candidates. Specifically, the amendments required firms to provide substantial evidence of a new drug's efficacy based on "adequate and well controlled trials." As a result, after 1962 the Food and Drug Administration (FDA) shifted from being an evaluator of evidence and research findings at the end of the R&D process to an active participant in the process itself (Grabowski and Vernon, 1983).

The effects of the 1962 law on innovative activities and market structure have been the subject of considerable debate.[5] The law certainly led to large increases in the resources devoted to obtaining approval of a new drug application (NDA), and it probably caused sharp increases in both R&D costs and in the gestation times for new chemical entities (NCEs). As a consequence, the annual rate of NCE introduction declined sharply and there was a lag in the introduction of significant new drugs therapies in the United States compared with Germany and the United Kingdom. However, the creation of a stringent drug approval process in the United States may have also helped reduce rates of entry into the industry and thus may have indirectly served to protect the margins that attracted further investment in research. Although the process of development and approval increased costs, it significantly increased barriers to imitation, even after patents expired. Until the Waxman-Hatch Act was passed in 1984, generic versions of drugs that had gone off patent still had to undergo extensive human clinical trials before they could be sold in the U.S. market, so that it might be years before a generic version appeared even after a key patent had expired. In 1980 generics held only 2 percent of the U.S. drug market.

The institutional environment surrounding drug approval in the United Kingdom was quite similar to that in the United States. Regulation of product safety, which began in 1964 and was tightened with passage of the Medicine Act in 1971, relied heavily from the beginning on formal academic medicine, in particular on well-controlled clinical trials, to demonstrate the safety and efficacy of new

---

[5]See, for example, Chien (1979) and Peltzman (1974).

drugs. Extensive documentation and high academic standards were required of all submissions. The Committee on Safety of Drugs (CSD), known as the Committee on Safety of Medicines (CSM) after 1971, comprised independent academic experts, voluntarily organized and supported by the industry. Based on strong cooperation among the regulatory body, the industry, and academe, the British system effectively imposed very high standards on the industry (Davies, 1967; Hancher, 1990; Thomas, 1994; Wardell, 1978). As in the United States, the introduction of a tougher regulatory environment in the United Kingdom in 1971 was followed by a sharp fall in the number of new drugs launched in Britain and a shakeout of the industry. Several smaller, weaker firms exited the market, and the proportion of minor local products launched into the British market shrunk significantly. The strongest British firms gradually reoriented their R&D activities toward the development of more ambitious, global products (Thomas, 1994).

Japan represented a very different case from either the United States or the United Kingdom. Before 1967 any drug approved for use in another country and listed in an accepted official pharmacopœia could be sold in Japan without going through additional clinical trials or regulatory approval. At the same time non-Japanese firms were prohibited from applying for drug approval. Thus Japanese firms were simultaneously protected from foreign competition and given strong incentives to license products that had been approved overseas. Under this regime the primary technology strategy for Japanese pharmaceutical companies became the identification of promising foreign products to license (Reich, 1990).

The Structure of the Health Care System and Systems of Reimbursement

Perhaps the biggest differences in institutional environments across countries was in the structure of the various health care systems. In the United States, pharmaceutical companies' rents from product innovation were further protected by the fragmented structure of health care markets and by the consequent low bargaining power of buyers. Moreover, the U.S. government does not regulate drug prices. Until the mid-1980s most U.S. companies marketed directly to physicians, who largely made the key purchasing decisions by deciding which drug to prescribe. The ultimate customers, patients, had little bargaining power, even in those instances where multiple drugs were available for the same condition. Because insurance companies generally did not cover prescription drugs,[6] they did not provide a major source of pricing leverage. Pharmaceutical companies were afforded a relatively high degree of pricing flexibility. This pricing flexibility, in turn, contributed to the profitability of investments in drug R&D.

Drug prices were also relatively high in other countries that did not have strong government intervention in prices, such as Germany and the Netherlands. In the United Kingdom, price regulation was framed as voluntary cooperation

---

[6]In 1960, only 4 percent of prescription drug expenditures were funded by third-party payers.

between the pharmaceutical industry and the Ministry of Health. This scheme let companies set their own prices, but the ministry negotiated a global profit margin with each firm designed to ensure each of them an appropriate return on capital investments, including research, made in the United Kingdom. The allowed rate of return was negotiated directly and was set higher for export-oriented firms. In general, this scheme tended to favor both British and foreign research-intensive companies that operated directly in the United Kingdom. Conversely, it tended to penalize weak, imitative firms as well as those foreign competitors, primarily the Germans, trying to enter the British market without direct innovative effort in loco (Burstall, 1985; Thomas, 1994).

In Japan the Ministry of Health and Welfare set the prices of all drugs, using suggestions from the manufacturer based on the drug's efficacy and the prices of comparable products. Once fixed, however, the price was not allowed to change over the life of the drug (Mitchell et al., 1995). Thus, whereas in many competitive contexts prices began to fall as a product matured, this was not the case in Japan. Because manufacturing costs often fall with cumulative experience, old drugs thus probably offered the highest profit margins for many Japanese companies, further curtailing the incentive to introduce new drugs. Moreover, generally high prices in the domestic market provided Japanese pharmaceutical companies with ample profits and little incentive to expand overseas.

Thus, by the time the revolution in molecular biology began to have its effect on the industry, differences in national policies across regions had already shaped industry structure to a very considerable degree. Across the world, the industry had fragmented into two groups—large, highly diversified firms that were tightly connected to the public sector and quick to take advantage of the latest scientific developments and smaller, more marketing-driven firms whose research was either governed by the older paradigm of "random" search or who concentrated on making improvements to existing therapies. A disproportionate number of American firms were of the former type. Their development was largely predicated on the complex mix of policies outlined above, and thus these policies continued to have a very significant influence in shaping the industry even as the revolution in molecular biology further transformed industry dynamics.

REVOLUTION IN MOLECULAR BIOLOGY AND CHANGING COMPETENCE IN DRUG R&D

If effective public policy was critical to the health of the U.S. pharmaceutical industry in the 1980s, the revolution in genetics and molecular biology that began 40 years ago with Watson and Crick's discovery of the double helix structure of DNA and continued with Cohen and Boyer's discovery of the techniques of genetic engineering made effective policy even more important.

The revolution had an enormous impact on the nature of pharmaceutical research and development and on the organizational capabilities required to intro-

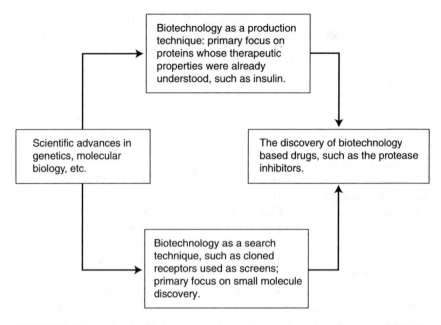

FIGURE 2 The molecular biology revolution and the trajectories of commercial R&D.

duce new drugs.[7] Application of the new techniques initially followed two relatively distinct technical trajectories (see Figure 2). One trajectory was rooted in the use of genetic engineering as a process technology to manufacture proteins *whose existing therapeutic qualities were already quite well understood* in large enough quantities to permit their development as therapeutic agents. The second trajectory used advances in genetics and molecular biology as tools to enhance the productivity of the discovery of conventional "small molecule" synthetic chemical drugs.

More recently, as the industry has gained experience with the new technologies, these two trajectories have converged, and contemporary efforts in biotechnology are largely focused on the search for large molecular weight drugs that must be produced using the tools of genetic engineering but whose therapeutic properties are not, as yet, fully understood.

Understanding the distinction between these two trajectories is of critical importance to understanding the role of public policy in the history of the industry because the two require quite different organizational competencies and have had quite different implications for industry structure and for the nature of competition across the world. In some regions, particularly the United States, the

---

[7]Biotechnology has also had far-ranging impacts on several other fields including diagnostics and agriculture. For the purposes of this paper we consider only its impact on human therapeutics.

ability to manufacture proteins in quantity triggered an explosion of entry into the industry and a proliferation of new firms. Although the success of the U.S. biotechnology industry undoubtedly had a multitude of causes, few observers doubt that a unique mix of publicly funded research and an institutional and financial climate that encouraged the formation of new firms was of key significance. The use of genetics as a tool for small molecule discovery, in contrast, appears to have reinforced the dominance of the large, global pharmaceutical firms at the expense of smaller regional players. Public policy appears to have been important in this case not only through the role that variation in access to publicly funded, leading edge research across the world has played in shaping the success of these large firms but also, and perhaps more important, through its influence as one of the factors that led to the emergence of these large firms in the first place.

Biotechnology as a Process Technology

Historically, most drugs have been derived from natural sources or synthesized through organic chemistry. Proteins, or molecules composed of long interlocking chains of amino acids, are simply too large and complex to synthesize feasibly through traditional synthetic chemical methods. Those proteins that were used historically as therapeutic agents, notably insulin, were extracted from natural sources or produced through traditional fermentation methods. But traditional fermentation processes, which were used to produce many antibiotics, could use only naturally occurring strains of bacteria, yeast, or fungi, so they were incapable of producing the vast majority of proteins. Cohen and Boyer's key contribution was the invention of a method for manipulating the genetics of a cell to induce it to produce a specific protein. This invention made it possible for the first time to produce a wide range of proteins and thus opened up an entirely new domain of search for new drugs, the vast store of proteins that the body uses to carry out a host of biological functions.

The human body produces approximately 500,000 different proteins, the vast majority of whose functions are not well understood. In principle Cohen and Boyers' discovery thus opened up an enormous new arena for research. The first firms to exploit the new technology chose, however, to focus on proteins such as insulin, human growth hormone, tissue plasminogen activator (tPA), and Factor VIII—proteins whose probable therapeutic effects were already relatively well understood. This knowledge greatly simplified both the process of research for the first biotechnology-based drugs and the process of gaining regulatory approval. Marketing these new drugs was also easier because their effects were well known and a preliminary patient population was already in place.

Thus for those firms choosing to exploit this route, the organizational capabilities most critical to success have been those of manufacturing and process development—learning to use the new recombinant DNA techniques as a production process to produce natural or modified human proteins. The development of this competence created significant challenges for nearly all of the established pharmaceutical firms because it required both the creation of an enormous body of new knowledge and a fundamental shift in the ways in which manufacturing process development was managed inside the firm.

The manufacture of small molecular weight drugs is essentially a problem in *chemical* process R&D. It draws primarily on chemistry and chemical engineering, disciplines in which there is a long history of basic scientific research. As a result much of the relevant theoretical knowledge has been codified in scientific journals and textbooks and, in searching for and selecting alternative chemical processes for the development of small molecular weight drugs, the pharmaceutical firm has at its disposal a wealth of scientific laws, principles, and models that describe the structure of relationships between different variables such as pressure, volume, and temperature. Thus process research chemists approaching the manufacture of a small molecular weight drug can often begin their work by deriving alternative feasible synthetic routes from theory.

The characteristics of the knowledge base underlying successful biotechnology process development are quite different. The major discovery underlying the field was made only in 1973, so biotechnology is in its infancy. Moreover, although basic scientific research has been extensive in molecular biology, cell biology, biochemistry, protein chemistry, and other relevant scientific disciplines, most of this work has been geared toward the problems of product "discovery" or to the identification of potentially important proteins rather than to their manufacture. Very little basic research has been conducted on the problems of engineering larger-scale biotechnology processes. Thus process developers in biotechnology have little theory to guide them in the development of new manufacturing processes.

Perhaps just as important, there is a long history of practical experience with chemical processes, whereas process developers in biotechnology had initially almost no practical experience to draw on. The chemical industry emerged in the eighteenth century, and chemical synthesis has been used to produce pharmaceuticals since the late 1800s. Through this experience a large body of heuristics have evolved that are widely used to guide process selection, scale-up, and plant design. Most pharmaceutical firms have also developed standard operating procedures for production activities such as quality assurance, process control, production scheduling, changeovers, and maintenance. Experience with these routines provides concrete starting points for development and guidance about the types of process techniques that are feasible within an actual production environment.

In contrast, some observers were initially skeptical that recombinantly engineered processes could be scaled up at all. Since 1982, when regulatory authorities approved recombinant insulin, the first biotechnology-based pharmaceutical to be manufactured at commercial scale, only about 25 biotechnology-based therapeutics have been approved for marketing. When a company develops and scales up a specific new biotechnology process, it is likely not only to be the company's first attempt, but also the first time *anyone* has attempted that process.

These differences imply that an organization developing a process for a protein molecule needs not only new technological or scientific capabilities but also organizational capabilities different from those required for developing a manufacturing process for a new small molecular weight compound. As Pisano (1996) put it, biotechnology process development requires the capability to "learn by doing" in the actual production environment because it is virtually impossible to "learn before doing" in the laboratory. In contrast, small molecule pharmaceutical process development requires the capability to exploit the rich theoretical and empirical knowledge base of chemistry through laboratory research.

Biotechnology as a Research Tool

The new techniques of genetic engineering have also had a significant impact on the organizational competencies required to be a successful player in the pharmaceutical industry through their effect on the competencies required to *discover* "conventional" small molecular weight drugs. However, although the adoption of biotechnology as a process technology was unambiguously competence destroying for incumbent pharmaceutical firms, adoption of biotechnology as a search tool was competence destroying only for those firms that had not made the transition from "random" to "guided" drug discovery.

Those firms that had made the transition initially used the tools of genetic engineering as another source of "screens" with which to search for new drugs. For example, genetic engineering techniques allow researchers to clone target receptors, so that firms can screen against a "pure" target rather than against, say, a pulverized solution of rat's brains that probably contain the receptor. The new techniques also permit the breeding of rats or mice that have been genetically altered to make them particularly sensitive to interference with a particular enzymatic pathway. Firms had to learn some new science, but *for those firms that had already made the transition to guided or science-driven drug discovery,* these techniques did not destroy existing competence in the way that the use of biotechnology as a process technology did.

The transition from random to guided drug discovery required the development of a large body of new knowledge and substantially new organizational capabilities in drug research. So-called random drug discovery drew on two core disciplines—medicinal chemistry and pharmacology. Successful firms employed battalions of skilled synthetic chemists and pharmacologists who managed smoothly running, large-scale screening operations. Although a working knowledge of current biomedical research might prove useful as a source of ideas about possible compounds to test or alternative screens to try, by and large firms did not need to employ researchers at the leading edge of their field or to sustain a tight connection to the publicly funded research community, and firms differed greatly in the degree to which they invested in advanced biomedical research.

The ability to take advantage of the techniques of "guided search," in contrast, required a very substantial extension of the range of scientific skills employed by the firm—a scientific workforce that was tightly connected to the larger scientific community and an organizational structure that supported a rich and rapid exchange of scientific knowledge across the firm (Gambardella, 1995; Henderson and Cockburn, 1994). The new techniques also significantly increased returns to the scope of the research effort (Henderson and Cockburn, 1996).

Managing the transition from random to guided drug discovery was thus not a straightforward matter. In general the larger organizations who had indulged a "taste" for science under the old regime were at a considerable advantage in adopting the new techniques, while smaller firms, firms that had been particularly successful in the older regime, and firms that were much less connected to the publicly funded research community were much slower to follow their lead (Cockburn et al., 1998; Gambardella, 1995).

These differences were critical in shaping responses to the use of biotechnology as a research tool. For those firms that had already made the transition to guided drug discovery, the adoption of the tools of genetic engineering as an additional resource in the search for small molecule drugs was a fairly natural extension of the existing competence base. Molecular geneticists could be hired as one additional scientific discipline among many, and the genetically engineered screens that they provided could be easily accommodated within the existing research procedures. The larger, more scientifically sophisticated firms were at an enormous advantage in employing biotechnology as a research tool in the search for small molecule drugs (Zucker and Darby, 1996, 1997), and this advantage shaped national responses to the biotechnology revolution. It continues to shape responses as the two trajectories have begun to converge.

PATTERNS OF INDUSTRY EVOLUTION

Thus the techniques of molecular biology had dramatic implications for both the discovery of new drugs, on the one hand, and for the ways in which they were manufactured, on the other. "Biotechnology" in the popular sense, has provided an important additional source of new drugs, but, as discussed above, it is by no means the only way in which these techniques have changed the industry. Each trajectory, biotechnology-based proteins and the use of genetics as a tool in the search for conventional drugs, has been associated with different organizational regimes and patterns of industry evolution across countries.

Tables 3 and 4 present some summary data that provide a preliminary picture of some of these differences. Table 3 shows the number of firms active in biotechnology across the world for the periods 1978-1986 and 1987-1993, as defined by their activity at the European patenting office. The United States clearly

TABLE 3 Patent Applications at the European Patent Office, 1978-1993

	World patentshares (%), 1978-1993	No. of firms 1978-1986	No. of firms 1987-1993
U.S.	36.5	213	303
Japan	19.5	108	185
U.K.	5.9	39	64
Germany	12.0	45	58
France	6.0	37	52
CH	4.2	11	19

Source: European Patent Office.

hosts the majority of firms, but the Japanese are also very highly represented. Table 4 illustrates the dramatic differences in institutional form. Newly founded firms are far more important in the United States and the United Kingdom than they are elsewhere, while the public sector plays a disproportionately important role in France. New firms play a negligible role in Japan, Switzerland, and Germany. Comprehensive data that would allow us to match trajectory to institution type is not available, but we believe that the vast majority of the new biotechnology firms initially pursued the first trajectory, or a focus on biotechnology as a

TABLE 4 Patent Activity in Genetic Engineering by Type of Institution

	Percent of patents filed at European patent office		
	NBFs	Established corporations	Universities and other research institutions
1978-1986			
U.S.	43.2	34.5	22.3
Japan	0.00	87.7	12.3
Germany	0.01	81.8	17.7
U.K.	27.3	49.1	23.6
France	18.7	21.5	59.8
Switzerland	0.00	92.9	7.1
Netherlands	12.7	56.4	30.9
Denmark	0.00	93.5	6.5
Italy	0.00	95.7	4.3
1987-1993			
U.S.	40.4	38.1	20.7
Japan	3.1	86.9	10.0
Germany	3.0	80.0	17.0
U.K.	23.7	44.7	31.6
France	16.7	35.0	48.3
Switzerland	4.7	89.0	6.3
Netherlands	20.0	62.5	17.5
Denmark	5.7	92.5	1.9

Source: European Patent Office.

process technology, while the established firms—with the important exception of the Japanese firms entering the industry from fermentation-related fields—largely pursued the second trajectory, or a focus on the use of biotechnology as a research tool in the search for small molecule drugs. Newly founded firms were initially far more successful than the established firms in bringing new biological entities to market. Zucker and Darby (1996) present an analysis of 21 new biological entities approved for the U.S. market by 1994: 7 were discovered by small independent firms, 12 by small firms that were subsequently acquired, and only 2 by established pharmaceutical firms acting "in their own right."

More recently, as the two trajectories have merged, intracompany agreements have proliferated, the majority between new biotechnology firms and the larger, established firms. Many companies that were initially slow to respond to the opportunities offered by the new science have attempted to "catch up" through joint research agreements or the outright purchase of promising new firms. For example, out of 95 biotechnology drugs that entered clinical trials in the United States between 1980 and 1988, 15 were developed solely by pharmaceutical firms, 36 were developed solely by biotechnology firms, and 44 were developed jointly by pharmaceutical and biotechnology firms (Bienz-Tadmore et al., 1992).

Below we explore these geographical differences in more detail as a prelude to our concluding discussion of the degree to which they can be explained by differences in the institutional structure and in the public policy regime surrounding the industry across the different regions of the world.

The United States

In the United States, the use of biotechnology as a process technology was the motive force behind the first large-scale entry into the pharmaceutical industry since the early postwar period. The first new biotechnology start-up was Genentech, founded in 1976 by Herbert Boyer, one of the scientists who developed the recombinant DNA technique, and Robert Swanson, a venture capitalist. Genentech became the model for a large number of new entrants. They were primarily university spin-offs, and they were usually formed through collaboration between scientists and professional managers, backed by venture capital. Their specific skills resided in the knowledge of the new techniques and in the research capabilities in that area. Their goal was to apply the new scientific discoveries to commercial drug development. Entry rates soared in 1980 and remained at a very high level at least until 1985. By the beginning of 1992, there were 48 publicly traded biotechnology companies specialized in pharmaceuticals and health care and several times this number still privately held.

Between 1982, when human insulin was approved, and 1992, 16 biotechnology drugs were approved for the U.S. market. As is the case for small molecular weight drugs, the distribution of sales of biotechnology products is highly skewed. Three products were major commercial successes: insulin (Genentech and Eli

Lilly), tPA (Genentech in 1987), and erythropoietin (Amgen and Ortho in 1989). By 1991 more than 100 biotechnology drugs were in clinical development, and applications for 21 biotechnology drugs had been submitted to the FDA (Grabowski and Vernon, 1994). This was roughly one-third of all drugs in clinical trials (Bienz-Tadmore et al., 1992). Sales of biotechnology-derived therapeutic drugs and vaccines had reached $2 billion, and two new biotechnology firms, Genentech and Amgen, had entered the club of the top eight major pharmaceutical innovators (Grabowski and Vernon, 1994).

Established pharmaceuticals initially played a less direct role in this application of biotechnology, at least in the United States. Zucker and Darby (1997) show that of all the firms in their sample—U.S. firms that either employed or were closely tied to "star" biotechnology scientists—taking out worldwide genetic-sequence patents between 1980 and 1990, 81 percent were dedicated biotechnology firms. Most of the major companies invested in biotechnology R&D through collaborative arrangements, R&D contracts, and joint ventures with the new biotechnology start-ups (Arora and Gambardella, 1990; Barbanfi et al., 1998; Pisano, 1991). As outlined above, the application of molecular biology to the development of protein-based drugs required a completely different set of competencies in both drug discovery and process development. Incumbents were thus poorly positioned to exploit the technical opportunities afforded by the new trajectory through in-house research or manufacturing. The competencies required for clinical development, regulatory approval, and marketing were essentially the same between biotechnology and traditional synthetic drugs, however, and new firms sought out incumbents as partners who could help commercialize the fruits of their R&D. Thus, during the 1970s and 1980s, a market for know-how emerged in biotechnology with the start-up firms positioned as upstream suppliers of technology and R&D services and established firms positioned as downstream buyers who could provide capital as well as access to complementary assets (Pisano and Mang, 1993).

Although newly founded firms pioneered the use of genetics as a source of large molecular weight drugs, established firms led the way in the use of genetic technology as a tool for the discovery of traditional or small molecular weight drugs. The speed with which the new techniques were adopted varied enormously, however. For those firms that were already heavily investing in fundamental research and in which participating in the broader scientific community was already recognized to be of value, the new knowledge presented itself as a natural extension of existing work. They might have been exploring the mechanisms of hypertension, for example. Knowledge of the genetic bases of these mechanism was a fairly easily accommodated "competence" and in general these firms moved quite quickly to adopt the new techniques (Gambardella, 1995; Zucker and Darby, 1997). Firms such as Merck, Pfizer, and SmithKline-Beecham, for example, made the transition relatively straightforwardly. Those firms that had been more firmly oriented toward the techniques of random drug design, however, found the transition much more difficult. Firms that had no history of publication or of investment in basic science often found it hard to recruit scientists of adequate caliber and to create the communication patterns that the new techniques required.

The new techniques probably also significantly increased returns to scope. As drug research came to rely increasingly on the insights of modern molecular biology, discoveries in one field often had implications for work in other areas, and firms that had the size and scope to capitalize on these opportunities for cross fertilization and the organizational mechanisms in place to take advantage of these opportunities reaped significant rewards. Thus one of the major impacts of the revolution in molecular biology has been to drive a wedge between those firms that have been able to absorb the new science into their research efforts and those that are still struggling to make the transition (Cockburn et al., 1997; Zucker and Darby, 1997).

Europe and Japan

In Europe and Japan the exploitation of genetics as a tool to produce proteins as drugs lagged considerably behind the effort in the United States and proceeded along different lines (Orsenigo, 1995). The most striking difference, of course, is the virtual lack of specialized biotechnology start-ups in Europe and Japan, with some exceptions in the United Kingdom and isolated cases elsewhere, at least until the late-1980s.

This difference is particularly striking because governments in Japan and most European countries at the community, national, and local government levels have devised a variety of measures to foster industry-university collaboration and the development of venture capital to favor the birth of new biotechnology ventures. To date the results of these policies have not been particularly impressive, although the increase in the rate of formation of new biotechnology-based firms in the 1990s may reflect the fact that these policies are now beginning to have an impact. Ernst and Young (1995) suggest that there are now approximately 380 biotechnology companies in Europe. Britain has the largest number of new biotechnology firms, followed by France, Germany, and the Netherlands (Escourrou, 1992; SERD,1996). Recent data, moreover, suggest a dramatic increase of new biological firms in Germany, with different sources estimating their number in the 400 to 500 range or as more than 600 (Coombs, 1995).

Very few of these companies resemble the American prototype, however. Many of the new European firms are not involved in drug research or development but are instead intermediaries commercializing products developed elsewhere or are active in diagnostics, the agricultural sector, or the provision of instrumentation and/or reagents (MERIT, 1996; SERD, 1996). Moreover, some of these companies, especially the most significant ones like Celltech and Transgene, have been founded through the direct support and involvement of governments and large pharmaceutical companies rather than through the venture capital market.

The contribution of this new breed of companies to the development of European biotechnology remains to be seen. They already seem to be suffering from the disadvantages of entering the market relatively late. Only the earliest entrants are significant innovators, and some of the most successful, like their American counterparts, have already been acquired or expect to be acquired shortly by U.S. companies.

In the absence of extensive new firm founding, most of the innovation in biotechnology in mainland Europe has occurred within established firms. In France there has been significant entry, largely from firms diversifying into biotechnology and from other research institutions, while in Germany there has been almost no entry at all. Thus in mainland Europe a few firms account for a large proportion of biotechnology patents, and innovation in biotechnology rests essentially on the activities of a relatively small and stable group of large established companies. However, in contrast to the majority of the established American firms that adopted the techniques of genetic engineering as a manufacturing tool primarily through acquisition and collaboration with the small American start-ups, the European firms showed considerable variation in the methods through which they acquired the technology.

The British (Glaxo, Wellcome, and to a lesser extent ICI) and the Swiss companies (particularly Hoffman La Roche, Ciba Geigy, and Sandoz) moved early and decisively in the direction pioneered by the large U.S. firms in collaborating with or acquiring American start-ups. Firms in the rest of Europe tended to establish a network of alliances with local research institutes, although German companies lagged somewhat behind. Hoechst signed a 10-year agreement with Massachusetts General Hospital as early as 1981, but Bayer did not enter seriously until 1985. In general the Germans made little progress in the field, and they are not now considered to be among the leaders in European biotechnology. In some countries such as Italy, the scientific community took the lead in the attempt to promote the commercial development of genetic engineering through the establishment of linkages and collaboration with the pharmaceutical industry. The biggest European innovators are a research institution, Institut Pasteur in France, and two companies that have not been traditional players in the pharmaceutical industry, Gist-Brocades and Novo Nordisk. While data are difficult to obtain, it appears that almost all of the established French, Italian, German, and Japanese companies have been slow to adopt the tools of biotechnology as an integral part of their own drug research efforts.

In Japan, the large food and chemical companies with strong capabilities in process technologies, such as Takeda, Kyowa Hakko, Ajinomoto, and Suntory, pioneered entry into biotechnology. Although these firms have strong competencies in process development, they generally lack capabilities in basic drug research. During the 1980s some U.S. observers expressed concern that biotechnology would be the next industry in which Japanese firms achieved dominance, but to date that has not occurred, and there has been only limited entry into the pharmaceutical industry through biotechnology.

NATIONAL SYSTEMS OF INNOVATION:
HOW DID PUBLIC POLICY MATTER?

This brief description of the impact of the revolution in molecular biology on the pharmaceutical industry highlights the diversity of responses across the world and suggests several "stylized facts" to be explored in examining the relationship between "national systems of innovation," or the entire set of public policies and institutional constraints that shaped the evolution of any particular firm and the evolution of the industry across the world. First, why was the use of molecular biology as a production tool pioneered in the United States by small, newly founded firms, in Japan by firms diversifying into the industry from other fields, and in Europe largely by established pharmaceutical firms? Why did new entrants play a much smaller role in the European context? Second, did national systems of innovation play a role in shaping the diffusion of the use of molecular biology as a research tool? This technology was pioneered by established pharmaceutical firms in almost every case, yet its rate of adoption varied widely across the world.

The Evolution of "Biotechnology"

Why the small, independently funded biotechnology start-up was initially an American phenomenon is an old question and a much discussed one. One of the reasons that it cannot be answered definitively is that the answer is to a large degree overdetermined; many factors were clearly at play, almost any one of which may have been sufficient. As the discussion has already suggested, the use of molecular biology as a production technology was a competence-destroying innovation for the vast majority of the established pharmaceutical firms. In the United States a combination of factors allowed small, newly founded firms to take advantage of the opportunity this created. These factors included a favorable financial climate, strong intellectual property protection, a scientific and medical establishment that could supplement the necessarily limited competencies of the new firms, a regulatory climate that did not restrict genetic experimentation, and, perhaps most importantly, a combination of a very strong local scientific base and academic norms that permitted the rapid translation of academic results into competitive enterprises. In Europe, apart from the United Kingdom, and in Japan many of these factors were not in place, and it was left to larger firms to exploit the new technology.

A Strong Scientific Base and Academic Norms

The majority of the American biotechnology start-ups were tightly linked to university departments, and the very strong state of American academic molecular biology clearly played an important role in facilitating the wave of start-ups that characterized the 1980s (Zucker et al. forthcoming). The strength of the local science base may also be responsible within Europe for the relative British advantage and the relative German and French delay. Similarly, the weakness of Japanese industry may partially reflect the weakness of Japanese science. There seems to be little question about the superiority of the American and British scientific systems in the field of molecular biology, and it is tempting to suggest that the strength of the local science base explains much of the regional differences in the speed with which molecular biology was exploited as a tool for the production of large molecular weight drugs.

Although this explanation might seem unsatisfying to the degree that academic science is rapidly published and thus, in principle, rapidly available across the world, the American lead appears to have been particularly important because the exploitation of biotechnology in the early years required the mastery of a considerable body of tacit knowledge that could not be easily acquired from the literature (Zucker et al., 1997; Pisano, 1996). Geographic proximity probability facilitated the transmission of this kind of tacit knowledge (Jaffe et al., 1993). In the case of biotechnology, however, several authors have suggested that the U.S. start-ups were not simply the result of geographic proximity (Zucker et al., 1997). These authors have suggested that the flexibility of the American academic system, the high mobility characteristic of the scientific labor market, and, in general, the social, institutional, and legal context that made it relatively straightforward for leading academic scientists to become deeply involved with commercial firms were also major factors in the health of the new industry.

The willingness to exploit the results of academic research commercially also distinguishes the U.S. environment from that in either Europe or Japan. This willingness has been strengthened since the late 1970s and the passage of the Bayh-Dole Act (see below), and the resulting role of universities as seedbeds of entrepreneurship has probably also been extremely important in the take-off the biotechnology industry.

In contrast, links between the academy and industry, especially the relatively free exchange of personnel, appear to have been much weaker in Europe and Japan. Indeed, the efforts of several European governments were targeted precisely toward strengthening industry-university collaboration, and it has been argued that the rigidities of the research system of continental Europe and the large role played in France and Germany by the public, nonacademic institutions have significantly hindered the development of biotechnology in those countries. That these kinds of factors, as distinct from the strength of the science base per se, were absolutely critical to the wave of new entry in biotechnology that occurred in America in the early 1980s is given further credibility by the rate at which the use of molecular biology diffused across the world.

Access to Capital

It is commonly believed that lack of venture capital restricted the start-up activity of biotechnology firms outside the United States. Clearly, venture capital, a largely American institution, played an enormous role in fueling the growth of the new biotechnology-based firms. Prospective start-ups in Europe, however, appear to have had many other sources of funds, usually through government programs. The results of several surveys also suggest that financial constraints did not constitute a significant obstacle for the founding of new biotechnology firms in Europe (Ernst and Young, 1995; MERIT, 1996; SERD, 1996).

In addition, although venture capital played a critical role in the founding of U.S. biotechnology firms, collaborations between the new firms and the larger, more established firms provided a potentially even more important source of capital. Why did prospective European or Japanese biotechnology start-ups not turn to established pharmaceutical firms as a source of capital? A plausible explanation focuses on the market for know-how in biotechnology. The evolution of that market created many opportunities for European and Japanese companies to collaborate with U.S. biotechnology firms. Although some U.S.-based new firms, such as Amgen, Biogen, Chiron, Genentech, and Genzyme, pursued a strategy of vertical integration from research through marketing in the U.S. market, most firms' strategies emphasized licensing product rights outside the U.S. to foreign partners. Thus to an even greater extent than many established U.S. pharmaceutical firms, European and Japanese firms were well positioned as partners for U.S. new biotechnology firms. Given the plethora of new U.S. firms in search of capital, European and Japanese firms interested in commercializing biotechnology had little incentive to invest in local biotechnology firms. Even in the absence of other institutional barriers to entrepreneurial ventures, start-ups in Europe or Japan might have been crowded out by the large number of U.S.-based firms anxious to trade non-U.S. marketing rights for capital.

Intellectual Property Rights

The establishment of clearly defined property rights also played a major role in making possible the explosion of new firm foundings in the United States, because the new firms, by definition, had few complementary assets that would have enabled them to appropriate returns from the new science in the absence of strong patent rights (Teece, 1986).

In the early years of biotechnology, considerable confusion surrounded the conditions under which patents could be obtained. In the first place, research in genetic engineering was on the borderline between basic and applied science. Much of it was conducted in universities or otherwise publicly funded, and the degree to which it was appropriate to patent the results of such research became the subject of bitter debate. Millstein and Kohler's groundbreaking discovery, hybridoma technology, was never patented, while Stanford University filed a patent for Boyer and Cohen's process in 1974. Boyer and Cohen renounced their own rights to the patent, but nevertheless they were strongly criticized for having being instrumental in patenting what many observers considered to be a basic technology. Similarly a growing tension emerged between publishing research results versus patenting them. The norms of the scientific community and the search for professional recognition had long stressed rapid publication, but patent laws prohibited the granting of a patent to an already published discovery (Merton, 1973; Kenney, 1986). In the second place the law surrounding the possibility of patenting life-formats and procedures relating to the modification of life-forms was not defined. This issue involved a variety of problems, but it essentially boiled down, first, to whether living things could be patented at all and, second, to the scope of the claims that could be granted to such a patent (Merges and Nelson, 1994).

These hurdles were gradually overcome. In 1980 Congress passed the Patent and Trademark Amendments of 1980 (Public Law 96-517). Also known as the Bayh-Dole Act, this law gave universities, and other nonprofit institutions and small businesses, the right to retain the property rights to inventions deriving from federally funded research. In 1984 Congress expanded the rights of universities further, by removing certain restrictions contained in Bayh-Dole regarding the kinds of inventions that universities could own and the right of universities to assign their property rights to other parties. In 1980 the U.S. Supreme Court ruled in favor of granting patent protection to living things (Diamond v. Chakrabarty); the case involved a scientist working for General Electric who had induced genetic modifications on a *Pseudomonas* bacterium that enhanced its ability to break down oil. In the same year the second reformulation of the Cohen and Boyer patent for the recombinant DNA process was approved. In subsequent years, a number of patents were granted establishing the right for very broad claims (Merges and Nelson, 1994). Finally, a one-year grace period was introduced for filing a patent after publication of the invention.

It is often stressed that the lack of adequate patent protection was a major obstacle to the development of the biotechnology industry in Europe.[8] First, the grace period available in the United States is not available in Europe; any discovery that has been published is not patentable. Second, the interpretation has prevailed that naturally occurring entities, whether cloned or uncloned, cannot be patented. As a consequence, the scope for broad claims on patents is greatly reduced, and usually process rather than product patents are granted. In 1994 the European Parliament *rejected* a draft directive from the European Commission that attempted to strengthen the protection offered to biotechnology.

Although it is clear that stronger intellectual property protection is not unambiguously advantageous, as the controversy surrounding NIH's decision to seek patents for human gene sequences clearly illustrated, in the early days of the industry, the United States probably reaped an advantage from its relatively stronger regime.

---

[8] See, for instance, Ernst and Young (1995).

Regulatory Climate

Although public opposition to genetic engineering was a significant phenomenon in the United States in the earliest years of the industry, it has quickly become less important, and in general the regulatory climate has been a favorable one (Kenney, 1986). In Europe, however, opposition to genetic engineering research by the "Green" parties is often cited as an important factor hindering the development of biotechnology, especially in Germany and other Northern European countries, and public opposition to biotechnology is said to have been a factor behind the decision of some European companies to establish research laboratories in the United States.

The Use of Molecular Biology as a Research Tool

Explaining variations in the rate of adoption of molecular biology as a research tool across the regions of the world is, in contrast, rather more difficult. In general the techniques were adopted first by the large, globally oriented U.S., British, and Swiss firms. Adoption by the other European firms, and by the Japanese, appears to have been a much slower process.

At first glance the relative strength of the local science base and the degree to which university research was connected to the industrial community appears to be as important an explanation here as it was in understanding the case of the diffusion of "biotechnology." Science in Japan and mainland Europe was arguably not as advanced as it was in the United States and Britain, a factor that slowed the adoption of the new techniques. Unfortunately this explanation is made much less plausible by the Swiss case. The Swiss companies established strong connections with the U.S. scientific system, suggesting that geographic proximity played a much less important role in the diffusion of molecular biology as a research tool.

A second possible explanation is that diffusion was shaped by the relative size and structure of the various national pharmaceutical industries. Henderson and Cockburn (1996) have shown that between 1960 and 1990 there were significant returns to size in pharmaceutical research, and that since 1975 these returns have come primarily from the exploitation of economies of scope. They interpret this finding as suggesting that the effective adoption of the techniques of guided search and more rational drug design placed a premium on the ability to integrate knowledge within the firm and thus that the larger, more experienced firms may have been at a significant advantage in the exploitation of the new techniques. To the degree that those firms that had already adopted the techniques of "rational" drug discovery were at a significant advantage in adopting molecular biology as a research tool, the pre-existence of a strong national pharmaceutical industry with some large internationalized companies may have been a fundamental prerequisite for the rapid adoption of molecular biology as a tool for product screening and design. The U.S. pharmaceutical industry has traditionally been internationally oriented and, at least since the early 1980s, open to international competition in the domestic market. But in many European countries, such as France and Italy, the pharmaceutical industry was highly fragmented into relatively small companies engaged essentially in the marketing of licensed products and in the development of minor products for the domestic markets.

Although size or global reach may have been a necessary condition, the failure of the largest German and Japanese firms to adopt these techniques suggests that it was not sufficient. The largest Japanese and German firms were arguably as international and as large as the Swiss.

The most plausible explanation is that institutional variables, particularly the stringency of the regulatory environment and the nature of patent regime, were also important. As mentioned earlier, there is now widespread recognition that the introduction of the Kefauver-Harris Amendments had a significant impact in inducing a deep transformation of the U.S. pharmaceutical industry, particularly through raising the cost and complexity of R&D. Partly as a result many U.S. firms were forced to upgrade their scientific capability.

Similarly the two European countries, the United Kingdom and Switzerland, whose leading firms did move more rapidly to adopt the new techniques appear to have actively encouraged a "harsher" competitive environment. The British system encouraged the entry of highly skilled foreign pharmaceutical firms, especially the American and the Swiss, and a stringent regulatory environment also facilitated a more rapid trend toward the adoption by British companies of institutional practices typical of the American and Swiss companies—in particular, product strategies based on high-priced patented molecules, strong links with universities, and aggressive marketing strategies focused on local doctors. The resulting change in the competitive environment in the home market induced British firms to pursue strategies that moved away from fragmenting innovative efforts into numerous minor products toward concentration on a few important products that could diffuse widely into the global market. By the 1970s the ensuing transformations of British firms had led to their increasing expansion into the world markets.

Lacy Glenn Thomas (1994) has suggested that the slowness with which the majority of the European firms, apart from British and Swiss firms, adopted the techniques of guided drug discovery reflected much weaker competitive pressures in their domestic markets. The Japanese experience also looks in many respects like that pursued in Europe outside Switzerland and the United Kingdom. In Japan legal and regulatory policies combined to frame a very "soft" competitive environment that appears to have seriously slowed the adoption of modern techniques by the Japanese pharmaceutical industry. As a result of the combination of patent laws, the policies surrounding drug licensing, and the drug reimbursement regime, Japanese pharmaceutical firms had little incentive to develop world-class product development capabilities, and in general they concentrated on finding novel processes for making existing foreign or domestically originated molecules (Mitchell et al., 1995). Moreover, Japanese firms were protected from foreign competition and simultaneously had strong incentives to license products that had been approved overseas. Under this regime the predominant technology strategy for Japanese pharmaceutical companies became the identification of promising foreign products to license.

Mitchell et al. (1995) have noted that some of these institutional factors are beginning to change and that these changes are starting to have effects on the R&D strategies and capabilities of some but not all firms participating in the Japanese pharmaceutical sector. After 1967 foreign-originated products required clinical testing in Japan before they could be approved for sale. After 1976 drug products could be patented. After 1981 pricing policy was changed so that prices for established drugs are reviewed periodically and compared with prices of newer drugs. Together these factors have combined to increase the incentives for original research. Recent evidence suggests that the share of new chemical entities approved in the United States that originate in Japan has increased substantially, from 4 percent in the 1970s to around 25 percent in 1988 (Mitchell et al., 1995). Nevertheless, because they lack a history of strong internal R&D, it is taking time for Japanese pharmaceutical companies to develop world class research capabilities.

Strong domestic competition, the existence of appropriate incentive mechanisms toward aggressive R&D strategies, and integration into the world markets thus appear to be important explanatory variables in analyzing variations in the diffusion of the new technologies in drug screening and design across *regions*. Note, however, that they appear to say little about variations in diffusion across *firms*.

Most of the firms that rapidly adopted the new techniques were large multinational or global companies, with a strong presence, at least as far research is concerned, in the United States and generally on the international markets. Zucker and Darby (1997) present some evidence that size alone is a reasonable predictor of adoption, at least in the United States. We suspect that this correlation reflects the fact that adoption is highly correlated with the degree to which firms have made the transition to guided drug discovery. By and large these were larger firms that had early developed a taste for science and that were able to build and sustain tight links to the public research community (Gambardella, 1995). Here institutional factors appear to have been a necessary but not sufficient condition. To the extent that the adoption of the new techniques also involved the successful adoption of particular, academic-like forms of organization of research within companies (Henderson, 1994), and this process was in turn influenced by the proximity and availability of first-rate scientific research in universities, it was much easier for American and to a lesser extent British firms to adopt them.

From this perspective, it is tempting to suggest that the origin of the American advantage in the use of biotechnology as a research tool as well as a process technique lies in the comparatively closer integration between industry and the academic community, compared with other countries. One might also speculate that this closer integration resulted in some degree from the strong scientific base of the American medical culture and from the adoption of tight scientific procedures in clinical trials. Through this mechanism, American companies might have come to develop earlier and stronger relationships with the biomedical community and with molecular biologists in particular. Segregation of the research system from both medical practice and from close contact with commercial firms (as in France and possibly in Germany) has been highlighted as a major factor hindering the transition to molecular biology in these two countries (see, for instance, Thomas, 1994).

CONCLUSION

Public policy plays a crucial role in shaping private sector productivity in many modern economies. The case of the pharmaceutical industry provides a particularly intriguing window into this process and into the importance of national systems of innovation in shaping industrial structure (Nelson, 1992).

Before the revolution in molecular biology, the U.S. pharmaceutical industry was shaped by public policy choices in a number of areas. Strong support for the life sciences provided both highly trained employees and a steady stream of knowledge that was a critical input to the industry. A tough regulatory environment and an intellectual property regime together increased the returns to fundamental innovation and further combined to create a cohort of large, diversified, highly skilled firms able to manage the transition from random to science-guided discovery effectively.

The revolution in molecular biology further reinforced the power of public policy in shaping the industry. In the first place, the revolution's extraordinary dependence on fundamental science meant that the large commitment the United States made to public funding became even more important. But the importance of public policy extends far beyond the simple provision of funding for research. In the case of biotechnology, or the use of molecular biology as a production technique, advances in basic science rendered obsolete several of the core competencies of existing firms, particularly those related to process development and manufacturing. In the United States, institutional flexibility on a wide range of dimensions led to the formation of specialized biotechnology firms that could provide these competencies and bridge the gap between basic university research, on the one hand, and clinical development of drugs on the other. Thus the new biotechnology-based firms were, in many ways, an institutional, or public policy-shaped, response to the technical opportunities created by new scientific know-how.

The case of biotechnology as a research tool presents a different but complementary picture. This trajectory was born within the confines of established pharmaceutical firms, and institutional factors appear to played a "necessary" rather than "sufficient" role in its diffusion. Pharmaceutical firms adopted biotechnology as a research tool as a way to use molecular biology to enhance the value and productivity of their existing assets and competencies, and in this sense biotechnology tools were "competence enhancing." But they were only competence enhancing for *some* pharmaceutical firms—those that were already oriented toward "high science" research and already firmly embedded in the global scientific community. Thus this case is one of existing institutional arrangements and structures shaping, rather than creating, the path of technical change. Forces facilitating institutional flexibility and responsiveness played a less prominent role in this domain, which may help to explain why Swiss and British firms have joined U.S. firms as leaders in the application of molecular biology to small molecule discovery.

We hesitate to draw any hard and fast conclusions about how public policy might best be shaped in the future to support the health of the industry. But this brief historical overview does raise several intriguing questions. First, it highlights the extraordinarily important role of publicly funded science in supporting the industry. Most important from a policy perspective, perhaps, it highlights the fact that *the ways in which this research is conducted may be as important as the level to which it is funded.* The published results of publicly funded research are, with some lag, widely diffused across the world, and this kind of "output" clearly had an important impact on the industry. But our discussion suggests American industry was able to gain extraordinary benefits from this research because of the fluid nature of the boundary between public and private research institutions in the field. In the case of the larger, more established firms, this led to the creation of several exceptionally creative and flexible research organizations that were heavily influenced by the norms of "open" science. In the case of biotechnology, it led to the foundation of an extraordinary number of new firms whose energy and creativity has been the envy of the world. To the extent that efforts to realize a direct return on public investments in research lead to a weakening of the culture and incentives of "open science," our results are consistent with the hypothesis that the productivity of the whole system of biomedical research may suffer.

REFERENCES

Arora, A., and A. Gambardella. (1990). "Complementarity and external linkage: the strategies of the large firms in biotechnology." *Journal of Industrial Economics* 37(4):361-379.

Barbanfi, P., A. Gambardella, and L. Orsenigo. (1998). "The evolution of the forms of organization of innovative activities in biotechnology." Forthcoming in *Biotechnology/International Journal of Technology Management.*

Bienz-Tadmore, B., P. Decerbo, G. Tadmore, and L. Lasagna. (1992). "Biopharmaceuticals and conventional drugs: Clinical Success Rates." *Bio/Technology* 10:521-525.

Burstall, M.L. (1985). "The Community's pharmaceutical industry," Bruxelles: Commission of the European Communities.

Chien, R.I. (1979). *Issues in pharmaceutical economics.* Lexington, MA: Lexington Books.

Cockburn, I., and R. Henderson. (1996). "Public-Private Interaction in Pharmaceutical Research" *Proceedings of the National Academy of Sciences* 93/23(November):12725-12730.

Cockburn, I., and R. Henderson. (1998). "Absorptive Capacity, Coauthoring Behavior, and the Organization of Research in Drug Discovery." *Journal of Industrial Economics* XLVI(2):157-182.

Cockburn, I., R. Henderson, and S. Stern. (1997). "Fixed Effects and the Diffusion of Organizational Practice in Pharmaceutical Research." MIT Mimeo.

Coombs, A. (1995). *The European Biotechnology Yearbook.*

Davies, W. (1967). *The Pharmaceutical Industry: A Personal Study.* Oxford: Pergamon Press.

Ernst and Young. (1995). "European Biotech 95: Gathering Momentum."

Escourrou, N. (1992). "Les societés de biotechnologie européennes: un reseau très imbriqué Biofutur." (July-August):40-42.

Gambardella, A. (1995). *Science and Innovation in the U.S. Pharmaceutical Industry.* Cambridge University Press: Cambridge.

Grabowski, H., and J. Vernon. (1994). "Innovation and structural change in pharmaceuticals and biotechnology." *Industrial and Corporate Change* 3(2):435-450.

Grabowski, H., and J. Vernon. (1983). *The regulation of pharmaceuticals.* Washington and London: American Enterprise Institute for Public Policy Research.

Griliches, Z. (1994). "Productivity, R&D and the Data Constraint." *The American Economic Review* 84(1):1-23.

Hancher, L. (1990). *Regulating for Competition: Government, Law and the Pharmaceutical Industry in the United Kingdom and France.* Oxford: Oxford University Press.

Henderson, R. (1994). "The Evolution of Integrative Competence: Innovation in Cardiovascular Drug Discovery." *Industrial and Corporate Change* 3(3):607-630.

Henderson, R., and I. Cockburn. (1994). "Measuring Competence? Exploring Firm Effects in Pharmaceutical Research." *Strategic Management Journal* 15(Winter):63-84.

Henderson, R., and I. Cockburn. (1996). "Scale, Scope and Spillovers: The Determinants of Research Productivity in Drug Discovery." *Rand Journal of Economics* 27(1):32-59

Irvine, J., B. Martin, and P. Isard. (1990). *Investing in the Future: An International Comparison of Government Funding of Academic and Related Research.* Aldershot, England: Edward Elgar Publishers.

Jaffe, A., M. Trajtenberg, and R. Henderson. (1993). "Geographic Localization of Knowledge Spillovers as Evidenced by Patent Citations." *Quarterly Journal of Economics* 434:578-598

Kenney, M. (1986). *Biotechnology: The Industry-University Complex.* Ithaca: Cornell University Press.

Maxwell, R., and S. Eckhardt. (1990). *Drug Discovery: A Case Book and Analysis.* Clifton, NJ: Humana Press.

Merges, R., and R. Nelson. (1994). "On limiting or encouraging rivalry in technical progress: The effect of patent scope decisions," *Journal of Economic Behavior and Organization* 25:1-24.

MERIT. (1996). "The Organization of Innovative Activities in the European Biotechnology Industry and its Implications for Future Competitiveness." Report for the European Commission. Maastricht.

Merton, D. (1973). *The Sociology of Science: Theoretical and Empirical Investigation.* Chicago: University of Chicago Press.

Mitchell, W., T. Roehl, and R. Slattery. (1995). "Influences on R&D Growth among Japanese Pharmaceutical Firms, 1975-1990." *Journal of High Technology Management Research* 6(1):17-31.

Nelson, R.R., ed. (1992). *National Systems of Innovation.* Oxford: Oxford University Press.

Orsenigo, L. (1995). *The Emergence of Biotechnology.* London: Pinter Publishers.

Peltzman, S. (1974). *Regulation of Pharmaceutical Innovation: The 1962 Amendments.* Washington, DC: American Enterprise Institute for Public Policy.

Pisano, G. (1991). "The Governance of Innovation: Vertical Integration and Collaborative Arrangements in the Biotechnology Industry." *Research Policy* 20:237-249.

Pisano, G. (1996). *The Development Factory. Unlocking the Potential of Process Innovation.* Boston: Harvard Business School Press.

Pisano, G., and P. Mang. (1993). "Collaborative Product Development and the Market for Know-How: Strategies and Structures in the Biotechnology Industry." In *Research on Technological Innovation, Management and Policy, Vol 5.* R. Rosenbloom and R. Burgelman, eds. Greenwich. CT: JAI Press.

Reich, M. (1990) "Why the Japanese Don't Export More Pharmaceuticals: Health Policy as Industrial Policy." *California Management Review* (Winter):124-150.

SERD. (1996). "The Role of SMEs in Technology Creation and Diffusion: Implications for European Competitiveness in Biotechnology." Report for the European Commission. Maastricht, Netherlands.

Statman, M. (1983). *Competition in the Pharmaceutical Industry: The Declining Profitability of Drug Innovation.* Washington, DC: American Enterprise Institute.

Teece, D.J. (1986). "Profiting from Technological Innovation: Implications for Integration, Collaboration, Licensing and Public Policy." *Research Policy* 15(6):185-219.

Thomas, L.G. III. (1994). "Implicit Industrial Policy: The Triumph of Britain and the Failure of France in Global Pharmaceuticals." *Industrial and Corporate Change* 3(2):451-489.

Ward, M., and D. Dranove. (1995). "The Vertical Chain of R&D in the Pharmaceutical Industry." *Economic Inquiry* 33(January):1-18.

Wardell, W. (1978). *Controlling the Use of Therapeutic Drugs: An International Comparison.* Washington, DC: American Enterprise Institute.

Zucker, L., and M. Darby. (1996). "Costly Information in Firm Transformation, Exit, or Persistent Failure." *American Behavioral Scientist* 39:959-974

Zucker, L., and M. Darby. (1997). "Present at the revolution: Transformation of technical identity for a large incumbent pharmaceutical firm." *Research Policy* 26:429-446.

Zucker, L., M. Darby, and M. Brewer. "Intellectual Human Capital and the Birth of U.S. Biotechnology Enterprises." Forthcoming, *American Economic Review.*

Index